T0296705

LONDON MATHEMATICAL SOCIETY LECTURE NOTE SERIES

Managing Editor: Professor N.J. Hitchin, Mathematical Institute,
University of Oxford, 24–29 St Giles, Oxford OX1 3LB, United Kingdom

The titles below are available from booksellers, or, in case of difficulty, from Cambridge University Press.

London Mathematical Society Lecture Note Series. 288

Rational Points on Curves over Finite Fields: Theory and Applications

Harald Niederreiter
National University of Singapore

Chaoping Xing
National University of Singapore

CAMBRIDGE
UNIVERSITY PRESS

CAMBRIDGE UNIVERSITY PRESS
Cambridge, New York, Melbourne, Madrid, Cape Town, Singapore,
São Paulo, Delhi, Dubai, Tokyo, Mexico City

Cambridge University Press
The Edinburgh Building, Cambridge CB2 8RU, UK

Published in the United States of America by Cambridge University Press, New York

www.cambridge.org
Information on this title: www.cambridge.org/9780521665438

© Cambridge University Press 2001

First published 2001
Reprinted 2002

A catalogue record for this publication is available from the British Library

ISBN 978-0-521-66543-8 Paperback

To Gerlinde and Youqun Shi

Contents

Preface

Algebraic curves over finite fields and their function fields have been and are still a source of great fascination for number theorists and geometers alike, ever since the seminal work of Hasse and Weil in the 1930s and 1940s. Many important and fruitful ideas have arisen out of this area, where number theory and algebraic geometry meet, and these developments have even spawned a new subject called arithmetic algebraic geometry which now has a broad appeal.

For a long time, the study of algebraic curves over finite fields and their function fields was the province of pure mathematicians. But then, in a series of three papers in the period 1977–1982, Goppa found stunning applications of algebraic curves over finite fields, and especially of those with many rational points, to coding theory. This created a much stronger interest in the area and attracted new groups of researchers such as coding theorists and algorithmically inclined mathematicians. An added incentive was provided by the invention of elliptic-curve cryptosystems in 1985. Algebraic geometry over finite fields is a flourishing subject nowadays which produces exciting research and is immensely relevant for applications.

There has been tremendous research activity focused on algebraic curves over finite fields and their function fields in the last five years. Important theoretical advances were achieved, such as new techniques of constructing algebraic curves over finite fields with many rational points, or equivalently global function fields with many rational places, and improved lower bounds on $A(q)$, the crucial quantity in the asymptotic theory of the number of $\mathbf{F}_q$-rational points on algebraic curves over the finite field $\mathbf{F}_q$ of order q. Explicit towers of global function fields meeting the Vlăduţ-Drinfeld bound were constructed for the first time. These and other results have a significant impact on coding theory since they lead, in particular, to sequences of algebraic-geometry codes of increasing length that beat the asymptotic Gilbert-Varshamov bound, the classical benchmark for families of good linear codes. Algebraic-geometry codes have received a further impetus from completely new methods of constructing linear codes from algebraic curves over finite fields that allow a much greater flexibility than Goppa's construction. It is equally important that entirely new areas of applications have opened up for algebraic curves over finite fields and their function fields in the last five years. These include stream ciphers, hash functions, and authentication schemes in cryptography as well as the construction of low-discrepancy sequences for quasi-Monte Carlo methods. In all these applications, the methods of algebraic geometry have been more successful than classical approaches.

The main aim of this book is to make interested graduate students and researchers conversant with these recent developments, by not only offering a unified exposition of the relevant results and techniques, but also providing the necessary background as far as possible in a limited space. An ideal preparation for reading this book will be the study of Stichtenoth's excellent monograph *Algebraic Function Fields and Codes*. A prior exposure to class field theory will also be helpful for the reader. Summaries of pertinent facts on algebraic function fields and class field theory can be found in Chapters 1 and 2 of the present book, together with appropriate references.

Just like Stichtenoth's book and most of the recent research papers on the topics of relevance here, our book favors the function-field viewpoint over the algebraic-geometry viewpoint. It is our experience that, particulary for students with a background in classical algebra and number theory, the language of global function fields is easier to master than that of projective curves over finite fields, mainly because of the close analogy between global function fields and algebraic number fields. The two viewpoints are, of course, equivalent, and a succinct discussion of the algebraic-geometry viewpoint, and in particular of the connections between global function fields and smooth projective curves over finite fields, is presented in Appendix A of this book.

As mentioned above, Chapter 1 on algebraic function fields and Chapter 2 on the class field theory of global function fields are of an introductory character. Chapter 3 surveys explicit global function fields that are useful for the following core chapter, in which recent work on the construction of global function fields with many rational places is discussed. In Chapter 4 we have emphasized methods that lead to general results and are not just *ad hoc* techniques. Another core part is Chapter 5 which studies the asymptotic behavior of the number of rational places of global function fields when the genus tends to infinity, again with a focus on recent research. The remaining three chapters are devoted to applications. Chapter 6 discusses the well-known use of algebraic-geometry methods in coding theory, but we go beyond the several textbooks on this subject by also covering results of the last few years. Chapters 7 and 8 present the recent applications to cryptography, respectively low-discrepancy sequences, that we mentioned earlier in this preface. In all three applications-oriented chapters we include some background on the underlying target area for a better appreciation of the material.

We express our gratitude to the Austrian Academy of Sciences and the National University of Singapore for sponsoring mutual visits of the authors in crucial periods of this book project. We are grateful also to Helmut Kopka for having given to the world the treasure trove of a book *LaTeX: Einführung* which proved invaluable in all TEXnical matters. It is a pleasure to thank the staff of Cambridge University Press, and in particular Roger Astley, for being so receptive to our idea of writing this book and for seeing through, and advising on, the project with their usual professionalism.

Singapore, October 2000   HARALD NIEDERREITER
                                                                CHAOPING XING

Chapter 1

Background on Function Fields

Some basic definitions and fundamental properties of algebraic function fields are introduced in this chapter. In particular, we focus on those concepts and results that are needed in the subsequent chapters, such as the Riemann-Roch theorem, divisor class groups, Galois extensions of algebraic function fields, ramification theory, constant field extensions, zeta functions, and the Hasse-Weil bound.

In later chapters, we will be interested only in algebraic function fields over finite fields, but a lot of the background material can be developed for arbitrary constant fields. In this chapter, we will not always state results in their most general form, but only in the form in which we need them. Most results will be presented here without proof since they are standard results from textbooks. The reader will find the books of Stichtenoth [152] and Weiss [173] particularly useful. We refer also to the books of Cassels and Fröhlich [13], Deuring [18], Koch [60], Moreno [82], Neukirch [85], and Stepanov [150], [151] for further background.

1.1 Riemann-Roch Theorem

In this section, we always assume that k is an arbitrary field. The field k will serve as the constant field of algebraic function fields.

An extension field F of k is called an **algebraic function field (of one variable) over** k if there exists an element z of F that is transcendental over k and such that F is a finite extension of the rational function field $k(z)$ over k. Furthermore, k is called the **full constant field** of F if k is algebraically closed in F, that is, if each element of F that is algebraic over k belongs to k. For brevity, we simply denote by F/k an algebraic function field (of one variable) with full constant field k. If the full constant field is clear from the context, we often write just F instead of F/k.

A **place** P of F is, by definition, the maximal ideal of some valuation ring of F. We denote by O_P the valuation ring corresponding to P. Later on we will also use the symbol M_P to stand for the ideal P.

A **normalized discrete valuation** of an algebraic function field F over k is a surjective map $\nu : F \longrightarrow \mathbf{Z} \cup \{\infty\}$ which satisfies:

(i) $\nu(x) = \infty$ if and only if $x = 0$;

(ii) $\nu(xy) = \nu(x) + \nu(y)$ for all $x, y \in F$;

(iii) $\nu(x + y) \geq \min(\nu(x), \nu(y))$ for all $x, y \in F$;

(iv) $\nu(a) = 0$ for any $a \in k^*$.

As a consequence of these axioms we get the following useful strengthening of (iii):

(iii') $\nu(x + y) = \min(\nu(x), \nu(y))$ for $x, y \in F$ with $\nu(x) \neq \nu(y)$.

There is a bijective correspondence between the places of F and the normalized discrete valuations of F.

For a place P of F, we write ν_P for the normalized discrete valuation of F corresponding to P. We denote by $\mathbf{P}_F$ the set of places of F. For a place P of F/k, its valuation ring

$$O_P = \{x \in F : \nu_P(x) \geq 0\}$$

is a local ring. Its maximal ideal is

$$M_P = \{x \in O_P : \nu_P(x) > 0\}.$$

The residue class field O_P/M_P, denoted by $\tilde{F}_P$, can be identified with a finite extension of k. The degree $[\tilde{F}_P : k]$ of this extension is called the **degree** of the place P. It is denoted by $\deg(P)$. A place of degree 1 is called **rational**.

Example 1.1.1 Consider the rational function field $F = k(x)$ over k, where x is transcendental over k. For a fixed monic irreducible polynomial $p(x)$ of $k[x]$, a normalized discrete valuation ν_P of F is uniquely determined by putting

$$\nu_P(f(x)) = r \quad \text{if } f(x) \in k[x] \setminus \{0\} \text{ and } p(x)^r || f(x).$$

It is easy to verify that

$$O_P = \left\{ \frac{f(x)}{g(x)} : f(x), g(x) \in k[x], \ p(x) \nmid g(x) \right\}$$

is a valuation ring with maximal ideal

$$M_P = \left\{ \frac{f(x)}{g(x)} : f(x), g(x) \in k[x], \ p(x) | f(x), \ p(x) \nmid g(x) \right\}.$$

Hence $p(x)$ yields a place P. Later on we will also use $p(x)$ to denote the place P. The residue class field corresponding to $p(x)$ is isomorphic to $k[x]/(p(x))$. Thus, the place P is rational if and only if $p(x)$ is a linear polynomial.

There is another normalized discrete valuation ν_∞ of F that is uniquely determined by putting

$$\nu_\infty(f(x)) = -\deg(f(x)) \quad \text{for } f(x) \in k[x] \setminus \{0\}.$$

This yields the place ∞ with the valuation ring

$$O_\infty = \left\{ \frac{f(x)}{g(x)} : f(x) \in k[x], \ g(x) \in k[x] \setminus \{0\}, \ \deg(f(x)) \leq \deg(g(x)) \right\}$$

of F and the maximal ideal

$$M_\infty = \left\{ \frac{f(x)}{g(x)} : f(x) \in k[x],\ g(x) \in k[x] \setminus \{0\},\ \deg(f(x)) < \deg(g(x)) \right\}.$$

We will sometimes call ∞ the infinite place of F. Since $O_\infty/M_\infty \simeq k$, it is a rational place of F. The places of the form $p(x)$, which are called the finite places of F, plus the infinite place ∞ are exactly all the places of F. In particular, F has altogether $q+1$ rational places if k is the finite field $\mathbf{F}_q$ of order q.

Recall that a **divisor** D of an algebraic function field F is a formal sum

$$D = \sum_{P \in \mathbf{P}_F} m_P P$$

with integer coefficients m_P and $m_P \neq 0$ for at most finitely many $P \in \mathbf{P}_F$. We often write $\nu_P(D)$ for the coefficient m_P of P. Then

$$D = \sum_{P \in \mathbf{P}_F} \nu_P(D)P.$$

The **support** $\operatorname{supp}(D)$ of D is the set

$$\operatorname{supp}(D) = \{P \in \mathbf{P}_F : \nu_P(D) \neq 0\}.$$

The **degree** $\deg(D)$ of D is defined by

$$\deg(D) = \sum_{P \in \operatorname{supp}(D)} \nu_P(D) \deg(P).$$

A divisor D of F is called **positive** if $\nu_P(D) \geq 0$ for all $P \in \mathbf{P}_F$. For two divisors D_1 and D_2 of F we write $D_1 \leq D_2$ if $\nu_P(D_1) \leq \nu_P(D_2)$ for all $P \in \mathbf{P}_F$.

Let P be a place of F and x a nonzero element of F. The place P is called a **zero** of x if $\nu_P(x) > 0$ and a **pole** of x if $\nu_P(x) < 0$. It is clear that a constant element in k has neither poles nor zeros. However, for $x \in F \setminus k$, x has at least one zero place and one pole place.

Let $x \in F^*$ and denote by $Z(x)$, respectively $N(x)$, the set of zero places, respectively pole places, of x. We define the **zero divisor** of $x \in F^*$ by

$$(x)_0 = \sum_{P \in Z(x)} \nu_P(x)P$$

and the **pole divisor** of x by

$$(x)_\infty = \sum_{P \in N(x)} (-\nu_P(x))P.$$

Then the **principal divisor** of x is given by

$$\operatorname{div}(x) = (x)_0 - (x)_\infty.$$

The degree of $\operatorname{div}(x)$ is equal to 0, i.e.,

$$\deg((x)_0) = \sum_{P \in Z(x)} \nu_P(x) \deg(P) = \sum_{P \in N(x)} (-\nu_P(x)) \deg(P) = \deg((x)_\infty).$$

For a divisor D of F we form the **Riemann-Roch space**

$$\mathcal{L}(D) = \{x \in F^* : \operatorname{div}(x) + D \geq 0\} \cup \{0\}.$$

Then $\mathcal{L}(D)$ is a finite-dimensional vector space over k, and we denote its dimension by $\ell(D)$. Obviously, we have two facts:
   (i) $\ell(0) = 1$;
   (ii) $\ell(D) = 0$ if $\deg(D) < 0$.
The dimension $\ell(D)$ is the subject of the Riemann-Roch theorem.

Theorem 1.1.2 (Riemann-Roch Theorem) *Let F be an algebraic function field of genus g. Then for any divisor D of F we have*

$$\ell(D) \geq \deg(D) + 1 - g,$$

and equality holds whenever $\deg(D) \geq 2g - 1$.

The Riemann-Roch theorem is often used to define the genus of F implicitly. Thus, we may define the **genus** of F as the integer

$$g := g(F) := \max_D \left(\deg(D) - \ell(D) + 1\right),$$

where the maximum is extended over all divisors D of F. By putting $D = 0$ in this definition, we see that the genus of F is nonnegative. For a rational function field F we have $g(F) = 0$. A function field of genus 1 is also called **elliptic**.

Definition 1.1.3 A divisor D of the algebraic function field F of genus g is called **nonspecial** if we have equality in the Riemann-Roch theorem, i.e., if

$$\ell(D) = \deg(D) + 1 - g.$$

If D is an arbitrary divisor of F and A is a positive divisor of F, then

$$\ell(D + A) \leq \ell(D) + \deg(A),$$

where the sum $D_1 + D_2$ of two divisors D_1 and D_2 of F is the divisor of F with

$$\nu_P(D_1 + D_2) = \nu_P(D_1) + \nu_P(D_2) \qquad \text{for all } P \in \mathbf{P}_F.$$

This result and the Riemann-Roch theorem yield the following consequence.

Corollary 1.1.4 *If D is a nonspecial divisor of F and G is a divisor of F with $G \geq D$, then G is nonspecial.*

Let P be a place of F. An integer $n > 0$ is called a **pole number** of P if there exists an element $x \in F^*$ with $(x)_\infty = nP$. Otherwise, n is called a **gap number** of P.

Corollary 1.1.5 (Weierstrass Gap Theorem) *Let F have genus $g \geq 1$ and let P be a rational place of F. Then there are exactly g gap numbers $i_1, \ldots, i_g$ of P and they satisfy*

$$1 = i_1 < \cdots < i_g \leq 2g - 1.$$

Proof. This is an immediate consequence of the following three facts:

(i) For any $i \geq 1$ we have $\mathcal{L}((i-1)P) \subseteq \mathcal{L}(iP)$, and $\mathcal{L}((i-1)P) = \mathcal{L}(iP)$ if and only if i is a gap number of P.

(ii) $\ell(iP) \leq \ell((i-1)P) + 1$.

(iii) $\ell(0 \cdot P) = 1$ and $\ell((2g-1)P) = g$. $\qquad\qquad\square$

For a place P of F, an element $t \in F$ with $\nu_P(t) = 1$ is called a **local parameter** at (or a **prime element** for) P.

For a place $P \in \mathbf{P}_F$ and a function $f \in F$ with $\nu_P(f) \geq 0$, we denote by $f(P)$ the residue class $f + M_P$ of f in $\tilde{F}_P$. Thus, $f(P) \in \tilde{F}_P$ can be viewed as an element of a finite extension of k.

Now we choose a sequence $\{t_r\}_{r=-\infty}^{\infty}$ of elements in F such that

$$\nu_P(t_r) = r$$

for all integers r. For a given function $f \in F$, we can find an integer v such that $\nu_P(f) \geq v$. Hence

$$\nu_P\left(\frac{f}{t_v}\right) \geq 0.$$

Put

$$a_v = \left(\frac{f}{t_v}\right)(P),$$

i.e., a_v is the value of the function f/t_v at P. Then a_v is an element of $\tilde{F}_P$. Note that the function $f/t_v - a_v$ satisfies

$$\nu_P\left(\frac{f}{t_v} - a_v\right) \geq 1,$$

hence we know that

$$\nu_P\left(\frac{f - a_v t_v}{t_{v+1}}\right) \geq 0.$$

Put

$$a_{v+1} = \left(\frac{f - a_v t_v}{t_{v+1}}\right)(P).$$

Then a_{v+1} belongs to $\tilde{F}_P$ and $\nu_P(f - a_v t_v - a_{v+1} t_{v+1}) \geq v + 2$.

Assume that we have obtained a sequence $\{a_r\}_{r=v}^{m}$ $(m > v)$ of elements of $\tilde{F}_P$ such that

$$\nu_P\left(f - \sum_{r=v}^{k} a_r t_r\right) \geq k + 1$$

for all $v \leq k \leq m$. Put

$$a_{m+1} = \left(\frac{f - \sum_{r=v}^{m} a_r t_r}{t_{m+1}} \right)(P).$$

Then $a_{m+1} \in \bar{F}_P$ and $\nu_P(f - \sum_{r=v}^{m+1} a_r t_r) \geq m+2$. In this way we continue our construction of the a_r. Then we obtain an infinite sequence $\{a_r\}_{r=v}^{\infty}$ of elements of $\bar{F}_P$ such that

$$\nu_P(f - \sum_{r=v}^{m} a_r t_r) \geq m + 1$$

for all $m \geq v$. We summarize the above construction in the formal expansion

$$f = \sum_{r=v}^{\infty} a_r t_r, \tag{1.1}$$

which is called the **local expansion** of f at P. A typical choice for the t_r is $t_r = t^r$ with t being a local parameter at P.

The above local expansion shows that for a given element $f \in F$ and a place P of F, there exists a sequence $\{f_n = \sum_{i=v}^{n} a_i t_i\}_{n=v}^{\infty}$ of special elements of F such that f_n tends to f at P, i.e., $\nu_P(f_n - f) \to \infty$ as $n \to \infty$. The following result shows approximation at several places.

Theorem 1.1.6 (Approximation Theorem) *Let S be a proper nonempty subset of $\mathbf{P}_F$ and $P_1, \ldots, P_r \in S$. Then for any given elements $x_1, \ldots, x_r \in F$ and integers $n_1, \ldots, n_r \in \mathbf{Z}$, there exists an element $x \in F$ such that $\nu_{P_i}(x - x_i) = n_i$ for all $i = 1, \ldots, r$ and $\nu_P(x) \geq 0$ for all $P \in S \setminus \{P_1, \ldots, P_r\}$.*

1.2 Divisor Class Groups and Ideal Class Groups

Throughout this section, the constant field k is finite. Thus, an algebraic function field F over k is now a **global function field**.

Denote the **divisor group** of F by $\mathrm{Div}(F)$, i.e., $\mathrm{Div}(F)$ is the free abelian group generated by all places of F:

$$\mathrm{Div}(F) = \left\{ \sum_{P \in S} m_P P : S \text{ is a finite subset of } \mathbf{P}_F \text{ and } m_P \in \mathbf{Z} \text{ for all } P \in S \right\}.$$

Let $\mathrm{Div}^0(F)$ be the subset of $\mathrm{Div}(F)$ consisting of all divisors of F of degree 0. Then $\mathrm{Div}^0(F)$ is a subgroup of $\mathrm{Div}(F)$. The group $\mathrm{Div}^0(F)$ is called the **divisor group of degree zero** of F.

Since $\deg(\mathrm{div}(x)) = 0$ for any $x \in F^*$, we have

$$\mathrm{Princ}(F) := \{\mathrm{div}(x) : x \in F^*\} \subseteq \mathrm{Div}^0(F).$$

Moreover, $\mathrm{Princ}(F)$ forms a subgroup of $\mathrm{Div}(F)$ which is called the **principal divisor group** of F. The factor group $\mathrm{Div}(F)/\mathrm{Princ}(F)$ is called the **divisor class group** of F.

The divisor class $[D] := D + \mathrm{Princ}(F)$ of $D \in \mathrm{Div}(F)$ consists of all divisors G of F that are **equivalent** to D, in the sense that $G = D + \mathrm{div}(x)$ for some $x \in F^*$. Now we consider the factor group

$$\mathrm{Div}^0(F)/\mathrm{Princ}(F).$$

It is a finite group and it is called the **divisor class group** (or the **group of divisor classes**) **of degree zero** of F, denoted by $\mathrm{Cl}(F)$. The cardinality of $\mathrm{Cl}(F)$ is called the **divisor class number** of F, denoted by $h(F)$.

Now choose a nonempty subset S of $\mathbf{P}_F$ with $S \neq \mathbf{P}_F$ and define the S-**integral ring** of F by

$$O_S = \{z \in F : \nu_P(z) \geq 0 \text{ for all } P \in S\}.$$

Then O_S is a Dedekind ring. In particular, we have $O_{\{P\}} = O_P$. Thus,

$$O_S = \bigcap_{P \in S} O_P.$$

A nonempty subset U of F is said to be a **fractional S-ideal** (or an S-**ideal**) of F if:

(i) $U \neq \{0\}$, and

(ii) U is an O_S-module, and

(iii) there exists an element $a \in F^*$ such that $aU \subseteq O_S$.

An S-ideal U is called **integral** if $U \subseteq O_S$. Thus, an integral S-ideal is an ordinary nonzero ideal of the ring O_S. It is trivial that F is not an S-ideal since $zF = F \neq O_S$ for any $z \in F^*$. We denote the set of all fractional S-ideals of F by $\mathrm{Fr}_S(F)$ or Fr_S.

For any S-ideals U and V of F, it is easy to check that

$$U + V = \{x + y : x \in U, \ y \in V\},$$

$$U \cdot V = \{\sum_{i=1}^{n} x_i y_i : x_i \in U, \ y_i \in V, \ n \in \mathbf{N}\},$$

and $U \cap V$ are S-ideals of F. Note that Fr_S forms an abelian group under the multiplication operation "$\cdot$". It is called the **fractional ideal group** of O_S or the **fractional S-ideal group** of F.

For any $z \in F^*$, the S-ideal zO_S is called a **principal S-ideal**. It is obvious that all principal S-ideals of F form a subgroup of Fr_S, denoted by $\mathrm{Princ}_S(F)$ or Princ_S. The factor group

$$\mathrm{Fr}_S/\mathrm{Princ}_S$$

is called the **fractional ideal class group** of O_S or the **fractional S-ideal class group** of F, and is denoted by $\mathrm{Cl}(O_S)$.

Proposition 1.2.1 *If $\mathbf{P}_F \setminus S$ is a finite nonempty set, then the fractional ideal class group $\mathrm{Cl}(O_S)$ is a finite abelian group.*

If $\mathbf{P}_F \setminus S$ is a finite nonempty set, we call the cardinality of $\mathrm{Cl}(O_S)$ the **fractional ideal class number** of O_S, denoted by $h(O_S)$.

Lemma 1.2.2 *For $P \in S$ and $U \in \mathrm{Fr}_S$, put*

$$\nu_P(U) = \min_{z \in U} \nu_P(z).$$

Then:

(i) $\nu_P(zO_S) = \nu_P(z)$ *for any* $z \in F^*$;

(ii) $\nu_P(U + V) = \min(\nu_P(U), \nu_P(V))$ *and* $\nu_P(UV) = \nu_P(U) + \nu_P(V)$ *for any two S-ideals* U *and* V;

(iii) U *is integral if and only if* $\nu_P(U) \geq 0$ *for all* $P \in S$.

Proof. (i) Since $1 \in O_S$, we have $\nu_P(zO_S) \leq \nu_P(z)$ by the definition. For any $zx \in zO_S$ with $x \in O_S$, we have $\nu_P(zx) = \nu_P(z) + \nu_P(x) \geq \nu_P(z)$, and so $\nu_P(zO_S) \geq \nu_P(z)$.

(ii) Note that

$$\begin{aligned}
\nu_P(U + V) &= \min_{x \in U,\, y \in V} \nu_P(x + y) \\
&= \min_{x \in U,\, y \in V} \min(\nu_P(x + 0), \nu_P(y + 0), \nu_P(x + y)) \\
&= \min_{x \in U,\, y \in V} \min(\nu_P(x), \nu_P(y)) \\
&= \min(\nu_P(U), \nu_P(V))
\end{aligned}$$

and

$$\begin{aligned}
\nu_P(UV) &= \min_{x_i \in U,\, y_i \in V} \nu_P\left(\sum x_i y_i\right) \\
&= \min_{x \in U,\, y \in V} \nu_P(xy) \\
&= \min_{x \in U} \nu_P(x) + \min_{y \in V} \nu_P(y) \\
&= \nu_P(U) + \nu_P(V).
\end{aligned}$$

(iii) This follows immediately from the definitions. □

For each $P \in S$, let us put

$$M_P(S) = \{z \in F : \nu_P(z) \geq 1,\ \nu_Q(z) \geq 0 \text{ for all } Q \in S \text{ with } Q \neq P\}.$$

Then $M_P(S)$ is called a **prime ideal** of O_S. It is clear that

$$M_P(S) = O_S \cap M_P \quad \text{and} \quad O_S / M_P(S) \simeq \tilde{F}_P.$$

Proposition 1.2.3 *Every S-ideal U has a unique decomposition into prime ideals*

$$U = \prod_{P \in S} M_P(S)^{\nu_P(U)}.$$

Its inverse is

$$U^{-1} = \prod_{P \in S} M_P(S)^{-\nu_P(U)}.$$

Corollary 1.2.4 (i) $zO_S = \prod_{P \in S} M_P(S)^{\nu_P(z)}$ *for any* $z \in F^*$.
(ii) $U + V = \prod_{P \in S} M_P(S)^{\min(\nu_P(U), \nu_P(V))}$ *for any two S-ideals* U *and* V.
(iii) $UV = \prod_{P \in S} M_P(S)^{\nu_P(U) + \nu_P(V)}$ *for any two S-ideals* U *and* V.

Proof. This follows directly from Lemma 1.2.2 and Proposition 1.2.3.   □

There is a close relationship between the divisor class group $\mathrm{Cl}(F)$ of degree zero of F and the fractional ideal class group $\mathrm{Cl}(O_S)$ of O_S. We present the result only for the case where $\mathbf{P}_F \setminus S$ consists of a single place (see Rosen [130] for a more general result).

Proposition 1.2.5 *If* $S = \mathbf{P}_F \setminus \{P\}$ *for some place* P, *then there exists an exact sequence*

$$0 \longrightarrow \mathrm{Cl}(F) \longrightarrow \mathrm{Cl}(O_S) \longrightarrow \mathbf{Z}/d\mathbf{Z} \longrightarrow 0,$$

where d *is the degree* $\deg(P)$ *of* P. *In particular,*

$$h(O_S) = dh(F).$$

Furthermore, $\mathrm{Cl}(F)$ *is isomorphic to* $\mathrm{Cl}(O_S)$ *if* P *is a rational place.*

Proof. Consider the homomorphism

$$\theta_1 : \mathrm{Div}^0(F) \longrightarrow \mathrm{Cl}(O_S), \quad \sum_Q m_Q Q \mapsto \left(\prod_{Q \neq P} M_Q(S)^{m_Q} \right) \mathrm{Princ}_S.$$

Then it is easy to check that $\ker(\theta_1) = \mathrm{Princ}(F)$, and so θ_1 induces an injective homomorphism

$$\theta : \mathrm{Cl}(F) \longrightarrow \mathrm{Cl}(O_S).$$

Define another homomorphism

$$\phi_1 : \mathrm{Fr}_S \longrightarrow \mathbf{Z}/d\mathbf{Z}, \quad \prod_{Q \neq P} M_Q(S)^{m_Q} \mapsto \sum_{Q \neq P} m_Q \deg(Q) + d\mathbf{Z}.$$

Note that ϕ_1 is surjective since the degree map $\mathrm{Div}(F) \longrightarrow \mathbf{Z}$ is surjective by [152, Corollary V.1.11]. Furthermore, it is easily seen that $\mathrm{Princ}_S \subseteq \ker(\phi_1)$, and so ϕ_1 induces a surjective homomorphism

$$\phi : \mathrm{Cl}(O_S) \longrightarrow \mathbf{Z}/d\mathbf{Z}.$$

It is straightforward to show that $\ker(\phi) = \mathrm{im}(\theta)$.   □

Let $D = \sum_P \nu_P(D)P$ be a positive divisor of F. If $x \in F^*$, we define

$$x \equiv 1 \pmod{D}$$

to mean that x satisfies the following condition:

if $P \in \mathrm{supp}(D)$, then x lies in the valuation ring O_P and $\nu_P(x - 1) \geq \nu_P(D)$.

Let S be a proper subset of $\mathbf{P}_F$ such that S contains $\operatorname{supp}(D)$ and $\mathbf{P}_F \setminus S$ is finite. Let $\operatorname{Fr}_{D,S}$ be the subgroup of Fr_S consisting of the S-ideals that are relatively prime to D, that is,

$$\operatorname{Fr}_{D,S} = \{\mathsf{U} \in \operatorname{Fr}_S : \nu_P(\mathsf{U}) = 0 \text{ for all } P \in \operatorname{supp}(\mathsf{D})\}.$$

Define the subgroup $\operatorname{Princ}_{D,S}$ of $\operatorname{Fr}_{D,S}$ by

$$\operatorname{Princ}_{D,S} = \{xO_S : x \in F^*, \ x \equiv 1 \,(\operatorname{mod} D)\}.$$

We also use $\operatorname{Princ}_D(O_S)$ to denote $\operatorname{Princ}_{D,S}$. The factor group

$$\operatorname{Fr}_{D,S}/\operatorname{Princ}_{D,S}$$

is called the S-**ray class group** modulo D. It is a finite group and denoted by $\operatorname{Cl}_D(O_S)$. If $D = 0$, then we obtain the fractional S-ideal class group $\operatorname{Cl}(O_S)$.

1.3   Algebraic Extensions and the Hurwitz Formula

Throughout this section, we assume that F'/k' and F/k are two algebraic function fields, that F'/F is a finite separable extension, and that $k' \supseteq k$. The constant field k is assumed to be finite, for simplicity, although the results hold also for more general k, e.g. for perfect fields k.

Let P be a place of F and P' a place of F' lying over P, that is, $P \subseteq P'$. We sometimes express this situation by $P'|P$. Choose a local parameter $t_P \in F$ at P, then the positive integer $\nu_{P'}(t_P)$ is called the **ramification index** of P' over P. It is independent of the choice of the local parameter t_P at P. We denote $\nu_{P'}(t_P)$ by $e_{P'}(F'/F)$ or $e(P'|P)$. We say that F'/F is **ramified** at P' if $e_{P'}(F'/F) > 1$ and **unramified** at P' if $e_{P'}(F'/F) = 1$. Alternatively, P' is called ramified, respectively unramified, in F'/F. Furthermore, we say that P' is **totally ramified** in F'/F if $e_{P'}(F'/F) = [F' : F]$. Let $\tilde{F}'_{P'}$ and $\tilde{F}_P$ be the residue class fields of P' and P, respectively. Then $\tilde{F}'_{P'}/\tilde{F}_P$ is a finite extension and the degree $[\tilde{F}'_{P'} : \tilde{F}_P]$ of this extension is called the **relative degree** of P' over P, denoted by $f_{P'}(F'/F)$ or $f(P'|P)$.

It follows from the definitions of ramification index and relative degree that we have the following tower formulas.

Proposition 1.3.1 *Suppose that $F \subseteq F' \subseteq F''$ is a field tower of finite separable extensions and that P, P', P'' are places of F, F', F'', respectively, with $P''|P'|P$. Then*

$$e_{P''}(F''/F) = e_{P''}(F''/F')e_{P'}(F'/F),$$

$$f_{P''}(F''/F) = f_{P''}(F''/F')f_{P'}(F'/F).$$

The following proposition expresses an important relationship between ramification indices, relative degrees, and the degree of the extension.

Proposition 1.3.2 *For a place P of F, let $P_1', \ldots, P_r'$ be all the distinct places of F' lying over P. Then*

$$\sum_{i=1}^{r} e(P_i'|P) f(P_i'|P) = [F' : F].$$

If $e(P'|P) = f(P'|P) = 1$ for all places $P'|P$, then we say that P **splits completely** in F'/F. Proposition 1.3.2 shows that in this case there are exactly $[F' : F]$ places of F' lying over P. If one of the places P_i' in Proposition 1.3.2 is totally ramified in F'/F, then it follows immediately from this proposition that we must have $r = 1$. Thus, in this situation it is meaningful to say that P is **totally ramified** in F'/F, or alternatively that F'/F is totally ramified at P.

A simple sufficient condition for totally ramified places is the Eisenstein criterion. A monic polynomial $f(x) = x^n + a_{n-1}x^{n-1} + \cdots + a_0 \in F[x]$ is called an **Eisenstein polynomial** at the place P if

$$\nu_P(a_0) = 1 \quad \text{and} \quad \nu_P(a_i) \geq 1 \quad \text{for } i = 1, \ldots, n-1.$$

Proposition 1.3.3 (Eisenstein Criterion) *Suppose that $f(x) \in F[x]$ is an Eisenstein polynomial at some place P of F. Then $f(x)$ is irreducible in $F[x]$. Moreover, if α is a root of $f(x)$ in some algebraic closure of F, then the place P is totally ramified in the simple extension $F(\alpha)/F$.*

In order to introduce the Hurwitz genus formula, we have to discuss the different exponent of a place in an extension.

First let us define norm and trace maps for a finite separable extension F'/F. For $\alpha \in F'$, we define the F-linear map

$$\phi_\alpha(z) = \alpha z$$

for all $z \in F'$. Let $\{\alpha_1, \ldots, \alpha_n\}$ be an ordered basis of F'/F and suppose that

$$(\alpha\alpha_1, \ldots, \alpha\alpha_n) = (\alpha_1, \ldots, \alpha_n)(a_{ij})$$

is the matrix representation of ϕ_α. Then we define the **norm** and the **trace** of α with respect to F'/F by

$$N_{F'/F}(\alpha) = \det(a_{ij}), \qquad \mathrm{Tr}_{F'/F}(\alpha) = \sum_{i=1}^{n} a_{ii},$$

respectively. If F'/F is a Galois extension and $\mathrm{Gal}(F'/F) = \{\sigma_1, \ldots, \sigma_n\}$, then

$$N_{F'/F}(\alpha) = \prod_{i=1}^{n} \sigma_i(\alpha), \qquad \mathrm{Tr}_{F'/F}(\alpha) = \sum_{i=1}^{n} \sigma_i(\alpha).$$

Let S be a nonempty proper subset of $\mathbf{P}_F$ and $\mathcal{T}$ the subset of $\mathbf{P}_{F'}$ that consists of all places of F' lying over places in S. Then $\mathcal{T}$ is called the **over-set** of S with respect to the extension F'/F. The integral closure of O_S in F' is given by

$$O_{\mathcal{T}} = \{z \in F' : \nu_{P'}(z) \geq 0 \text{ for all } P' \in \mathcal{T}\}.$$

Now let S consist of a single place $P \in \mathbf{P}_F$ and let T again be the over-set of S with respect to F'/F. A basis $\{\alpha_1, \ldots, \alpha_n\}$ of F'/F is called P-**integral** if

$$O_T = \alpha_1 O_P + \cdots + \alpha_n O_P.$$

For any basis $\{\alpha_1, \ldots, \alpha_n\}$ of F'/F we define the **discriminant**

$$\Delta(\alpha_1, \ldots, \alpha_n) = \det\left(\mathrm{Tr}_{F'/F}(\alpha_i \alpha_j)\right) = (\det(\sigma_i(\alpha_j)))^2,$$

where $\sigma_1, \ldots, \sigma_n$ are all F-embeddings of F' into an algebraic closure of F. Note that $\Delta(\alpha_1, \ldots, \alpha_n) \neq 0$ since the extension F'/F is separable.

Lemma 1.3.4 *Suppose that S is a nonempty proper subset of $\mathbf{P}_F$ and T the over-set of S with respect to F'/F. Define the complementary set of O_T by*

$$\mathrm{co}(O_T) = \{z \in F' : \mathrm{Tr}_{F'/F}(zO_T) \subseteq O_S\}.$$

Then:

(i) $\mathrm{co}(O_T)$ *is a T-ideal of F' containing O_T;*
(ii) $(\mathrm{co}(O_T))^{-1}$ *is an integral ideal of O_T.*

Definition 1.3.5 The **different** of O_T with respect to O_S is defined by

$$\mathsf{D}_S(F'/F) = (\mathrm{co}(O_T))^{-1}.$$

If S is a set consisting of one place P of F, we denote $\mathsf{D}_S(F'/F)$ simply by $\mathsf{D}_P(F'/F)$.

Proposition 1.3.6 *For a tower $F \subseteq F' \subseteq F''$ of finite separable extensions we have*

$$\mathsf{D}_S(F''/F) = \mathsf{D}_T(F''/F') \cdot \mathsf{D}_S(F'/F).$$

Definition 1.3.7 Let the place P' of F' lie over the place P of F. Then the **different exponent** of P' over P is defined by

$$d(P'|P) := \nu_{P'}(\mathsf{D}_P(F'/F)).$$

Proposition 1.3.8 *For any nonempty proper subset S of $\mathbf{P}_F$ we have*

$$\mathsf{D}_S(F'/F) = \prod_{P \in S} \prod_{P'|P} \mathsf{M}_{P'}(T)^{d(P'|P)}.$$

The different exponent $d(P'|P)$ is always a nonnegative integer and we have $d(P'|P) = 0$ for all but finitely many places P' of F'. Therefore the following definition is meaningful.

Definition 1.3.9 The **global different divisor** of F'/F is defined by

$$\text{Diff}(F'/F) = \sum_{P \in \mathbf{P}_F} \sum_{P'|P} d(P'|P)P'.$$

Observe that the global different divisor $\text{Diff}(F'/F)$ is a positive divisor of F'. We are now ready to state the Hurwitz genus formula.

Theorem 1.3.10 (Hurwitz Genus Formula) *Suppose that F'/k' is a finite separable extension of F/k. Then*

$$2g(F') - 2 = \frac{[F' : F]}{[k' : k]}(2g(F) - 2) + \deg(\text{Diff}(F'/F)),$$

where $g(F')$ and $g(F)$ are the genera of F' and F, respectively.

According to the Hurwitz genus formula, in order to calculate the genus of F' it is important to compute different exponents for all places of F'. The rest of this section is devoted to the investigation of different exponents.

First of all, we have the following tower formula for different exponents.

Proposition 1.3.11 *For a field tower $F \subseteq F' \subseteq F''$ and a corresponding place tower $P''|P'|P$ as in Proposition 1.3.1, we have*

$$d(P''|P) = e(P''|P')d(P'|P) + d(P''|P').$$

Proof. Let $\mathcal{T}$ be the set of places of F' lying over P. Then we have $\nu_{P''}(D_{\mathcal{T}}(F''/F')) = d(P''|P')$ by Proposition 1.3.8. By the tower formula for differents in Proposition 1.3.6, we get

$$\begin{aligned}
d(P''|P) &= \nu_{P''}(D_P(F''/F)) \\
&= \nu_{P''}(D_P(F'/F)) + \nu_{P''}(D_{\mathcal{T}}(F''/F')) \\
&= e(P''|P')d(P'|P) + d(P''|P'),
\end{aligned}$$

which is the desired result.   □

There is a close relationship between ramification index and different exponent that is given by the following proposition.

Proposition 1.3.12 *For places $P'|P$ we have:*

(i) $d(P'|P) \geq e(P'|P) - 1$;

(ii) $d(P'|P) = e(P'|P) - 1$ *if and only if $e(P'|P)$ is relatively prime to the characteristic of k.*

Note that Proposition 1.3.12(ii) shows, in particular, that $d(P'|P) = 0$ whenever P' is unramified in F'/F. We say that the extension F'/F is **unramified** if all places of F'

are unramified in F'/F. In this case we have $\mathrm{Diff}(F'/F) = 0$, and so the Hurwitz genus formula attains a particularly simple form.

Proposition 1.3.12(ii) also suggests the following definitions. If P' is a place of F' lying over the place P of F, then we say that P' is **tamely ramified** in F'/F if $e(P'|P) > 1$ and $e(P'|P)$ is not divisible by the characteristic of k. Furthermore, P' is called **wildly ramified** in F'/F if $e(P'|P)$ is divisible by the characteristic of k.

Proposition 1.3.3 guarantees that a finite extension $F' = F(\alpha)$ of F is totally ramified at a place P if α satisfies an Eisenstein polynomial at P. The following proposition provides an efficient method to compute the different exponents of totally ramified places in simple extensions.

Proposition 1.3.13 *Let $F' = F(\alpha)$ be a finite, simple, separable extension of F and let P' be a place of F' lying over a place P of F. Suppose that P' is totally ramified in F'/F and α is a local parameter at P'. Then*

$$d(P'|P) = \nu_{P'}(f'(\alpha)),$$

where $f(x) \in F[x]$ is the minimal polynomial of α over F and $f'(x)$ is the derivative of $f(x)$.

1.4 Ramification Theory of Galois Extensions

Throughout this section, we always assume that F'/k' is a finite Galois extension of F/k with $k' \supseteq k$, and we denote by $\mathrm{Gal}(F'/F)$ the Galois group of F'/F. The constant field k is again assumed to be finite, for simplicity.

For any automorphism $\sigma \in \mathrm{Gal}(F'/F)$ and a place P' of F', let P be the place of F lying under P'. Consider the set

$$\sigma(P') := \{\sigma(x) : x \in P'\} = \{\sigma(x) : x \in \mathsf{M}_{P'}\}.$$

Then it is easy to verify that $\sigma(P')$ is also a place of F' lying over P. The place $\sigma(P')$ is called a **conjugate place** of P'. It is also easy to show that for any $z \in F'$ we have

$$\nu_{\sigma(P')}(\sigma(z)) = \nu_{P'}(z). \tag{1.2}$$

We have seen that σ permutes all places of F' lying over P. Furthermore, $\mathrm{Gal}(F'/F)$ acts transitively on the set of places of F' lying over P, as we prove in the following proposition.

Proposition 1.4.1 *Let Q_1 and Q_2 be two places of F' lying over a place P of F. Then there exists an automorphism $\sigma \in \mathrm{Gal}(F'/F)$ such that $\sigma(Q_1) = Q_2$.*

Proof. Suppose that $\sigma(Q_1) \neq Q_2$ for all $\sigma \in G := \mathrm{Gal}(F'/F)$. It follows from the approximation theorem (see Theorem 1.1.6) that there exists an element $x \in F'$ such that $\nu_{\sigma(Q_1)}(x) = 0$ and $\nu_{\sigma(Q_2)}(x) > 0$ for all $\sigma \in G$. Then for $i = 1, 2$ we get

$$
\begin{aligned}
\nu_P(\mathrm{N}_{F'/F}(x)) &= \frac{1}{e(Q_i|P)}\nu_{Q_i}(\mathrm{N}_{F'/F}(x)) \\
&= \frac{1}{e(Q_i|P)}\sum_{\sigma\in G}\nu_{Q_i}(\sigma(x)) \\
&= \frac{1}{e(Q_i|P)}\sum_{\sigma\in G}\nu_{Q_i}(\sigma^{-1}(x)) \\
&= \frac{1}{e(Q_i|P)}\sum_{\sigma\in G}\nu_{\sigma(Q_i)}(x),
\end{aligned}
$$

where we used (1.2) in the final step. The last expression is 0 for $i = 1$ and positive for $i = 2$. This contradiction proves the proposition. $\qquad\square$

From the above transitivity, we can obtain some very important identities for ramification indices, relative degrees, and different exponents among the places of F' lying over a fixed place of F.

Theorem 1.4.2 *Suppose that F'/F is a finite Galois extension. Let P be a place of F and $Q_1,\ldots,Q_r$ all places of F' lying over P. Then for $1\le i,j\le r$ we have*

$$
e(Q_i|P) = e(Q_j|P), \quad f(Q_i|P) = f(Q_j|P), \quad d(Q_i|P) = d(Q_j|P).
$$

Proof. Fix i and j. Let $t \in F$ be a local parameter at P and $\sigma \in \mathrm{Gal}(F'/F)$ with $\sigma(Q_i) = Q_j$. Then

$$
e(Q_i|P) = \nu_{Q_i}(t) = \nu_{\sigma(Q_i)}(\sigma(t)) = \nu_{Q_j}(t) = e(Q_j|P).
$$

By the definition of $\sigma(Q_i)$ we have

$$
O_{\sigma(Q_i)} = \sigma(O_{Q_i}), \quad M_{\sigma(Q_i)} = \sigma(M_{Q_i}).
$$

Hence

$$
O_{Q_j}/M_{Q_j} = O_{\sigma(Q_i)}/M_{\sigma(Q_i)} = \sigma(O_{Q_i})/\sigma(M_{Q_i}) \simeq O_{Q_i}/M_{Q_i}.
$$

This implies $f(Q_i|P) = f(Q_j|P)$.

Let $\mathcal{T} = \{Q_1,\ldots,Q_r\}$ be the set of places of F' lying over P. Then $\mathsf{D}_P(F'/F)$ has the following decomposition in $O_{\mathcal{T}}$:

$$
\mathsf{D}_P(F'/F) = \prod_{h=1}^{r} \mathsf{M}_{Q_h}(\mathcal{T})^{d(Q_h|P)}.
$$

Thus,

$$
\sigma\left(\mathsf{D}_P(F'/F)\right) = \prod_{h=1}^{r} \sigma(\mathsf{M}_{Q_h}(\mathcal{T}))^{d(Q_h|P)} = \prod_{h=1}^{r} \mathsf{M}_{\sigma(Q_h)}(\mathcal{T})^{d(Q_h|P)}.
$$

By the definition of complementary sets, we immediately see that

$$
\sigma\left(\mathsf{D}_P(F'/F)\right) = \mathsf{D}_P(F'/F).
$$

From the uniqueness of decomposition of ideals we obtain

$$d(Q_j|P) = d(\sigma(Q_i)|P)) = d(Q_i|P),$$

and so all identities in the theorem are shown.   □

Since $e(Q_i|P), f(Q_i|P)$, and $d(Q_i|P)$ are independent of Q_i by the above theorem, we also denote them by $e_P(F'/F)$, $f_P(F'/F)$, and $d_P(F'/F)$, respectively. Then Proposition 1.3.2 can be written as follows.

Corollary 1.4.3 *Suppose that F'/F is a finite Galois extension and that there are exactly r places of F' lying over $P \in \mathbf{P}_F$. Then*

$$e_P(F'/F)f_P(F'/F)r = [F' : F].$$

In particular, both $e_P(F'/F)$ and $f_P(F'/F)$ divide $[F' : F]$.

If $e_P(F'/F) > 1$, we say also that F'/F is **ramified** at P (or P is ramified in F'/F), and if $e_P(F'/F) = 1$, we say also that F'/F is **unramified** at P (or P is unramified in F'/F). Furthermore, if $e_P(F'/F) > 1$, then P is called **tamely ramified** in F'/F provided that $e_P(F'/F)$ is not divisible by the characteristic of k, and P is called **wildly ramified** in F'/F if $e_P(F'/F)$ is divisible by the characteristic of k.

Let Q be a place of F' lying over $P \in \mathbf{P}_F$. For every integer $i \geq -1$, we define the ith **ramification group** by

$$
\begin{aligned}
G_i(Q|P) = G_i(Q, F'/F) &= \{\sigma \in \mathrm{Gal}(F'/F) : \nu_Q(\sigma(x) - x) \geq i + 1 \text{ for all } x \in O_Q\} \\
&= \{\sigma \in \mathrm{Gal}(F'/F) : (\sigma - \mathrm{id})(O_Q) \subseteq \mathsf{M}_Q^{i+1}\}.
\end{aligned}
$$

The subfield F_i of F'/F fixed by $G_i(Q|P)$ is called the ith **ramification field**.

The groups $G_{-1}(Q|P)$ and $G_0(Q|P)$ are particularly interesting; they are called the **decomposition group** and the **inertia group** of Q over P, respectively. The groups $G_{-1}(Q|P)$ and $G_0(Q|P)$ are also denoted by $G_Z(Q|P)$ and $G_T(Q|P)$, respectively. The corresponding fields F_{-1} and F_0 are called the **decomposition field** and the **inertia field** of Q over P, respectively. Note that the decomposition group $G_Z(Q|P)$ can be described also as

$$G_Z(Q|P) = \{\sigma \in \mathrm{Gal}(F'/F) : \sigma(Q) = Q\}.$$

Proposition 1.4.4 *If Q_1 and Q_2 are two places of F' lying over $P \in \mathbf{P}_F$, then $G_i(Q_1|P)$ and $G_i(Q_2|P)$ are conjugate for all $i \geq -1$. More precisely,*

$$G_i(\sigma(Q)|P) = \sigma G_i(Q|P)\sigma^{-1}$$

for any $\sigma \in \mathrm{Gal}(F'/F)$ and $Q|P$.

Proof. Note that

$$
\begin{aligned}
\tau \in G_i(\sigma(Q)|P) &\iff (\tau - \mathrm{id})\left(O_{\sigma(Q)}\right) \subseteq \mathsf{M}_{\sigma(Q)}^{i+1} \\
&\iff (\tau - \mathrm{id})\sigma\left(O_Q\right) \subseteq \sigma\left(\mathsf{M}_Q^{i+1}\right) \\
&\iff (\sigma^{-1}\tau\sigma - \mathrm{id})\left(O_Q\right) \subseteq \mathsf{M}_Q^{i+1} \\
&\iff \sigma^{-1}\tau\sigma \in G_i(Q|P) \\
&\iff \tau \in \sigma G_i(Q|P)\sigma^{-1}.
\end{aligned}
$$

$\square$

Proposition 1.4.5 *For a place Q of F' lying over $P \in \mathbf{P}_F$, let Z be the decomposition field of Q over P. Let $R \in \mathbf{P}_Z$ be the place lying under Q. Then we have:*
(i) $[F' : Z] = |G_Z(Q|P)| = e(Q|P)f(Q|P)$;
(ii) $e(R|P) = f(R|P) = 1$, *that is, P splits completely in Z/F;*
(iii) *Q is the unique place of F' lying over R.*

Proof. (i) Since $\mathrm{Gal}(F'/F)$ acts transitively on the set of places of F' lying over P, we can find $\sigma_1,\ldots,\sigma_r \in \mathrm{Gal}(F'/F)$ such that $\{\sigma_1(Q),\ldots,\sigma_r(Q)\}$ is the set of all places of F' lying over P. By the definition, $\sigma(Q) = Q$ for all $\sigma \in G_Z(Q|P)$. Hence $\sigma_1,\ldots,\sigma_r$ are in distinct left cosets of $G_Z(Q|P)$. For any $\tau \in \mathrm{Gal}(F'/F)$, we have $\tau(Q) = \sigma_i(Q)$ for some $1 \le i \le r$. Then $\tau^{-1}\sigma_i(Q) = Q$. It follows from the definition of the decomposition group that $\tau^{-1}\sigma_i$ is an element of $G_Z(Q|P)$, i.e., τ and σ_i are in the same left coset of $G_Z(Q|P)$. Thus, we have proved that

$$\frac{|\mathrm{Gal}(F'/F)|}{|G_Z(Q|P)|} = r,$$

that is,

$$|G_Z(Q|P)| = \frac{[F' : F]}{r} = e(Q|P)f(Q|P).$$

(ii) The decomposition group of Q over R is obviously equal to $\mathrm{Gal}(F'/Z)$. By (i) we have

$$e(Q|R)f(Q|R) = e(Q|P)f(Q|P).$$

Hence

$$e(R|P)f(R|P) = \frac{e(Q|P)}{e(Q|R)} \cdot \frac{f(Q|P)}{f(Q|R)} = 1.$$

This is equivalent to $e(R|P) = f(R|P) = 1$.
(iii) This is a direct consequence of (i) and (ii). $\square$

We recall that the constant field k is finite. Note that then for any place P of F, any finite extension of the residue class field of P is a Galois extension.

By the proof of Theorem 1.4.2, each $\sigma \in \mathrm{Gal}(F'/F)$ provides an isomorphism

$$O_Q/\mathsf{M}_Q \simeq O_{\sigma(Q)}/\mathsf{M}_{\sigma(Q)}$$

for any place Q of F'. More precisely, we have an isomorphism

$$\bar{\sigma}: \quad \tilde{F}'_Q \longrightarrow \tilde{F}'_{\sigma(Q)}, \quad \bar{\sigma}(\bar{z}) = \overline{\sigma(z)} \quad \text{for all } z \in O_Q.$$

Of course, $\bar{\sigma}$ leaves $\tilde{F}_P$ pointwise fixed, so that $\sigma \mapsto \bar{\sigma}$ is a map from $\mathrm{Gal}(F'/F)$ to $\mathrm{Gal}(\tilde{F}'_Q/\tilde{F}_P)$ if we identify $\tilde{F}'_Q$ with $\tilde{F}'_{\sigma(Q)}$. It is easily seen that this map is a group homomorphism.

Proposition 1.4.6 *Suppose that Q is a place of F' lying over $P \in \mathbf{P}_F$ and Z is the decomposition field of Q over P. Then the group homomorphism*

$$\mathrm{Gal}(F'/Z) \longrightarrow \mathrm{Gal}(\tilde{F}'_Q/\tilde{F}_P), \quad \sigma \mapsto \bar{\sigma},$$

is surjective. Furthermore, $G_T(Q|P)$ is the kernel of this homomorphism and thus a normal subgroup of $G_Z(Q|P)$.

Proof. Let $z \in O_Q$ satisfy $\tilde{F}'_Q = \tilde{F}_P(\bar{z})$. Let $R \in \mathbf{P}_Z$ be the place lying under Q. Then O_Q is the integral closure of O_R in F' since Q is the only place of F' lying over R by Proposition 1.4.5. Thus, there exists a monic polynomial $f(x) \in O_R[x]$ such that $f(x)$ is the minimal polynomial of z over Z.

Any $\mu \in \mathrm{Gal}(\tilde{F}'_Q/\tilde{F}_P)$ is uniquely determined by $\mu(\bar{z})$. As $\mu(\bar{z})$ is a root of $\overline{f}(x) \in \tilde{Z}_R[x] = \tilde{F}_P[x]$, there exists a $\sigma \in \mathrm{Gal}(F'/Z)$ such that $\overline{\sigma(z)} = \mu(\bar{z})$, i.e., $\bar{\sigma}(\bar{z}) = \mu(\bar{z})$, which implies $\bar{\sigma} = \mu$. This shows that our homomorphism is surjective.

Furthermore, for $\sigma \in \mathrm{Gal}(F'/Z)$ we have

$$\bar{\sigma} = \overline{\mathrm{id}} \Longleftrightarrow \bar{\sigma}(\bar{z}) = \bar{z} \quad \text{for all } z \in O_Q$$
$$\Longleftrightarrow \sigma(z) - z \in M_Q \quad \text{for all } z \in O_Q$$
$$\Longleftrightarrow (\sigma - \mathrm{id})(O_Q) \subseteq M_Q \Longleftrightarrow \sigma \in G_T(Q|P).$$

This shows that $G_T(Q|P)$ is the kernel of this homomorphism, so $G_T(Q|P)$ is a normal subgroup of $\mathrm{Gal}(F'/Z) = G_Z(Q|P)$. $\qquad\qquad\square$

Theorem 1.4.7 *Suppose that F'/F is a finite Galois extension and let Q be a place of F' lying over $P \in \mathbf{P}_F$. Let Z be the decomposition field and T the inertia field of Q over P. If $Q_Z \in \mathbf{P}_Z$ and $Q_T \in \mathbf{P}_T$ are two places lying under Q, then:*

(i) $[F' : T] = e(Q|P)$, $[T : Z] = f(Q|P)$.

(ii) $e(Q|Q_T) = e(Q|P) = [F' : T]$, $f(Q|Q_T) = 1$.

(iii) $f(Q_T|Q_Z) = f(Q|P) = [T : Z]$, $e(Q_T|Q_Z) = 1$.

(iv) $G_{-1}(Q|P) \supseteq G_0(Q|P) \supseteq G_1(Q|P) \supseteq \ldots$ *and* $G_m(Q|P) = \{\mathrm{id}\}$ *for m sufficiently large.*

(v) $G_1(Q|P)$ *is a normal subgroup of* $G_0(Q|P)$. *The order of $G_1(Q|P)$ is a power of p and the factor group $G_0(Q|P)/G_1(Q|P)$ is cyclic of order relatively prime to p, where p is the characteristic of k.*

(vi) $G_{i+1}(Q|P)$ *is a normal subgroup of $G_i(Q|P)$ for all $i \geq 1$ and $G_i(Q|P)/G_{i+1}(Q|P)$ is an elementary abelian p-group.*

Corollary 1.4.8 *Let K be an intermediate field of F'/F. Let Q be a place of F' lying over $P \in \mathbf{P}_F$ and let $R \in \mathbf{P}_K$ be the place lying under Q. Denote by T the inertia field and by Z the decomposition field of Q over P. Then we have:*

(i) *P is unramified in K/F if and only if $K \subseteq T$;*

(ii) *R is totally ramified in F'/K if and only if $K \supseteq T$;*

(iii) *P splits completely in K/F if and only if $K \subseteq Z$;*

(iv) *Q is the only place of F' lying over R if and only if $K \supseteq Z$.*

Proof. We prove only (i). The proofs of the other parts are similar. If $K \subseteq T$, then $e(R|P)$ is a divisor of $e_P(T/F) = 1$. Hence $e(R|P)$ is also equal to 1.

Conversely, if $e(R|P) = 1$, then $e(Q|P) = e(Q|R)e(R|P) = e(Q|R)$. From the definitions it is obvious that $G_T(Q|R) \subseteq G_T(Q|P)$, and since $e(Q|R) = e(Q|P)$, Theorem 1.4.7(i) shows that $G_T(Q|R) = G_T(Q|P)$. Thus,

$$\mathrm{Gal}(F'/T) = G_T(Q|P) = G_T(Q|R) \subseteq \mathrm{Gal}(F'/K),$$

and so $K \subseteq T$. $\qquad\qquad\square$

Corollary 1.4.9 *Suppose that K_1/F and K_2/F are two finite separable extensions and K is the composite field of K_1 and K_2. Then for a place $P \in \mathbf{P}_F$ we have:*

(i) *P splits completely in K/F if and only if P splits completely in both K_1/F and K_2/F;*

(ii) *P is unramified in K/F if and only if P is unramified in both K_1/F and K_2/F.*

The different exponent at a place is determined by the orders of its ramification groups.

Theorem 1.4.10 (Hilbert Different Formula) *Let F'/F be a finite Galois extension and $G_i(Q|P)$ the ith ramification group of a place Q of F' over $P \in \mathbf{P}_F$. Then the different exponent of Q over P is given by*

$$d(Q|P) = \sum_{i=0}^{\infty}(|G_i(Q|P)| - 1).$$

Suppose that $P \in \mathbf{P}_F$ is unramified in F'/F and Q is a place of F' lying over P. Then we know by Proposition 1.4.6 that there is an isomorphism

$$\mathrm{Gal}(F'/Z) \simeq \mathrm{Gal}(\tilde{F}'_Q/\tilde{F}_P).$$

The residue class fields $\tilde{F}_P$ and $\tilde{F}'_Q$ are both finite fields by assumption, so $\mathrm{Gal}(\tilde{F}'_Q/\tilde{F}_P)$ is a cyclic group. Thus, there exists a unique $\sigma \in \mathrm{Gal}(F'/Z)$ such that $\overline{\sigma}$ is the canonical generator of $\mathrm{Gal}(\tilde{F}'_Q/\tilde{F}_P)$, in other words, $\overline{\sigma} : a \mapsto a^r$ for any $a \in \tilde{F}'_Q$ if $|\tilde{F}_P| = r$. Obviously, σ satisfies the property that

$$\sigma(z) \equiv z^r \pmod{\mathsf{M}_Q} \qquad \text{for all } z \in O_Q. \qquad (1.3)$$

We denote this unique σ by $\left[\frac{F'/F}{Q}\right]$ and call it the **Frobenius automorphism** or **Frobenius symbol** of Q over P. Clearly, $\left[\frac{F'/F}{Q}\right]$ is characterized by (1.3). The Frobenius symbol has the following properties.

Theorem 1.4.11 *Suppose that the finite Galois extension F'/F is unramified at $P \in \mathbf{P}_F$ and let Q be a place of F' lying over P. Assume that K is an intermediate field of F'/F and $R \in \mathbf{P}_K$ is the place lying under Q. Then we have:*

(i) *For any $\tau \in \mathrm{Gal}(F'/F)$,*

$$\left[\frac{F'/F}{\tau(Q)}\right] = \tau \left[\frac{F'/F}{Q}\right] \tau^{-1}.$$

(ii)

$$\left[\frac{F'/K}{Q}\right] = \left[\frac{F'/F}{Q}\right]^{f(R|P)}.$$

(iii) *If K/F is a Galois extension, then the restriction of $\left[\frac{F'/F}{Q}\right]$ to K is equal to $\left[\frac{K/F}{R}\right]$.*

If F'/F is an abelian extension, then it follows from Theorem 1.4.11(i) that the Frobenius symbol $\left[\frac{F'/F}{Q}\right]$ does not depend on Q, but only on the place P of F lying under Q. It is thus unambiguous to write the Frobenius symbol as $\left[\frac{F'/F}{P}\right]$, which is called the **Artin symbol** of P in F'/F.

Proposition 1.4.12 *Let F'/F be a finite abelian extension and let E be a subfield of F'/F. Suppose that F'/F is unramified at $P \in \mathbf{P}_F$. Then P splits completely in E/F if and only if the Artin symbol $\left[\frac{F'/F}{P}\right]$ belongs to $\mathrm{Gal}(F'/E)$.*

1.5   Constant Field Extensions

Throughout this section, we assume that:

(i) $F/\mathbf{F}_q$ is an algebraic function field with full constant field $\mathbf{F}_q$;

(ii) $F' = F \cdot \mathbf{F}_{q^n}$ is a constant field extension of F, where n is a positive integer.

Let $\mathbf{F}_{q^n} = \mathbf{F}_q(\alpha)$, then we have $F' = F(\alpha)$. The Galois group $\mathrm{Gal}(\mathbf{F}_{q^n}/\mathbf{F}_q)$ is a cyclic group with a canonical generator

$$\sigma : \beta \mapsto \beta^q \qquad \text{for any } \beta \in \mathbf{F}_{q^n}.$$

Lemma 1.5.1 *With the notation above, we have:*

(i) F'/F *is a cyclic extension with $[F' : F] = n$ and*

$$\mathrm{Gal}(F'/F) \simeq \mathrm{Gal}(\mathbf{F}_{q^n}/\mathbf{F}_q).$$

(ii) *The full constant field of F' is $\mathbf{F}_{q^n}$.*

(iii) *Let $\{\alpha_1, \ldots, \alpha_n\}$ be a basis of $\mathbf{F}_{q^n}$ over $\mathbf{F}_q$. Then for any $P \in \mathbf{P}_F$, $\{\alpha_1, \ldots, \alpha_n\}$ is a P-integral basis of F'/F.*

Proof. (i) Let $\mathbf{F}_{q^n} = \mathbf{F}_q(\alpha)$ and $f(t) \in \mathbf{F}_q[t]$ be the minimal polynomial of α over $\mathbf{F}_q$. We will prove that f is irreducible over F. Suppose, on the contrary, that f has a factorization

$f = gh$ over F with g and h monic and $\deg(g) \geq 1$, $\deg(h) \geq 1$. It is obvious that all roots of g and h are elements of $\mathbf{F}_{q^n}$. Hence from the fact that the coefficients of a monic polynomial are polynomial expressions of its roots, it follows that g and h are polynomials in $\mathbf{F}_{q^n}[t]$. In particular, all coefficients of g and h are algebraic over $\mathbf{F}_q$. Thus, each coefficient of g and h is an element of $\mathbf{F}_q$ since $\mathbf{F}_q$ is algebraically closed in F, and we obtain a contradiction to f being irreducible over $\mathbf{F}_q$. This shows that $[F' : F] = n$ and $\mathrm{Gal}(F'/F) \simeq \mathrm{Gal}(\mathbf{F}_{q^n}/\mathbf{F}_q)$.

(ii) It is trivial that the full constant field of F' contains $\mathbf{F}_{q^n}$ since $\mathbf{F}_{q^n}/\mathbf{F}_q$ is a finite extension. Let $z \in F'$ be an algebraic element over $\mathbf{F}_{q^n}$, then $\mathbf{F}_{q^n}(z)/\mathbf{F}_q$ is a finite extension. From (i) we get

$$n = [F' : F] = [F \cdot \mathbf{F}_{q^n}(z) : F] = [\mathbf{F}_{q^n}(z) : \mathbf{F}_q].$$

Hence $\mathbf{F}_{q^n} = \mathbf{F}_{q^n}(z)$, i.e., $z \in \mathbf{F}_{q^n}$. This means that $\mathbf{F}_{q^n}$ is the full constant field of F'.

(iii) It is obvious that $\alpha_1, \ldots, \alpha_n$ are elements of $O_{\mathcal{T}}$, where $\mathcal{T}$ is the over-set of $S = \{P\}$ with respect to F'/F. The discriminant $\Delta(\alpha_1, \ldots, \alpha_n)$ of $\{\alpha_1, \ldots, \alpha_n\}$ is an element of $\mathbf{F}_q^*$. Thus, we have

$$\nu_P(\Delta(\alpha_1, \ldots, \alpha_n)) = 0.$$

The desired conclusion follows now from [173, Proposition 4-8-8]. $\qquad\square$

By Lemma 1.5.1, we can identify $\mathrm{Gal}(F'/F)$ with $\mathrm{Gal}(\mathbf{F}_{q^n}/\mathbf{F}_q)$ in the following way. Let $\{\alpha_1 = 1, \alpha_2, \ldots, \alpha_n\}$ be a basis of $\mathbf{F}_{q^n}$ over $\mathbf{F}_q$. Then for any $\sigma \in \mathrm{Gal}(\mathbf{F}_{q^n}/\mathbf{F}_q)$ and $x = \sum_{i=1}^n \alpha_i x_i \in F'$ with all $x_i \in F$, put

$$\sigma(x) = \sum_{i=1}^n \sigma(\alpha_i) x_i.$$

Then σ is a Galois automorphism of F'/F and all elements of $\mathrm{Gal}(F'/F)$ are obtained in this way.

Theorem 1.5.2 *For a constant field extension $F' = F \cdot \mathbf{F}_{q^n}$ of F we have:*

(i) *F'/F is unramified at all places of F.*

(ii) *$F'/\mathbf{F}_{q^n}$ has the same genus as $F/\mathbf{F}_q$.*

(iii) *For a place $P \in \mathbf{P}_F$ and a place Q of F' lying over P, the following holds:*

(a) *$\deg(Q) = d/\gcd(d,n)$, where $d = \deg(P)$ is the degree of P;*

(b) *$f(Q|P) = n/\gcd(d,n)$;*

(c) *there are exactly $\gcd(d,n)$ places of F' lying over P.*

Proof. (i) Let P be a place of F and $\{\alpha_1, \ldots, \alpha_n\}$ a basis of $\mathbf{F}_{q^n}$ over $\mathbf{F}_q$. Then by Lemma 1.5.1(iii) and its proof, $\nu_P(\Delta(\alpha_1, \ldots, \alpha_n)) = 0$ and $\{\alpha_1, \ldots, \alpha_n\}$ is a P-integral basis of F'/F. It follows now from Dedekind's discriminant theorem (see [173, Theorem 4-8-14]) that P is unramified in F'/F.

(ii) This is an immediate consequence of (i) and the Hurwitz genus formula.

(iii) We will first prove that $\tilde{F}_Q'$ is (up to an isomorphism) the composite field $\tilde{F}_P \cdot \mathbf{F}_{q^n}$. It is obvious that $\tilde{F}_P \cdot \mathbf{F}_{q^n}$ can be viewed as a subfield of $\tilde{F}_Q'$.

Now let $\mathcal{T}$ be the over-set of $\{P\}$ with respect to F'/F. Let $\{\alpha_1, \ldots, \alpha_n\}$ be a basis of $\mathbf{F}_{q^n}$ over $\mathbf{F}_q$. For any element $x(Q) \in \tilde{F}'_Q$ with some $x \in O_\mathcal{T}$ (by the approximation theorem), there exist n elements $x_1, \ldots, x_n \in O_P$ such that

$$x = \sum_{i=1}^{n} x_i \alpha_i$$

since $\{\alpha_1, \ldots, \alpha_n\}$ is a P-integral basis of F'/F. Hence

$$x(Q) = \sum_{i=1}^{n} x_i(Q)\alpha_i = \sum_{i=1}^{n} x_i(P)\alpha_i \in \tilde{F}_P \cdot \mathbf{F}_{q^n},$$

showing that $\tilde{F}'_Q = \tilde{F}_P \cdot \mathbf{F}_{q^n}$. This yields

$$\deg(Q) = [\tilde{F}'_Q : \mathbf{F}_{q^n}] = [\tilde{F}_P \cdot \mathbf{F}_{q^n} : \mathbf{F}_{q^n}] = [\mathbf{F}_{q^d} \cdot \mathbf{F}_{q^n} : \mathbf{F}_{q^n}] = \frac{d}{\gcd(d, n)}.$$

Furthermore, we have

$$f(Q|P) = [\tilde{F}'_Q : \tilde{F}_P] = [\tilde{F}_P \cdot \mathbf{F}_{q^n} : \tilde{F}_P] = [\mathbf{F}_{q^d} \cdot \mathbf{F}_{q^n} : \mathbf{F}_{q^d}] = \frac{n}{\gcd(d, n)}.$$

Since $e(Q|P) = 1$ by (i), it is now clear that there are exactly $\gcd(d, n)$ places of F' lying over P. $\square$

It follows from Theorem 1.5.2 that for a place $P \in \mathbf{P}_F$ and a place Q of F' with $Q|P$, the decomposition group $G_Z(Q|P)$ is isomorphic to $\mathrm{Gal}(\mathbf{F}_{q^d} \cdot \mathbf{F}_{q^n}/\mathbf{F}_{q^d})$ and the ith ramification group of Q over P is trivial for any $i \geq 0$.

Let S be a proper subset of $\mathbf{P}_F$ containing P and $\mathcal{T}$ the over-set of S with respect to F'/F. Then we have the prime-ideal decomposition

$$\mathsf{M}_P(S)O_\mathcal{T} = \prod_{i=1}^{r} \mathsf{M}_{Q_i}(\mathcal{T}),$$

where $r = \gcd(\deg(P), n)$ and $\{Q_1, \ldots, Q_r\}$ is the over-set of $\{P\}$ with respect to F'/F. We express this fact by the map

$$i_{F'/F}: \quad P \mapsto \sum_{i=1}^{r} Q_i.$$

Extend $i_{F'/F}$ to a map from the divisor group $\mathrm{Div}(F)$ of F to the divisor group $\mathrm{Div}(F')$ of F' by letting

$$\sum_P m_P P \mapsto \sum_P m_P i_{F'/F}(P) = \sum_P \sum_{Q|P} m_P Q.$$

Then $i_{F'/F}$ is a group homomorphism from $\mathrm{Div}(F)$ to $\mathrm{Div}(F')$. Moreover,

$$\deg\left(i_{F'/F}(\sum_P m_P P)\right) = \deg\left(\sum_P \sum_{Q|P} m_P Q\right)$$

$$= \sum_P \sum_{Q|P} m_P \deg(Q)$$

$$= \sum_P m_P \gcd(\deg(P), n) \cdot \frac{\deg(P)}{\gcd(\deg(P), n)}$$

$$= \sum_P m_P \deg(P) = \deg(\sum_P m_P P).$$

Hence if $i_{F'/F}$ is restricted to the subgroup $\mathrm{Div}^0(F)$, then the image $i_{F'/F}(\mathrm{Div}^0(F))$ is a subgroup of $\mathrm{Div}^0(F')$. We still denote by $i_{F'/F}$ the restricted homomorphism from $\mathrm{Div}^0(F)$ to $\mathrm{Div}^0(F')$.

Theorem 1.5.3 *The map $i_{F'/F}$ on $\mathrm{Div}^0(F)$ induces an injective homomorphism from the group $\mathrm{Cl}(F)$ of divisor classes of degree zero of F to the group $\mathrm{Cl}(F')$ of divisor classes of degree zero of F' by*

$$\sum_P m_P P + \mathrm{Princ}(F) \in \mathrm{Cl}(F) \mapsto \sum_P \sum_{Q|P} m_P Q + \mathrm{Princ}(F') \in \mathrm{Cl}(F').$$

Proof. Compare with [152, Theorem III.6.3(f)]. □

Let D be a positive divisor of F and $\mathcal{S}$ a proper subset of $\mathbf{P}_F$ such that $\mathcal{S}$ contains $\mathrm{supp}(D)$ and $\mathbf{P}_F \setminus \mathcal{S}$ is finite. Let $\mathcal{T}$ be the over-set of $\mathcal{S}$ with respect to F'/F. Then $i_{F'/F}$ also induces a map from $\mathrm{Fr}_{D,\mathcal{S}}$ to $\mathrm{Fr}_{D',\mathcal{T}}$ and $i_{F'/F}(\mathrm{Princ}_{D,\mathcal{S}})$ is a subset of $\mathrm{Princ}_{D',\mathcal{T}}$, where $D' = i_{F'/F}(D)$. In analogy with Theorem 1.5.3 we have the following result.

Theorem 1.5.4 *The map $i_{F'/F}$ induces an injective homomorphism from the $\mathcal{S}$-ray class group $\mathrm{Cl}_D(O_\mathcal{S})$ to the $\mathcal{T}$-ray class group $\mathrm{Cl}_{D'}(O_\mathcal{T})$ by*

$$\left(\prod_P \mathbf{M}_P(\mathcal{S})^{m_P} \right) \mathrm{Princ}_{D,\mathcal{S}} \mapsto \left(\prod_P \prod_{Q|P} \mathbf{M}_Q(\mathcal{T})^{m_P} \right) \mathrm{Princ}_{D',\mathcal{T}}.$$

Definition 1.5.5 A divisor $D = \sum_Q m_Q Q \in \mathrm{Div}(F')$ is called $\mathbf{F}_q$-**rational** if $\sigma(D) = \sum_Q m_Q \sigma(Q) = D$ for any $\sigma \in \mathrm{Gal}(F'/F)$.

Theorem 1.5.6 *A divisor $D = \sum_Q m_Q Q \in \mathrm{Div}(F')$ is $\mathbf{F}_q$-rational if and only if D is in the image $i_{F'/F}(\mathrm{Div}(F))$.*

Proof. It is obvious that all divisors in $i_{F'/F}(\mathrm{Div}(F))$ are $\mathbf{F}_q$-rational. To prove the converse, let $D = \sum_Q m_Q Q$ be an $\mathbf{F}_q$-rational divisor of F' and let $\mathcal{S}$ consist of the places of F lying under the places in the support of D. We rewrite D in the form

$$D = \sum_{P \in \mathcal{S}} \sum_{Q|P} m_Q Q.$$

Then

$$\sum_{P\in S}\sum_{Q|P} m_Q \sigma(Q) = \sigma(D) = D = \sum_{P\in S}\sum_{Q|P} m_Q Q$$

for any $\sigma \in \mathrm{Gal}(F'/F)$, i.e.,

$$\sum_{Q|P} m_Q \sigma(Q) = \sum_{Q|P} m_Q Q$$

for any $P \in S$ and $\sigma \in \mathrm{Gal}(F'/F)$. This is equivalent to

$$m_{\sigma^{-1}(Q)} = m_Q$$

for all $\sigma \in \mathrm{Gal}(F'/F)$ and $Q|P$. Now $\mathrm{Gal}(F'/F)$ acts transitively on the set of places of F' lying over P. Thus, m_Q is a constant for all places Q of F' lying over P. Define m_P to be m_Q for all places Q of F' lying over P. Then

$$D = \sum_{P\in S}\sum_{Q|P} m_P Q = i_{F'/F}\left(\sum_{P\in S} m_P P\right),$$

i.e., $D \in i_{F'/F}(\mathrm{Div}(F))$. $\qquad\square$

For an $\mathbf{F}_q$-rational divisor $D \in \mathrm{Div}(F')$, we have the Riemann-Roch space $\mathcal{L}(D)$ over $\mathbf{F}_{q^n}$ and the Riemann-Roch space $\mathcal{L}(i_{F'/F}^{-1}(D))$ over $\mathbf{F}_q$. We denote $\mathcal{L}(D)$ and $\mathcal{L}(i_{F'/F}^{-1}(D))$ by $\mathcal{L}_{\mathbf{F}_{q^n}}(D)$ and $\mathcal{L}_{\mathbf{F}_q}(D)$, respectively.

Theorem 1.5.7 *If $D \in \mathrm{Div}(F')$ is $\mathbf{F}_q$-rational, then any $\mathbf{F}_q$-basis of $\mathcal{L}_{\mathbf{F}_q}(D)$ is an $\mathbf{F}_{q^n}$-basis of $\mathcal{L}_{\mathbf{F}_{q^n}}(D)$. In particular,*

$$\dim_{\mathbf{F}_{q^n}}\left(\mathcal{L}_{\mathbf{F}_{q^n}}(D)\right) = \dim_{\mathbf{F}_q}\left(\mathcal{L}_{\mathbf{F}_q}(D)\right)$$

and

$$\mathcal{L}_{\mathbf{F}_q}(D) = \mathcal{L}_{\mathbf{F}_{q^n}}(D) \cap F.$$

We will still use $\ell(D)$ to denote both $\dim_{\mathbf{F}_{q^n}}\left(\mathcal{L}_{\mathbf{F}_{q^n}}(D)\right)$ and $\dim_{\mathbf{F}_q}\left(\mathcal{L}_{\mathbf{F}_q}(D)\right)$.

Let E/F be a finite Galois extension with $\mathbf{F}_q$ being the full constant field of both E and F. Consider the constant field extensions $F' = F \cdot \mathbf{F}_{q^n}$ and $E' = E \cdot \mathbf{F}_{q^n}$.

Proposition 1.5.8 *Let the notation be the same as above. Then we have:*
(i) $F = F' \cap E$;
(ii) E'/F *is a Galois extension and*

$$\mathrm{Gal}(E'/F) \simeq \mathrm{Gal}(F'/F) \times \mathrm{Gal}(E/F) \simeq \mathrm{Gal}(\mathbf{F}_{q^n}/\mathbf{F}_q) \times \mathrm{Gal}(E/F).$$

Proof. (ii) is a direct consequence of (i). Hence it suffices to prove $F = F' \cap E$. The full constant field of $F' \cap E$ is $\mathbf{F}_q$ since $F' \cap E$ is a subfield of the extension E/F. By Lemma 1.5.1,

$$[F' : F' \cap E] = [(F' \cap E) \cdot \mathbf{F}_{q^n} : F' \cap E] = [\mathbf{F}_{q^n} : \mathbf{F}_q] = [F' : F].$$

This implies that $F = F' \cap E$. □

The above proposition and Lemma 1.5.1 show that we have isomorphisms

$$\mathrm{Gal}(E'/F') \simeq \mathrm{Gal}(E/F) \tag{1.4}$$

and

$$\mathrm{Gal}(E'/E) \simeq \mathrm{Gal}(F'/F) \simeq \mathrm{Gal}(\mathbf{F}_{q^n}/\mathbf{F}_q). \tag{1.5}$$

The isomorphisms in (1.5) are quite clear. There are different ways to define an isomorphism in (1.4). Here we give a canonical isomorphism between $\mathrm{Gal}(E/F)$ and $\mathrm{Gal}(E'/F')$. Fix a basis $\{\alpha_1, \ldots, \alpha_n\}$ of $\mathbf{F}_{q^n}$ over $\mathbf{F}_q$. For an automorphism $\sigma \in \mathrm{Gal}(E/F)$ and an element $\sum_{i=1}^n x_i \alpha_i \in E'$ with $x_i \in E$, put

$$\sigma'\left(\sum_{i=1}^n x_i \alpha_i\right) = \sum_{i=1}^n \sigma(x_i) \alpha_i.$$

Then it is easy to verify that σ' is a Galois automorphism of E'/F'. A canonical isomorphism between $\mathrm{Gal}(E/F)$ and $\mathrm{Gal}(E'/F')$ is given by

$$\mathrm{Gal}(E/F) \simeq \mathrm{Gal}(E'/F'), \quad \sigma \mapsto \sigma'. \tag{1.6}$$

Theorem 1.5.9 *Let E/F be a finite Galois extension with $\mathbf{F}_q$ being the full constant field of both E and F. Define the constant field extensions $F' = F \cdot \mathbf{F}_{q^n}$ and $E' = E \cdot \mathbf{F}_{q^n}$. Let Q' be a place of E'. Denote by Q, P', and P the places of E, F', and F, respectively, lying under Q'. If the degree of Q divides $d(n/\gcd(d,n) - 1)$ with $\deg(P) = d$, then the Frobenius automorphism of Q' over P' corresponds to the Frobenius automorphism of Q over P under the canonical isomorphism* (1.6).

Proof. Let $\sigma \in \mathrm{Gal}(E/F)$ be the Frobenius automorphism of Q over P, i.e.,

$$(\sigma(x) - x^{q^d})(Q) = 0$$

for all $x \in O_Q$. Let $\mathcal{T}$ be the over-set of $\{Q\}$ with respect to the extension E'/E and let $\{\alpha_1, \ldots, \alpha_n\}$ be a basis of $\mathbf{F}_{q^n}$ over $\mathbf{F}_q$. For any $z \in O_\mathcal{T}$, there exist $x_1, \ldots, x_n \in O_Q$ such that $z = \sum_{i=1}^n x_i \alpha_i$. Put $l = d/\gcd(d,n) = \deg(P')$. Then we obtain

$$
\begin{aligned}
(\sigma'(z) - z^{q^{n\,\deg(P')}})(Q') &= (\sigma'(z) - z^{q^{nl}})(Q') \\
&= \left(\sum_{i=1}^n \sigma(x_i)\alpha_i - \sum_{i=1}^n x_i^{q^{nl}} \alpha_i^{q^{nl}}\right)(Q') \\
&= \sum_{i=1}^n (\sigma(x_i) - x_i^{q^{nl}})(Q')\alpha_i \\
&= \sum_{i=1}^n (\sigma(x_i)(Q) - x_i^{q^{nl}}(Q))\alpha_i
\end{aligned}
$$

$$= \sum_{i=1}^{n} (x_i(Q)^{q^d} - x_i(Q)^{q^{nl}}) \alpha_i$$

$$= \sum_{i=1}^{n} (x_i(Q) - x_i(Q)^{q^{nl-d}})^{q^d} \alpha_i = 0,$$

where in the last step we used that $\deg(Q)$ divides $nl - d$ by assumption. This shows that σ' is the Frobenius automorphism of Q' over P'. $\square$

1.6   Zeta Functions and Rational Places

In order to define the zeta function of a global function field, we need to prove the following proposition.

Proposition 1.6.1 *A global function field $F/\mathbf{F}_q$ has only finitely many rational places.*

Proof. Choose an element $x \in F \setminus \mathbf{F}_q$. Then $F/\mathbf{F}_q(x)$ is a finite extension. All rational places of F lie over rational places of $\mathbf{F}_q(x)$. For each rational place P of $\mathbf{F}_q(x)$, there are at most $[F : \mathbf{F}_q(x)]$ rational places of F lying over P. Moreover, there are only $q + 1$ rational places of $\mathbf{F}_q(x)$ by Example 1.1.1. Hence the number of rational places of F is at most $(q + 1)[F : \mathbf{F}_q(x)]$. $\square$

Let $F/\mathbf{F}_q$ be a global function field. For each integer $n \geq 1$, consider the constant field extension

$$F_n = F \cdot \mathbf{F}_{q^n}.$$

Denote by N_n the number of rational places of $F_n/\mathbf{F}_{q^n}$, which is a finite number by Proposition 1.6.1. A rational place of $F_n/\mathbf{F}_{q^n}$ is called an $\mathbf{F}_{q^n}$-**rational place** of F. For each integer $d \geq 1$, denote the number of places of F of degree d by B_d. By Theorem 1.5.2, a place of F of degree d splits into rational places of F_n if and only if d divides n. In the case where d divides n, a place of F of degree d splits into exactly d rational places of F_n. Therefore we get the relation

$$N_n = \sum_{d|n} dB_d.$$

By the Möbius inversion formula, we obtain

$$B_n = \frac{1}{n} \sum_{d|n} \mu(\frac{n}{d}) N_d, \tag{1.7}$$

where μ is the Möbius function. For N_1, i.e., for the number of rational places of $F/\mathbf{F}_q$, we will often write $N(F)$, or $N(F/\mathbf{F}_q)$ if we want to emphasize the full constant field $\mathbf{F}_q$.

The **zeta function** of $F/\mathbf{F}_q$ is defined to be the formal power series

$$Z(F, t) := Z(t) := \exp\left(\sum_{n=1}^{\infty} \frac{N_n}{n} t^n \right) \in \mathbf{C}[[t]].$$

Example 1.6.2 Let us compute the zeta function of the rational function field over $\mathbf{F}_q$. The number of $\mathbf{F}_{q^n}$-rational places of the rational function field is equal to $q^n + 1$ for all $n \geq 1$. Hence we get

$$
\begin{aligned}
\log(Z(t)) &= \sum_{n=1}^{\infty} \frac{q^n + 1}{n} t^n = \sum_{n=1}^{\infty} \frac{(qt)^n}{n} + \sum_{n=1}^{\infty} \frac{t^n}{n} \\
&= -\log(1 - qt) - \log(1 - t) = \log \frac{1}{(1-t)(1-qt)},
\end{aligned}
$$

that is,

$$
Z(t) = \frac{1}{(1-t)(1-qt)}.
$$

Theorem 1.6.3 *The zeta function $Z(t)$ of $F/\mathbf{F}_q$ can be represented also by the following two expressions:*

(i)
$$
Z(t) = \prod_{P \in \mathbf{P}_F} (1 - t^{\deg(P)})^{-1};
$$

(ii)
$$
Z(t) = \sum_{n=0}^{\infty} A_n t^n,
$$

where A_n is the number of positive divisors of F of degree n.

Proof. (i) We have

$$
\begin{aligned}
\log \left(\prod_{P \in \mathbf{P}_F} (1 - t^{\deg(P)})^{-1} \right) &= \log \left(\prod_{d=1}^{\infty} (1 - t^d)^{-B_d} \right) \\
&= \sum_{d=1}^{\infty} (-B_d) \log(1 - t^d) = \sum_{d=1}^{\infty} B_d \sum_{n=1}^{\infty} \frac{t^{dn}}{n} \\
&= \sum_{n=1}^{\infty} \sum_{d|n} \frac{d B_d t^n}{n} = \sum_{n=1}^{\infty} \frac{N_n}{n} t^n.
\end{aligned}
$$

Therefore

$$
Z(t) = \exp \left(\sum_{n=1}^{\infty} \frac{N_n}{n} t^n \right) = \prod_{P \in \mathbf{P}_F} (1 - t^{\deg(P)})^{-1}.
$$

(ii) From (i) we get

$$
\begin{aligned}
Z(t) &= \prod_{P \in \mathbf{P}_F} (1 - t^{\deg(P)})^{-1} = \prod_{P \in \mathbf{P}_F} \sum_{n=0}^{\infty} t^{\deg(nP)} \\
&= \sum_{A \in \mathrm{Div}(F),\ A \geq 0} t^{\deg(A)} = \sum_{n=0}^{\infty} A_n t^n,
\end{aligned}
$$

by collecting terms appropriately in the last step.   □

Definition 1.6.4 Let $F/\mathbf{F}_q$ be a global function field and n a positive integer. The nth **zeta function** $Z_n(t)$ of $F/\mathbf{F}_q$ is defined to be the zeta function of $F \cdot \mathbf{F}_{q^n}/\mathbf{F}_{q^n}$.

There is a close relationship between $Z_n(t)$ and $Z(t) = Z_1(t)$.

Proposition 1.6.5 *For every positive integer n we have*

$$Z_n(t^n) = \prod_{\zeta^n=1} Z(\zeta t),$$

where the product is extended over all complex nth roots of unity ζ.

Proof. Put $F_n = F \cdot \mathbf{F}_{q^n}$. Then by Theorem 1.6.3,

$$
\begin{aligned}
Z_n(t^n) &= \prod_{Q\in \mathbf{P}_{F_n}} (1 - t^{n \deg(Q)})^{-1} \\
&= \prod_{P\in \mathbf{P}_F} \prod_{Q|P}(1 - t^{n \deg(Q)})^{-1}.
\end{aligned}
$$

For a fixed place $P \in \mathbf{P}_F$ of degree d, we know by Theorem 1.5.2 that the degree of the places of F_n lying over P is $l := d/\gcd(d,n)$ and there are exactly d/l places of F_n lying over P. Hence

$$
\begin{aligned}
\prod_{Q|P}(1 - t^{n \deg(Q)})^{-1} &= (1 - t^{nl})^{-d/l} = \prod_{\zeta^n=1} (1 - (\zeta t)^d)^{-1} \\
&= \prod_{\zeta^n=1} (1 - (\zeta t)^{\deg(P)})^{-1}.
\end{aligned}
$$

This yields
$$Z_n(t^n) = \prod_{\zeta^n=1} \prod_{P\in \mathbf{P}_F} (1 - (\zeta t)^{\deg(P)})^{-1} = \prod_{\zeta^n=1} Z(\zeta t).$$

□

Without proof, we state the following important theorem on zeta functions which was established by Weil.

Theorem 1.6.6 *Let $F/\mathbf{F}_q$ be a global function field of genus g. Then:*
(i) $Z(F,t)$ is a rational function of the form

$$Z(F,t) = \frac{L(F,t)}{(1 - t)(1 - qt)},$$

where $L(F,t) \in \mathbf{Z}[t]$ is a polynomial of degree $2g$ with integral coefficients and $L(F,0) = 1$. Moreover, $L(F,1)$ is equal to the divisor class number $h(F)$ of F.

(ii) *Factor $L(F, t)$ into the form*

$$L(F, t) = \prod_{i=1}^{2g}(1 - \omega_i t) \in \mathbf{C}[t],$$

then $|\omega_i| = q^{1/2}$ for all $1 \leq i \leq 2g$.

There are several approaches to prove the above theorem. Bombieri's elementary proof is presented in [152, Chapter V]. The l-adic cohomology method can be found in [45, Appendix C].

Definition 1.6.7 The polynomial $L(F, t) := L(t) := (1 - t)(1 - qt)Z(F, t)$ is called the **L-polynomial** of $F/\mathbf{F}_q$. For a positive integer n, the L-polynomial of $F \cdot \mathbf{F}_{q^n}/\mathbf{F}_{q^n}$ is called the nth **L-polynomial** of $F/\mathbf{F}_q$, denoted by $L_n(t)$, i.e., $L_n(t) = (1 - t)(1 - q^n t)Z_n(t)$.

Corollary 1.6.8 *The following results on the L-polynomial of $F/\mathbf{F}_q$ hold:*
(i)

$$L(t) = q^g t^{2g} L(\frac{1}{qt}).$$

(ii) *Write $L(t) = a_0 + a_1 t + \cdots + a_{2g}t^{2g} \in \mathbf{Z}[t]$, then $a_{2g-i} = q^{g-i}a_i$ for all $0 \leq i \leq g$.*
(iii) $L_n(t) = \prod_{i=1}^{2g}(1 - \omega_i^n t)$ *if* $L(t) = \prod_{i=1}^{2g}(1 - \omega_i t)$.

Proof. (i) Write

$$L(t) = \prod_{i=1}^{2g}(1 - \omega_i t),$$

then the q/ω_i are complex conjugates of the ω_i and therefore are reciprocal roots of $L(t)$ for all $1 \leq i \leq 2g$. Hence

$$L(t) = \prod_{i=1}^{2g}(1 - \frac{q}{\omega_i}t) = q^g t^{2g} L(\frac{1}{qt}).$$

(ii) From (i) we get

$$L(t) = q^g t^{2g} L(\frac{1}{qt}) = \frac{a_{2g}}{q^g} + \frac{a_{2g-1}}{q^{g-1}}t + \cdots + q^g a_0 t^{2g}.$$

The result follows from the above equality.
(iii) By Proposition 1.6.5 we obtain

$$
\begin{aligned}
L_n(t^n) &= (1 - t^n)(1 - q^n t^n)Z_n(t^n) \\
&= (1 - t^n)(1 - q^n t^n) \prod_{\zeta^n=1} Z(\zeta t) \\
&= (1 - t^n)(1 - q^n t^n) \prod_{\zeta^n=1} \frac{L(\zeta t)}{(1 - \zeta t)(1 - \zeta qt)}
\end{aligned}
$$

$$= \prod_{\zeta^n=1} L(\zeta t) = \prod_{\zeta^n=1} \prod_{i=1}^{2g}(1 - \omega_i \zeta t)$$

$$= \prod_{i=1}^{2g} \prod_{\zeta^n=1} (1 - \omega_i \zeta t) = \prod_{i=1}^{2g}(1 - \omega_i^n t^n).$$

Hence

$$L_n(t) = \prod_{i=1}^{2g}(1 - \omega_i^n t).$$

□

Corollary 1.6.9 *Let* $1/\omega_1, \ldots, 1/\omega_{2g}$ *be the roots of the L-polynomial* $L(t) = \sum_{i=0}^{2g} a_i t^i$ *of* $F/\mathbf{F}_q$. *Then the number* $N(F)$ *of rational places of* $F/\mathbf{F}_q$ *is equal to*

$$N(F) = N_1 = q + 1 - \sum_{i=1}^{2g} \omega_i = q + 1 + a_1.$$

More generally, the number N_n *of* $\mathbf{F}_{q^n}$*-rational places of* $F/\mathbf{F}_q$ *is equal to*

$$N_n = q^n + 1 - \sum_{i=1}^{2g} \omega_i^n$$

for all $n \geq 1$.

Proof. By the definition of $Z_n(t)$ we obtain

$$N_n = \frac{d(\log(Z_n(t)))}{dt} \Big|_{t=0}. \tag{1.8}$$

On the other hand, by Theorem 1.6.6(i) we get

$$\frac{d(\log(Z(t)))}{dt} \Big|_{t=0} = \left(\frac{L'(t)}{L(t)} + \frac{1}{1-t} + \frac{q}{1-qt} \right) \Big|_{t=0} = a_1 + 1 + q. \tag{1.9}$$

Combining (1.8) with (1.9) yields

$$N_1 = q + 1 + a_1 = q + 1 - \sum_{i=1}^{2g} \omega_i.$$

By Corollary 1.6.8(iii), $\omega_1^n, \ldots, \omega_{2g}^n$ are the reciprocal roots of the nth L-polynomial $L_n(t)$. Hence

$$N_n = q^n + 1 - \sum_{i=1}^{2g} \omega_i^n.$$

□

Theorem 1.6.10 (Hasse-Weil Bound) *Let* $F/\mathbf{F}_q$ *be a global function field of genus g. Then the number* $N(F)$ *of rational places of* $F/\mathbf{F}_q$ *satisfies*

$$|N(F) - (q + 1)| \leq 2g q^{1/2}.$$

Proof. From Corollary 1.6.9 and Theorem 1.6.6(ii) we obtain

$$|N(F) - (q+1)| = |\sum_{i=1}^{2g} \omega_i| \leq \sum_{i=1}^{2g} |\omega_i| = 2gq^{1/2}.$$

□

Remark 1.6.11 Similarly, the number N_n of $\mathbf{F}_{q^n}$-rational places of $F/\mathbf{F}_q$ satisfies

$$|N_n - (q^n + 1)| \leq 2gq^{n/2}$$

for all $n \geq 1$.

The Hasse-Weil bound is sharp for relatively small genera with respect to q. For relatively large genera with respect to q, this bound is quite bad. There are many different improvements on the Hasse-Weil bound. We just state Serre's improvement (see Serre [141], [144]) for which the proof will be indicated in Remark 1.6.17 below.

Proposition 1.6.12 (Serre Bound) *Let $F/\mathbf{F}_q$ be a global function field of genus g. Then the number $N(F)$ of rational places of $F/\mathbf{F}_q$ satisfies*

$$|N(F) - (q+1)| \leq g\lfloor 2q^{1/2} \rfloor.$$

The following is an easy sufficient condition for the existence of at least one place of F of given degree $d \geq 2$.

Lemma 1.6.13 *Let $F/\mathbf{F}_q$ be a global function field of genus g. Suppose that for some integer $d \geq 2$ we have*

$$q^d - 2gq^{d/2} > \sum_{r|d, r<d} (q^r + 2gq^{r/2}).$$

Then there exists at least one place of F of degree d.

Proof. Let $\omega_1, \ldots, \omega_{2g}$ be all reciprocal roots of the L-polynomial of F. For $r \geq 1$ let B_r be the number of places of F of degree r and $s_r = \sum_{i=1}^{2g} \omega_i^r$. Then by (1.7) and Corollary 1.6.9, we obtain the identity

$$B_d = \frac{1}{d} \sum_{r|d} \mu(\frac{d}{r})(q^r - s_r) \qquad \text{for } d \geq 2.$$

Hence using Theorem 1.6.6(ii), we get

$$B_d \geq \frac{1}{d} \left(q^d - 2gq^{d/2} - \sum_{r|d, r<d} (q^r + 2gq^{r/2}) \right) > 0.$$

□

From the Hasse-Weil bound we know that for fixed g and q, the number of rational places of a global function field $F/\mathbf{F}_q$ of genus g is at most $q + 1 + 2gq^{1/2}$. Hence the following definition makes sense.

Definition 1.6.14 For a fixed prime power q and an integer $g \geq 0$, let $N_q(g)$ denote the maximum number of rational places that a global function field $F/\mathbf{F}_q$ of genus g can have.

Note that $N_q(g)$ is also the maximum number of $\mathbf{F}_q$-rational points that a smooth, projective, absolutely irreducible algebraic curve over $\mathbf{F}_q$ of genus g can have (compare with Appendix A). Obviously, $N_q(g)$ satisfies

$$N_q(g) \leq q + 1 + g\lfloor 2q^{1/2}\rfloor \qquad (1.10)$$

according to the Serre bound.

Definition 1.6.15 A global function field $F/\mathbf{F}_q$ of genus g is called **optimal** if its number $N(F)$ of rational places is equal to $N_q(g)$.

The bound (1.10) is not sharp if g is sufficiently large relative to q, according to the following theorem.

Theorem 1.6.16 *Put* $m = \lfloor 2q^{1/2}\rfloor$. *If*

$$g > \frac{q^2 - q}{m^2 + m - 2q},$$

then

$$N_q(g) < q + 1 + gm.$$

Proof. It is sufficient to prove that $g \leq (q^2 - q)/(m^2 + m - 2q)$ if $N_q(g) = q + 1 + gm$. Let $F/\mathbf{F}_q$ be a global function field of genus g with $N(F) = N_q(g) = q + 1 + mg$. Let $\omega_1, \ldots, \omega_{2g}$ be the reciprocal roots of the L-polynomial of $F/\mathbf{F}_q$, arranged in such a way that $\omega_{g+i} = \overline{\omega_i}$ for all $i = 1, \ldots, g$. Put

$$\gamma_i = \omega_i + \omega_{g+i} + m + 1 = \omega_i + \overline{\omega_i} + m + 1.$$

Then $\gamma_i \geq m + 1 - |\omega_i + \overline{\omega_i}| \geq m + 1 - 2q^{1/2} > 0$ for all $i = 1, \ldots, g$. Since $\omega_1, \ldots, \omega_{2g}$ are all roots of the monic polynomial $t^{2g}L(1/t) \in \mathbf{Z}[t]$, they are algebraic integers, and so $\prod_{i=1}^{g} \gamma_i > 0$ is an integer. Thus

$$\prod_{i=1}^{g} \gamma_i \geq 1.$$

Therefore

$$1 = 1 + m - \frac{1}{g}(N(F) - q - 1) = \frac{1}{g}(g + gm + \sum_{i=1}^{g}(\omega_i + \overline{\omega_i})) = \frac{1}{g}\sum_{i=1}^{g}\gamma_i \geq (\prod_{i=1}^{g}\gamma_i)^{1/g} \geq 1.$$

Thus $\gamma_1 = \gamma_2 = \cdots = \gamma_g = 1$, i.e., $\omega_i + \overline{\omega_i} = -m$ for all $i = 1, \ldots, g$.

By Corollary 1.6.9, the number N_2 of $\mathbf{F}_{q^2}$-rational places of $F/\mathbf{F}_q$ is given by

$$\begin{aligned} N_2 &= q^2 + 1 - \sum_{i=1}^{2g}\omega_i^2 = q^2 + 1 - \sum_{i=1}^{g}(\omega_i^2 + \overline{\omega_i}^2) \\ &= q^2 + 1 - \sum_{i=1}^{g}((\omega_i + \overline{\omega_i})^2 - 2\omega_i\overline{\omega_i}) = q^2 + 1 - gm^2 + 2gq. \end{aligned}$$

From the fact that the number of rational places of $F/\mathbf{F}_q$ is not greater than the number of $\mathbf{F}_{q^2}$-rational places, we obtain

$$g \leq \frac{q^2 - q}{m^2 + m - 2q}.$$

The desired result is proved. □

Remark 1.6.17 The proof of Theorem 1.6.16 shows that for any global function field $F/\mathbf{F}_q$ of genus g we have

$$1 + m - \frac{1}{g}(N(F) - q - 1) \geq 1,$$

hence $N(F) \leq q + 1 + gm$, which is the Serre upper bound. The lower bound is proved similarly, by working with $\delta_i = m + 1 - (\omega_i + \overline{\omega_i})$ instead of γ_i.

We present the method of "explicit Weil formulas" due to Serre [141] for improved upper bounds on $N_q(g)$ if g is relatively large with respect to q.

Theorem 1.6.18 *For $r \geq 1$, suppose that $c_1, \ldots, c_r$ are r nonnegative real numbers such that at least one of them is not equal to zero and the inequality*

$$1 + \lambda_r(t) + \lambda_r(t^{-1}) \geq 0$$

holds for all $t \in \mathbf{C}$ with $|t| = 1$, where $\lambda_r(t) = \sum_{n=1}^r c_n t^n$. Then we have

$$N_q(g) \leq \frac{g}{\lambda_r(q^{-1/2})} + \frac{\lambda_r(q^{1/2})}{\lambda_r(q^{-1/2})} + 1$$

for any q and g.

Proof. Let $F/\mathbf{F}_q$ be an arbitrary global function field of genus g and let $\omega_1, \ldots, \omega_{2g}$ be as in the proof of Theorem 1.6.16. With $\alpha_i = \omega_i q^{-1/2}$ we have $|\alpha_i| = 1$ for $1 \leq i \leq 2g$ and $\alpha_{g+i} = \alpha_i^{-1}$ for $1 \leq i \leq g$. Then Corollary 1.6.9 yields

$$N(F)q^{-n/2} \leq N_n q^{-n/2} = q^{n/2} + q^{-n/2} - \sum_{i=1}^{g} (\alpha_i^n + \alpha_i^{-n})$$

for any positive integer n. By forming an appropriate linear combination of these inequalities, we get

$$\begin{aligned} N(F)\lambda_r(q^{-1/2}) &\leq \lambda_r(q^{1/2}) + \lambda_r(q^{-1/2}) - \sum_{i=1}^{g}\left(\lambda_r(\alpha_i) + \lambda_r(\alpha_i^{-1})\right) \\ &\leq \lambda_r(q^{1/2}) + \lambda_r(q^{-1/2}) + g, \end{aligned}$$

where we used the condition on λ_r in the second step. Now the desired result follows from the definition of $N_q(g)$. □

Example 1.6.19 A systematic search for polynomials λ_r which yield a good bound for $N_q(g)$ in Theorem 1.6.18 was carried out by Oesterlé (see Serre [144]). We mention one particular example. Take $q = 2, r = 6$, and

$$c_1 = \frac{184}{203}, \quad c_2 = \frac{20}{29}, \quad c_3 = \frac{90}{203}, \quad c_4 = \frac{89}{406}, \quad c_5 = \frac{2}{29}, \quad c_6 = \frac{2}{203}.$$

Then the condition on λ_6 in Theorem 1.6.18 is satisfied since it is easily checked that

$$1 + \lambda_6(t) + \lambda_6(t^{-1}) = \frac{1}{406}\left(f(t) + f(t^{-1})\right)^2$$

with

$$f(t) = 2t^3 + 7t^2 + 10t + 5.$$

After a simple calculation, Theorem 1.6.18 yields the bound

$$N_2(g) \le (0.83)g + 5.35 \qquad \text{for all } g \ge 0.$$

Let us now turn to a brief discussion of the asymptotic behavior of $N_q(g)$ for fixed q and $g \to \infty$. The following important quantity was introduced by Ihara [56].

Definition 1.6.20 For any prime power q define

$$A(q) = \limsup_{g \to \infty} \frac{N_q(g)}{g}.$$

It follows from (1.10) that $A(q) \le \lfloor 2q^{1/2} \rfloor$ for all q. Some improvements on this bound were obtained by Ihara [56] and Manin [78], and somewhat later the following better bound was established in [167]. It will be seen in Section 5.4 that, at least for prime powers q that are squares, this bound is best possible.

Theorem 1.6.21 (Vlăduţ-Drinfeld Bound) *For every prime power q we have*

$$A(q) \le q^{1/2} - 1.$$

Proof. For any positive integer r consider the polynomial

$$\lambda_r(t) = \sum_{n=1}^{r}\left(1 - \frac{n}{r+1}\right)t^n.$$

For any $t \in \mathbf{C}$ with $|t| = 1$ we have

$$1 + \lambda_r(t) + \lambda_r(t^{-1}) = \sum_{n=-r}^{r}\left(1 - \frac{|n|}{r+1}\right)t^n = \frac{1}{r+1}\sum_{j,k=0}^{r}t^{j-k} = \frac{1}{r+1}\left|\sum_{n=0}^{r}t^n\right|^2 \ge 0,$$

and so λ_r can be used in Theorem 1.6.18. For $g > 0$ this yields

$$\frac{N_q(g)}{g} \le \frac{1}{\lambda_r(q^{-1/2})} + \frac{\lambda_r(q^{1/2})}{\lambda_r(q^{-1/2})g} + \frac{1}{g}. \qquad (1.11)$$

If $0 \leq t < 1$, then

$$\sum_{n=1}^{r} t^n \geq \lambda_r(t) \geq \sum_{n=1}^{r} t^n - \frac{1}{r+1} \sum_{n=1}^{\infty} n t^n,$$

and so

$$\lim_{r \to \infty} \lambda_r(q^{-1/2}) = \sum_{n=1}^{\infty} q^{-n/2} = \frac{1}{q^{1/2} - 1}.$$

Furthermore,

$$\lambda_r(q^{1/2}) \leq \sum_{n=1}^{r} q^{n/2} < \frac{q^{(r+1)/2}}{q^{1/2} - 1}.$$

If we now let g tend to infinity in (1.11) in such a way that $N_q(g)/g \to A(q)$ and put $r = \lfloor \log_q g \rfloor$, where $\log_q$ denotes the logarithm to the base q, then we obtain the desired bound on $A(q)$. □

In Chapter 5 we will establish lower bounds on $A(q)$. We have $A(q) > 0$ for all prime powers q, and there are, in fact, more specific results that bound $A(q)$ away from 0.

Chapter 2

Class Field Theory

For a global function field $F/\mathbf{F}_q$, all finite abelian extensions of F can be described by class field theory. For our purposes, we want to construct global function fields with many rational places by considering finite abelian extensions of a given global function field. In this chapter, we discuss some important results from class field theory and focus on ray class fields, narrow ray class fields, and class field towers. Some well-known results are presented without proof. The books of Cassels and Fröhlich [13], Koch [60], Neukirch [85], Serre [140], Weil [172], and Weiss [173] are suitable references for this chapter.

2.1 Local Fields

In order to obtain information about algebraic function fields over finite fields, it is often useful to study their various completions. It is a common technique in algebraic number theory and algebraic geometry to reduce problems to local fields.

Definition 2.1.1 A **discrete valuation** of a field E is a surjective map $\nu : E \longrightarrow G \cup \{\infty\}$, where G is a nonzero discrete subgroup of $(\mathbf{R}, +)$ and where ν satisfies:

(i) $\nu(x) = \infty$ if and only if $x = 0$;

(ii) $\nu(xy) = \nu(x) + \nu(y)$ for all $x, y \in E$;

(iii) $\nu(x + y) \geq \min(\nu(x), \nu(y))$ for all $x, y \in E$.

The field E together with the discrete valuation ν, or more precisely the ordered pair (E, ν), is called a **valued field**. If $\nu(E^*) = \mathbf{Z}$, then the discrete valuation ν is called **normalized**.

It follows from this definition that we have the strict triangle inequality

$$\nu(x + y) = \min(\nu(x), \nu(y)) \quad \text{if } x, y \in E \text{ and } \nu(x) \neq \nu(y).$$

For every discrete valuation ν there exists a uniquely determined equivalent valuation, i.e., a suitable scalar multiple of the map ν, which is normalized.

We say that a sequence $x_1, x_2, \ldots$ of elements of the valued field E is **convergent** (with respect to ν) if there exists an element $x \in E$ with the property that for any $N \in \mathbf{N}$ there

36

is an index n_0 such that $\nu(x_n - x) > N$ whenever $n \geq n_0$. This is also expressed by saying that the sequence $x_1, x_2, \ldots$ **converges** to x (with respect to ν). The sequence $x_1, x_2, \ldots$ is a **Cauchy sequence** (with respect to ν) if for any $N \in \mathbf{N}$ there exists an index n_0 such that $\nu(x_n - x_m) > N$ whenever $n, m \geq n_0$. It can be proved, as in real analysis, that any convergent sequence is a Cauchy sequence. However, the converse is not true in general. This leads to a further definition.

Definition 2.1.2 A valued field (K, ν) is called **complete** (with respect to ν) if any Cauchy sequence (with respect to ν) of elements of K is convergent with respect to ν.

Definition 2.1.3 For a valued field E with the discrete valuation ν_E, a **completion** of E is an extension field K of E with a discrete valuation ν_K such that:
  (i) ν_E is the restriction of ν_K to E;
  (ii) K is complete with respect to ν_K;
  (iii) E is dense in K, i.e., for any element $x \in K$ there exists a sequence of elements of E which converges to x with respect to ν_K.

Proposition 2.1.4 (i) *For any valued field (E, ν_E), there exists a completion (K, ν_K) and it is unique up to isomorphisms.*
  (ii) *If (K, ν) is a complete valued field, then ν has a unique extension to any algebraic extension field of K.*

Let (K, ν_K) be a complete valued field with ν_K normalized and let L/K be a finite extension. Consider the unique extension of ν_K to L and let ν_L be the equivalent normalized discrete valuation of L. Then there exists a positive integer e such that $\nu_L(x) = e\nu_K(x)$ for all $x \in K$. The integer e plays the role of a ramification index of ν_L over ν_K.

The **valuation ring** of a complete valued field (K, ν) is defined by

$$O_K = \{x \in K : \nu(x) \geq 0\}.$$

The valuation ring O_K has the unique maximal ideal

$$M_K = \{x \in K : \nu(x) > 0\}.$$

The group of units of the valuation ring O_K is given by

$$U_K = \{x \in K : \nu(x) = 0\}.$$

The factor ring O_K/M_K is called the **residue class field** of K.

Now let us restrict ourselves to global function fields. Let $F/\mathbf{F}_q$ be a global function field, let P be a place of F, and let ν_P be the corresponding normalized discrete valuation. Then (F, ν_P) forms a valued field.

Definition 2.1.5 The completion of the global function field F with respect to the valuation ν_P is called the P-**adic completion** of F. We denote this completion by F_P, whereas ν_P still denotes the corresponding discrete valuation of F_P.

According to Proposition 2.1.4, there exists (up to isomorphisms) a unique P-adic completion for each place P of F. As $O_P/M_P = \tilde{F}_P$ is a finite field, we can choose $s := |\tilde{F}_P|$ elements $x_1, \ldots, x_s$ from O_P such that $x_1 + M_P, \ldots, x_s + M_P$ are s different elements of $\tilde{F}_P$ and $x_i = 0$ if $x_i \equiv 0 \pmod{M_P}$. The set $\{x_1, \ldots, x_s\}$ is called a **complete system of representatives** for $\tilde{F}_P$. The following result is established in a similar way as the local expansion (1.1).

Theorem 2.1.6 *Let F_P be the P-adic completion of F and let $\mathcal{R}$ be a complete system of representatives for $\tilde{F}_P$. Suppose that the elements $t_n \in F$ satisfy $\nu_P(t_n) = n$ for all integers n. Then every nonzero element x of F_P can be expressed uniquely in the form*

$$x = \sum_{n=r}^{\infty} a_n t_n,$$

where $r \in \mathbf{Z}$, $a_n \in \mathcal{R}$ for all $n \geq r$, and $a_r \neq 0$. Moreover, every such series represents a nonzero element of F_P for which $\nu_P(x) = r$.

Corollary 2.1.7 *If $t \in F$ is a local parameter at P, then every element x of F_P has a unique expansion as a formal Laurent series*

$$x = \sum_{n=r}^{\infty} a_n t^n,$$

where the coefficients a_n come from a complete system of representatives for $\tilde{F}_P$ and $a_r \neq 0$ if $x \neq 0$.

It follows from Corollary 2.1.7 that the residue class field of the P-adic completion F_P of F is isomorphic to the residue class field $\tilde{F}_P$.

2.2   Newton Polygons

The Newton polygon provides a useful method to investigate factorization of polynomials over complete valued fields. Let (K, ν) be a complete valued field. Consider a polynomial $f(z) = a_0 + a_1 z + \cdots + a_n z^n \in K[z]$ and assume $a_0 a_n \neq 0$. Let us associate a polygon in 2-dimensional Euclidean space with the polynomial $f(z)$. To each term $a_i z^i$ of $f(z)$ with $a_i \neq 0$ we assign the point $(i, \nu(a_i)) \in \mathbf{R} \times \mathbf{R}$. After having assigned points to all terms of $f(z)$ with nonzero coefficient, we form the lower convex hull of the set of points $\{(i, \nu(a_i))\}$. The polygon determined by the above procedure is called the **Newton polygon** of $f(z)$. Let us give a more precise definition of the Newton polygon in terms of the vertices on it. All vertices on the Newton polygon are given by points

$$(0, \nu(a_0)), (i_1, \nu(a_{i_1})), \ldots, (i_h, \nu(a_{i_h})), (n, \nu(a_n))$$

with $0 < i_1 < i_2 < \cdots < i_h < n$ such that:

$$\text{(i)} \quad \frac{\nu(a_{i_{j+1}}) - \nu(a_{i_j})}{i_{j+1} - i_j} \geq \frac{\nu(a_{i_j}) - \nu(a_{i_{j-1}})}{i_j - i_{j-1}}$$

for $j = 1, 2, \ldots, h$, where $i_0 = 0$, $i_{h+1} = n$; and

$$\text{(ii)} \quad \frac{\nu(a_k) - \nu(a_{i_j})}{k - i_j} \geq \frac{\nu(a_{i_{j+1}}) - \nu(a_{i_j})}{i_{j+1} - i_j}$$

for any $0 \leq j \leq h$ and $i_j < k \leq i_{j+1}$.

Proposition 2.2.1 *Let (K, ν) be a complete valued field and let $f(z) = a_0 + a_1 z + \cdots + a_n z^n$ with $a_0 a_n \neq 0$ be a polynomial in $K[z]$. Suppose that $(r, \nu(a_r)) \leftrightarrow (s, \nu(a_s))$ with $s > r$ is any segment of the Newton polygon of $f(z)$ and that its slope is $-m$. Then $f(z)$ has exactly $s - r$ roots $\alpha_1, \ldots, \alpha_{s-r}$ with*

$$\nu_L(\alpha_1) = \cdots = \nu_L(\alpha_{s-r}) = m$$

in its splitting field L over K, where ν_L is the unique extension of ν to L. Moreover,

$$f_m(z) = \prod_{i=1}^{s-r} (z - \alpha_i)$$

is in $K[z]$ and $f_m(z)$ divides $f(z)$.

2.3 Ramification Groups and Conductors

Let K be a valued field which is complete with respect to a normalized discrete valuation ν_K. For a finite Galois extension L/K, consider the unique extension of ν_K to L and let ν_L be the equivalent normalized discrete valuation of L. For each integer $i \geq -1$, let $G_i(L/K)$ be the group

$$G_i(L/K) = \{\sigma \in \text{Gal}(L/K) : \nu_L(\sigma(a) - a) \geq i + 1 \text{ for all } a \in O_L\},$$

where $O_L = \{x \in L : \nu_L(x) \geq 0\}$ is the valuation ring of L. The group $G_i(L/K)$ is called the ith **ramification group** of L/K. Note that $G_{-1}(L/K)$ is just the Galois group $\text{Gal}(L/K)$. The group $G_0(L/K)$ is also called the **inertia group** of L/K.

Now let us consider a global function field $F/\mathbf{F}_q$ and a place P of F. Let E/F be a finite Galois extension and Q a place of E lying over P. As usual, we denote the P-adic completion of F by F_P and the Q-adic completion of E by E_Q. For an element $\sigma \in \text{Gal}(E/F)$, we have

$$\sigma(O_Q) = O_{\sigma(Q)}.$$

Thus, σ induces an F_P-isomorphism

$$E_Q \longrightarrow E_{\sigma(Q)}.$$

If σ is an element of the decomposition group $G_Z(Q|P)$ of Q over P, then $\sigma(Q) = Q$. Thus, σ induces an automorphism σ_Q of E_Q. Note also that σ_Q is the identity on F_P because σ is the identity on F, and so $\sigma_Q \in \text{Aut}(E_Q/F_P)$. It is clear that the group homomorphism

$$\eta: G_Z(Q|P) \longrightarrow \text{Aut}(E_Q/F_P), \quad \sigma \mapsto \sigma_Q,$$

is injective.

Theorem 2.3.1 (i) *If E/F is a finite Galois extension, then E_Q/F_P is also a finite Galois extension and the injection $\eta: G_Z(Q|P) \longrightarrow \mathrm{Gal}(E_Q/F_P)$ is an isomorphism.*

(ii) *The ith ramification group $G_i(Q|P)$ of Q over P is isomorphic to the ith ramification group $G_i(E_Q/F_P)$ for all $i \geq -1$.*

Proof. (i) It is obvious that

$$|G_Z(Q|P)| \leq |\mathrm{Aut}(E_Q/F_P)| \leq [E_Q : F_P],$$

and these inequalities are equalities if and only if (i) is true. Let $r = [\mathrm{Gal}(E/F) : G_Z(Q|P)]$, i.e., there are exactly r places of E lying over P. Let $\sigma_1(Q), \ldots, \sigma_r(Q)$ be all distinct places of E lying over P with $\sigma_i \in \mathrm{Gal}(E/F)$ for $1 \leq i \leq r$. Then

$$|\mathrm{Gal}(E/F)| = r|G_Z(Q|P)| = \sum_{i=1}^{r} |G_Z(\sigma_i(Q)|P)|$$

$$\leq \sum_{i=1}^{r} [E_{\sigma_i(Q)} : F_P] = [E : F] = |\mathrm{Gal}(E/F)|.$$

So we have equalities throughout.

(ii) This follows easily from (i). $\qquad\qquad\qquad\qquad\qquad\qquad\qquad\qquad\qquad\qquad\qquad$ $\square$

In view of the above result, we can identify $G_i(E_Q/F_P)$ with $G_i(Q|P)$. For a real number $u \geq -1$, we define G_u to be the group $G_{\lceil u \rceil}(E_Q/F_P)$. Let g_i be the order of $G_i(E_Q/F_P)$, i.e., $g_i = |G_i(E_Q/F_P)|$. Put

$$\varphi(u) = \begin{cases} \frac{1}{g_0}(g_1 + g_2 + \cdots + g_{\lfloor u \rfloor} + (u - \lfloor u \rfloor)g_{\lfloor u \rfloor+1}) & \text{if } u > 0, \\ u & \text{if } -1 \leq u \leq 0. \end{cases}$$

In particular,

$$\varphi(m) + 1 = \frac{1}{g_0} \sum_{i=0}^{m} g_i$$

for any integer $m \geq -1$, with an empty sum being 0 as usual.

The function φ is continuous, piecewise linear, strictly increasing, and concave on $[-1, \infty)$. Therefore there exists the inverse function ψ of φ. The function ψ is continuous, piecewise linear, strictly increasing, and convex on $[-1, \infty)$.

Lemma 2.3.2 *If v is an integer ≥ -1, then so is $u = \psi(v)$.*

Proof. We can assume that $u > 0$. By the definition of φ, we have

$$g_0\varphi(u) = g_0 v = g_0 + g_1 + \cdots + g_{\lfloor u \rfloor} + (u - \lfloor u \rfloor)g_{\lfloor u \rfloor+1}.$$

Since $G_{\lfloor u \rfloor+1}$ is a subgroup of G_i for all $0 \leq i \leq \lfloor u \rfloor$, we conclude that

$$u - \lfloor u \rfloor = \frac{g_0 v - \sum_{i=0}^{\lfloor u \rfloor} g_i}{g_{\lfloor u \rfloor+1}}$$

is an integer, i.e., u is an integer. □

For all real numbers $v \geq -1$ we define the **upper ramification groups** of E_Q/F_P by

$$G^v := G^v(E_Q/F_P) := G_{\psi(v)},$$

or equivalently

$$G^{\varphi(u)} := G^{\varphi(u)}(E_Q/F_P) = G_u.$$

Then we have $G^{-1} = \text{Gal}(E_Q/F_P)$, $G^0 = G_0(E_Q/F_P)$, and $G^v = \{\text{id}\}$ for sufficiently large v. Knowledge of the G^v is equivalent to that of the G_u. As we identify $\text{Gal}(E_Q/F_P)$ with $G_Z(Q|P)$, G^v and G_u are subgroups of $G_Z(Q|P)$. Thus, we can speak also of the upper ramification groups $G^v(Q|P) := G^v(E_Q/F_P)$ of Q over P.

Proposition 2.3.3 (Hasse-Arf Theorem) *Suppose that E_Q/F_P is a finite abelian extension. Then $\varphi(i)$ is an integer whenever $G_i \neq G_{i+1}$.*

From now on, we assume in this section that E/F is a finite abelian extension. Let P be a place of F. We can then speak of the ramification index $e_P(E/F)$ of P in E/F and of the different exponent $d_P(E/F)$ of P in E/F, as in any finite Galois extension (compare with Section 1.4). Furthermore, in the abelian case it follows from Proposition 1.4.4 that the ramification groups $G_i(Q|P)$, and so also the upper ramification groups of Q over P, are independent of the place Q of E lying over P. Thus, we can write $G_i(P, E/F) := G_i(Q|P)$ and $G^i(P, E/F) := G^i(Q|P)$ for the ith ramification group, respectively the ith upper ramification group, of P in E/F. We use the abbreviations G_u and G^v as before. Let $a_P(E/F)$ be the least integer $k \geq 0$ such that $|G_i(P, E/F)| = 1$ for all $i \geq k$. Define the **conductor exponent** $c_P(E/F)$ of P in E/F by

$$c_P(E/F) = \frac{d_P(E/F) + a_P(E/F)}{e_P(E/F)}.$$

Theorem 2.3.4 *Let E/F be a finite abelian extension of global function fields and let P be a place of F. Then:*

(i) *$c_P(E/F)$ is a nonnegative integer;*

(ii) *P is unramified in E/F if and only if $c_P(E/F) = 0$, and P is tamely ramified in E/F if and only if $c_P(E/F) = 1$;*

(iii) *$c_P(E/F)$ is the least integer $k \geq 0$ such that $G^v = \{\text{id}\}$ for all $v \geq k$.*

Proof. (i) Put $a = a_P(E/F)$, where we can assume $a \geq 1$. By the definition of $c_P(E/F)$ and the Hilbert different formula (see Theorem 1.4.10), we obtain

$$\begin{aligned}
c_P(E/F) &= \frac{d_P(E/F) + a_P(E/F)}{e_P(E/F)} \\
&= \frac{\sum_{i=0}^{a-1}(g_i - 1) + a}{g_0} = \frac{1}{g_0}\sum_{i=0}^{a-1} g_i = \varphi(a-1) + 1.
\end{aligned}$$

According to the definition of $a_P(E/F)$, we have $G_{a-1}(P, E/F) \neq G_a(P, E/F)$. It follows from the Hasse-Arf theorem that $\varphi(a-1)$ is an integer, i.e., $c_P(E/F)$ is a nonnegative integer.

(ii) P is unramified in E/F if and only if $g_0 = |G_0(P, E/F)| = e_P(E/F) = 1$. This is equivalent to $a_P(E/F) = 0$, i.e., $c_P(E/F) = 0$. The second part is shown similarly, using Proposition 1.3.12.

(iii) We can assume $c_P(E/F) \geq 1$. With the notation in part (i) of the proof we have

$$\psi(c_P(E/F)) = \psi(\varphi(a-1) + 1) > \psi(\varphi(a-1)) = a - 1.$$

By Lemma 2.3.2, $\psi(c_P(E/F))$ is an integer as $c_P(E/F)$ is an integer. Hence

$$\psi(c_P(E/F)) \geq a = a_P(E/F).$$

Therefore

$$G^v \subseteq G^{c_P(E/F)} = G_{\psi(c_P(E/F))} \subseteq G_a = \{\mathrm{id}\}$$

for all $v \geq c_P(E/F)$. On the other hand,

$$G^{c_P(E/F)-1} = G^{\varphi(a-1)} = G_{a-1} \neq \{\mathrm{id}\}.$$

This completes the proof.   □

The results of Theorem 2.3.4 make it meaningful to define the conductor of a finite abelian extension of global function fields in the following way.

Definition 2.3.5 Let E/F be a finite abelian extension of global function fields. The **conductor** of E/F is the positive divisor of F defined by

$$\mathrm{Cond}(E/F) = \sum_{P \in \mathbf{P}_F} c_P(E/F)P.$$

It follows from Theorem 2.3.4(ii) that the support of the divisor $\mathrm{Cond}(E/F)$ consists exactly of all places P of F that are ramified in E/F.

We use ramification theory to derive a formula due to Niederreiter and Xing [117] for the different exponent $d_P(E/F)$ in the case where $c_P(E/F) \leq 2$. An interesting aspect of this formula is that it depends only on the value of $e_P(E/F)$. The following proof determines not only $d_P(E/F)$, but also the orders of all ramification groups of P in E/F.

Theorem 2.3.6 Let E/F be a finite abelian extension of global function fields and let P be a place of F with conductor exponent $c_P(E/F) \leq 2$. Write $e_P(E/F) = bp^l$, where p is the characteristic of F, b is an integer with $\gcd(b, p) = 1$, and l is a nonnegative integer. Then

$$d_P(E/F) = 2e_P(E/F) - b - 1.$$

Proof. The case $l = 0$ of tame ramification or no ramification is trivial, and so we can assume $l \geq 1$. We use the abbreviations $a = a_P(E/F), c = c_P(E/F), d = d_P(E/F), e = e_P(E/F)$, and $g_i = |G_i(P, E/F)|$ for $i \geq 0$. Since $d \geq e$ and $a \geq 1$, we have $c = 2$. Ramification theory yields $g_0 = e = bp^l$ and $g_1 = p^l$ (see Theorem 1.4.7). For $1 \leq r \leq l$ let n_r be the number of $i \geq 1$ with $g_i = p^{l-r+1}$. The Hasse-Arf theorem shows that g_0 divides $\sum_{i=1}^{n_1} g_i = n_1 p^l$, and so b divides n_1. Furthermore, we have

$$a = 1 + \sum_{r=1}^{l} n_r,$$

$$d = e - 1 + \sum_{r=1}^{l} n_r \left(p^{l-r+1} - 1\right),$$

the latter by the Hilbert different formula. From $c = 2$ we get

$$2e = d + a = e + \sum_{r=1}^{l} n_r p^{l-r+1},$$

hence

$$\sum_{r=1}^{l} n_r p^{l-r+1} = bp^l.$$

Since b divides n_1, we must have $n_1 = b, n_2 = \ldots = n_l = 0$, thus

$$d = e - 1 + b(p^l - 1) = 2e - b - 1,$$

which is the desired result. $\qquad\square$

The following general lemma on conductor exponents will be useful later on.

Lemma 2.3.7 *Let $F \subseteq K \subseteq E$ be three global function fields with E/F being a finite abelian extension. Let P be a place of F and R a place of K lying over P. Then*

$$c_P(K/F) \leq c_P(E/F) \leq \max(c_P(K/F), c_R(E/K)).$$

Proof. We will freely use results on ramification theory from Serre [140, Chapter IV]. Put $c = c_P(E/F)$ and $G = \text{Gal}(E/F)$. Thus $G^c = \{\text{id}\}$ by Theorem 2.3.4(iii). With $H = \text{Gal}(E/K)$ we get $(G/H)^c = (G^c H)/H = \{\text{id}\}$, and so $c_P(K/F) \leq c$ again by Theorem 2.3.4(iii).

For the proof of the second inequality we can assume that $e_P(E/F) > 1$. If $H = \text{Gal}(E/K)$ as above, then we have

$$G_i(R, E/K) = G_i(P, E/F) \cap H \qquad \text{for all } i \geq 0. \tag{2.1}$$

Let $c = c_P(E/F) \geq 1$ as above and put $a = a_P(E/F) \geq 1$ and

$$m = \max(c_P(K/F), c_R(E/K)).$$

Suppose first that

$$|G_{a-1}(R, E/K)| > 1.$$

Then $a - 1 \leq a_R(E/K) - 1$, and so with φ being the function defined earlier in this section and indexed by the extension to which it belongs, we get

$$
\begin{aligned}
c &= \varphi_{E/F}(a - 1) + 1 \leq \varphi_{E/F}(a_R(E/K) - 1) + 1 \\
  &= \varphi_{K/F}(\varphi_{E/K}(a_R(E/K) - 1)) + 1,
\end{aligned}
$$

where we used the transitivity of φ in the last step. Therefore

$$c \leq \varphi_{K/F}(c_R(E/K) - 1) + 1 \leq \varphi_{K/F}(m - 1) + 1 \leq m.$$

In the remaining case we have

$$|G_{a-1}(R, E/K)| = 1.$$

Then by Herbrand's theorem with $j = \varphi_{E/K}(a - 1)$ we get

$$
\begin{aligned}
G_j(P, K/F) &\simeq (G_{a-1}(P, E/F)H)/H \\
            &\simeq G_{a-1}(P, E/F)/(G_{a-1}(P, E/F) \cap H) = G_{a-1}(P, E/F),
\end{aligned}
$$

where we used (2.1) in the last step. From $|G_{a-1}(P, E/F)| > 1$ it follows that

$$\varphi_{E/K}(a - 1) = j \leq a_P(K/F) - 1.$$

By again applying the transitivity of φ, we conclude that

$$
\begin{aligned}
c &= \varphi_{E/F}(a - 1) + 1 = \varphi_{K/F}(\varphi_{E/K}(a - 1)) + 1 \\
  &\leq \varphi_{K/F}(a_P(K/F) - 1) + 1 = c_P(K/F) \leq m,
\end{aligned}
$$

and so the second inequality is shown.   □

2.4   Global Fields

There are two kinds of global fields, namely algebraic number fields and algebraic function fields of one variable over finite constant fields. Note that, as in any global field, for a global function field $F/\mathbf{F}_q$ and any nonzero element x of F we have:

(i) $\nu_P(x) = 0$ for all but finitely many $P \in \mathbf{P}_F$;

(ii) $\sum_{P \in \mathbf{P}_F} \nu_P(x) \deg(P) = 0$.

For a global function field $F/\mathbf{F}_q$, we consider the ring direct product $\prod_{P \in \mathbf{P}_F} F_P = \prod_P F_P$ with componentwise operations, where F_P denotes the P-adic competion of F. By making use of the canonical embeddings $F \hookrightarrow F_P$, the field F may be embedded isomorphically into $\prod_P F_P$ along the diagonal. We identify F with its image in F_P, and thus we may write

$a = (a)_P \in \prod_P F_P$ for all $a \in F$. We shall use O_{F_P}, M_{F_P}, and U_{F_P} for the valuation ring of F_P, the maximal ideal of O_{F_P}, and the group of units of O_{F_P}, respectively. Although $\prod_P F_P$ contains all the information on F and its completions F_P, it is too big to be useful. Let us restrict ourselves to a subset of it.

An **adèle** of $\prod_P F_P$ is an element $(x_P)_{P\in\mathbf{P}_F} = (x_P) \in \prod_P F_P$ satisfying $x_P \in O_{F_P}$ for all but finitely many places $P \in \mathbf{P}_F$. It is easy to verify that all adèles form a subring of $\prod_P F_P$ with identity. This ring is called the **adèle ring** of F and denoted by A_F. It is obvious that each element of F is an adèle, so F is viewed as a subring of A_F. In this context, the elements of F are called **principal adèles** of F.

The units of the ring A_F are called **idèles** of F. The set of idèles is denoted by J_F. It is a multiplicative group consisting of all elements $(x_P) \in \prod_P F_P^*$ with $x_P \in U_{F_P}$ for all but finitely many $P \in \mathbf{P}_F$. The group J_F is called the **idèle group** of F. Obviously, we have $F^* \subset J_F \subset A_F$. The factor group $C_F := J_F/F^*$ is called the **idèle class group** of F.

The idèle group admits a natural surjective homomorphism

$$J_F \longrightarrow \mathrm{Div}(F), \quad (x_P) \mapsto \sum_P \nu_P(x_P)P.$$

The kernel of the above homomorphism is denoted by U_F. It consists of the idèles that have valuation 0 at all places. For $a \in F^*$, a belongs to U_F if and only if $\nu_P(a) = 0$ for all $P \in \mathbf{P}_F$. Hence $U_F \cap F^* = \mathbf{F}_q^*$.

Theorem 2.4.1 *For a global function field $F/\mathbf{F}_q$, we have the following diagram with exact rows and columns:*

$$
\begin{array}{ccccccc}
 & 1 & & 1 & & 0 & \\
 & \downarrow & & \downarrow & & \downarrow & \\
1 \to & \mathbf{F}_q^* & \to & F^* & \to & \mathrm{Princ}(F) & \to 0 \\
 & \downarrow & & \downarrow & & \downarrow & \\
1 \to & U_F & \to & J_F & \to & \mathrm{Div}(F) & \to 0 \\
 & \downarrow & & \downarrow & & \downarrow & \\
1 \to & U_F/\mathbf{F}_q^* & \to & C_F & \to & \mathrm{Div}(F)/\mathrm{Princ}(F) & \to 0 \\
 & \downarrow & & \downarrow & & \downarrow & \\
 & 1 & & 1 & & 0 &
\end{array}
$$

Proof. All exact sequences are obvious. □

Let S be a proper subset of $\mathbf{P}_F$ such that $\mathbf{P}_F \setminus S$ is finite. Consider

$$
\begin{aligned}
A_S &= \prod_{P \notin S} F_P \times \prod_{P \in S} O_{F_P} \\
&= \{(x_P) \in A_F : \nu_P(x_P) \geq 0 \text{ for all } P \in S\}.
\end{aligned}
$$

It is a subring of A_F containing $1 = (1)_P$. We call it the **S-integral ring** of A_F. The group of units of the ring A_S is known as the **S-idèle group** of F and denoted by J_S. It is immediate that

$$J_S = \prod_{P \notin S} F_P^* \times \prod_{P \in S} U_{F_P}.$$

The objects F_S and F_S^* are defined by

$$F_S = F \cap A_S, \qquad F_S^* = F^* \cap J_S.$$

Define the S-idèle class group to be the factor group $C_S := J_S/F_S^*$. Note that $C_S \simeq (F^* \cdot J_S)/F^*$, and so the following result is obvious.

Proposition 2.4.2 *We have a canonical embedding*

$$C_S = J_S/F_S^* \hookrightarrow C_F = J_F/F^*.$$

There are some close relationships between ray class groups and idèle classes. For a place P of F and an integer $n \geq 1$, we define the nth unit group

$$U_{F_P}^{(n)} = \{x \in U_{F_P} : \nu_P(x-1) \geq n\}.$$

In order to make $U_{F_P}^{(0)}$ meaningful, we define $U_{F_P}^{(0)}$ to be the unit group U_{F_P}. Let $D = \sum_P m_P P$ be a positive divisor of F with $\mathrm{supp}(D) \subseteq S$. We define a subgroup J_S^D of J_S by

$$J_S^D = \prod_{P \notin S} F_P^* \times \prod_{P \in S} U_{F_P}^{(m_P)}.$$

It is called the S-congruence subgroup modulo D. Its class group is defined by

$$C_S^D = \left(F^* \cdot J_S^D\right)/F^*.$$

Proposition 2.4.3 *We have isomorphisms*

$$C_F/C_S^D \simeq J_F/\left(F^* \cdot J_S^D\right) \simeq \mathrm{Cl}_D(O_S),$$

where $\mathrm{Cl}_D(O_S)$ is the S-ray class group modulo D defined in Section 1.2. In particular, C_F/C_S^D is a finite group.

Proof. The first isomorphism is obvious by the third isomorphism theorem for groups. Now we put

$$J^D = \{(x_P) \in J_F : x_P \in U_{F_P}^{(m_P)} \text{ for all } P \in \mathrm{supp}(D)\}.$$

We note that $J_F = F^* \cdot J^D$, since for any idèle $(x_P) \in J_F$ we can use the approximation theorem to obtain a $z \in F^*$ such that

$$\nu_P(x_P z - 1) \geq m_P \qquad \text{for all } P \in \mathrm{supp}(D),$$

hence $(x_P z) \in J^D$. We define the map

$$\phi : C_F = J_F/F^* \longrightarrow \mathrm{Cl}_D(O_S) = \mathrm{Fr}_{D,S}/\mathrm{Princ}_{D,S}$$

in the following way. Take any idèle $(x_P) \in J_F$, write it as $(x_P) = (y)(z_P)$ with $y \in F^*$ and $(z_P) \in J^D$, and set

$$\phi((x_P)F^*) = \left(\prod_{P \in \mathcal{T}} \mathsf{M}_P(\mathcal{S})^{\nu_P(z_P)} \right) \mathrm{Princ}_{D,\mathcal{S}},$$

where $\mathcal{T} = \mathcal{S} \setminus \mathrm{supp}(D)$. It is easily checked that ϕ is well defined, if we observe that the elements of $J^D \cap F^*$ are exactly those generating the principal $\mathcal{S}$-ideals in $\mathrm{Princ}_{D,\mathcal{S}}$. It is obvious that ϕ is a surjective group homomorphism and that $C_{\mathcal{S}}^D = \left(F^* \cdot J_{\mathcal{S}}^D \right) / F^*$ is contained in $\ker(\phi)$. On the other hand, by what we have shown above, any element of $\ker(\phi)$, like any element of J_F/F^*, can be written as $(z_P)F^*$ with $(z_P) \in J^D$. Furthermore, $(z_P)F^* \in \ker(\phi)$ implies that for some $u \in J^D \cap F^*$ we have

$$\nu_P(z_P) = \nu_P(u) \qquad \text{for all } P \in \mathcal{T}.$$

Consider the idèle $(w_P) = (z_P)(u^{-1})$. We have $\nu_P(w_P) = 0$ for all $P \in \mathcal{T}$. For $P \in \mathrm{supp}(D)$ we have $z_P, u \in U_{F_P}^{(m_P)}$, and so $w_P \in U_{F_P}^{(m_P)}$, hence $\nu_P(w_P) = 0$. Therefore $(w_P) \in J_{\mathcal{S}}$. But also $(w_P) \in J^D$, and so $(w_P) \in J_{\mathcal{S}} \cap J^D = J_{\mathcal{S}}^D$. Consequently, we get $(z_P)F^* = (w_P)F^* \in C_{\mathcal{S}}^D$, hence $\ker(\phi) = C_{\mathcal{S}}^D$ and $C_F/C_{\mathcal{S}}^D \simeq \mathrm{Cl}_D(O_{\mathcal{S}})$. □

Proposition 2.4.4 *The sequence*

$$F_{\mathcal{S}}^* \to J_{\mathcal{S}}/J_{\mathcal{S}}^D \to J_F/\left(F^* \cdot J_{\mathcal{S}}^D \right) \to J_F/(F^* \cdot J_{\mathcal{S}}) \to 1$$

is exact.

Proof. All homomorphisms are obtained in a canonical and obvious manner. □

2.5 Ray Class Fields and Hilbert Class Fields

In this section we first state some results from class field theory without proof. Then by applying these results we obtain ray class fields. For background and more results on class field theory, we refer to the books of Artin and Tate [4], Cassels and Fröhlich [13], Neukirch [84], Serre [140], and Weil [172].

We first review a central result of local class field theory. Suppose that K is the completion of a global function field with respect to one of its normalized discrete valuations, let ν_K be the corresponding normalized discrete valuation of K, and let L/K be a finite abelian extension. Then there exists a **local Artin reciprocity map** $\theta_{L/K} : K^* \longrightarrow \mathrm{Gal}(L/K)$ such that:

   (i) $\ker(\theta_{L/K}) = \mathrm{N}_{L/K}(L^*)$, $\quad \mathrm{im}(\theta_{L/K}) = \mathrm{Gal}(L/K)$.
   (ii) If L/K is unramified, i.e., $G_0(L/K) = \{\mathrm{id}\}$, then $\theta_{L/K}(x) = \pi^{\nu_K(x)}$ for all $x \in K^*$, where π is the Frobenius automorphism.

(iii) $\theta_{L/K}$ maps the vth unit group $U_K^{(v)}$ of K onto the upper ramification group $G^v(L/K)$ for all integers $v \geq 0$.

From the above we see that L/K is unramified if and only if $\theta_{L/K}(U_K) = \{\mathrm{id}\}$.

Now we turn to global class field theory. Let $F/\mathbf{F}_q$ be a global function field and E/F a finite abelian extension. We define the **global Artin reciprocity map** to be the product of local Artin reciprocity maps:

$$\theta_{E/F} := \prod_P \theta_P : J_F \longrightarrow \mathrm{Gal}(E/F),$$

where $\theta_P = \theta_{E_Q/F_P}$ is the local Artin reciprocity map $F_P^* \to \mathrm{Gal}(E_Q/F_P) \simeq G_Z(Q|P) \hookrightarrow \mathrm{Gal}(E/F)$ for a place Q of E lying over P. Since E/F is abelian, $G_Z(Q|P)$ remains unchanged for any places Q of E lying over P. Hence θ_P is well defined. For $x = (x_P) \in J_F$ we have

$$\theta_{E/F}(x) = \prod_P \theta_P(x_P).$$

This product is a proper definition since $\theta_P(x_P) = \pi_P^{\nu_P(x_P)}$ (π_P is the Frobenius of P) when P is unramified, and $\nu_P(x_P) = 0$ if $x_P \in U_{F_P}$. So $\theta_P(x_P) = 1$ for all but finitely many P.

The global Artin reciprocity map $\theta_{E/F}$ is a surjective homomorphism with kernel $F^* \cdot N_{E/F}(J_E)$, where $N_{E/F} : J_E \longrightarrow J_F$ is the canonical extension of the norm from E to F. Thus, we can define a surjective homomorphism

$$(\cdot, E/F) : C_F \longrightarrow \mathrm{Gal}(E/F)$$

induced by $\theta_{E/F}$. The map $(\cdot, E/F)$ is called the **norm residue symbol** of E/F. Its kernel is

$$(F^* \cdot N_{E/F}(J_E))/F^* =: \mathcal{N}_{E/F}.$$

Theorem 2.5.1 (i) (*Artin Reciprocity*) *For any finite abelian extension E/F of global function fields, there is a canonical isomorphism*

$$C_F/\mathcal{N}_{E/F} \simeq \mathrm{Gal}(E/F)$$

induced by the norm residue symbol $(\cdot, E/F)$.

(ii) (*Existence Theorem*) *For every (open) subgroup M of C_F of finite index, there exists a unique finite abelian extension E of F such that E is contained in the abelian closure F^{ab} of F and $\mathcal{N}_{E/F} = M$.*

(iii) *Given two finite abelian extensions E_1/F and E_2/F in F^{ab}, we have $E_1 \subseteq E_2$ if and only if $\mathcal{N}_{E_1/F} \supseteq \mathcal{N}_{E_2/F}$, i.e., if and only if $F^* \cdot N_{E_1/F}(J_{E_1}) \supseteq F^* \cdot N_{E_2/F}(J_{E_2})$.*

Remark 2.5.2 The abelian closure F^{ab} of F is the union of all finite abelian extensions of F in a fixed algebraic closure F^{ac} of F. Note that F^{ab} is contained in the separable closure of F inside F^{ac}, and this separable closure of F can be used for the same purpose as F^{ab} in the following. For the existence theorem, we require that M be an open subgroup of C_F in a suitable topology. Since in this book, all subgroups of C_F considered in the applications of Theorem 2.5.1(ii) will be open, we can avoid concepts from topology, though it is not difficult to introduce these definitions. Thus, in the following it is tacitly assumed that any subgroup of C_F of finite index that occurs is open.

Proposition 2.5.3 *Let M be a subgroup of C_F of finite index and let $E \subseteq F^{\text{ab}}$ be the finite abelian extension of F such that $\mathcal{N}_{E/F} = M$. Let H be the subgroup of J_F containing F^* such that $M = H/F^*$. For a place P of F, we have:*

(i) *P is unramified in E/F if and only if $U_{F_P} \subseteq H$;*

(ii) *P splits completely in E/F if and only if $F_P^* \subseteq H$.*

Proof. (i) Let $\theta_{E/F}$ be the global Artin reciprocity map. Then

$$\theta_{E/F}(U_{F_P}) = \theta_P(U_{F_P}) \times \prod_{R \neq P} \theta_R(\{1\}) = G_0(P, E/F).$$

We have $U_{F_P} \subseteq H$ if and only if $\theta_{E/F}(U_{F_P}) = \{\text{id}\}$. This is equivalent to $G_0(P, E/F) = \{\text{id}\}$, i.e., to P being unramified in E/F.

(ii) P splits completely in E/F if and only if $G_{-1}(P, E/F) = \{\text{id}\}$. This is equivalent to the Frobenius π_P of P being equal to the identity. We know that for an unramified place P in E/F we have

$$\theta_{E/F}(F_P^*) = \{\pi_P^{\nu_P(x)} : x \in F_P^*\}.$$

Thus, $\pi_P = \text{id}$ if and only if $\theta_{E/F}(F_P^*) = \{\text{id}\}$, which is in turn equivalent to $F_P^* \subseteq H$. $\square$

For a positive divisor D of F and a proper subset S of $\mathbf{P}_F$ such that $\text{supp}(D) \subseteq S$ and $\mathbf{P}_F \setminus S$ is finite, we know from Proposition 2.4.3 that C_S^D is a subgroup of C_F of finite index. Thus, by the existence theorem in Theorem 2.5.1, there exists a unique extension F_S^D/F with $F_S^D \subseteq F^{\text{ab}}$ such that

$$\mathcal{N}_{F_S^D/F} = C_S^D.$$

By the definition of C_S^D and Proposition 2.5.3(ii), all places in $\mathbf{P}_F \setminus S$ split completely in F_S^D/F. The field F_S^D is called a **ray class field** with modulus D.

Theorem 2.5.4 (Conductor Theorem) *Let E/F be a finite abelian extension of global function fields with $E \subseteq F^{\text{ab}}$ and let S be a set of places of F such that $\mathbf{P}_F \setminus S$ is nonempty and finite and all places in $\mathbf{P}_F \setminus S$ split completely in E/F. Denote the conductor $\text{Cond}(E/F)$ by C. Then:*

(i) *E is a subfield of F_S^C.*

(ii) *If D is a positive divisor of F with $\text{supp}(D) \subseteq S$ and $E \subseteq F_S^D$, then $D \geq C$.*

Proof. For a positive divisor D of F with $\text{supp}(D) \subseteq S$, by Theorem 2.5.1(iii) we have $E \subseteq F_S^D$ if and only if $\mathcal{N}_{E/F} \supseteq C_S^D$. This is equivalent to $F^* \cdot \text{N}_{E/F}(J_E) \supseteq F^* \cdot J_S^D$. Let $\theta_{E/F}$ be the global Artin reciprocity map from J_F to $\text{Gal}(E/F)$. For a place $P \in S$, we have $\{\mathrm{id}\} = \theta_{E/F}(U_{F_P}^{(n)}) = \theta_P(U_{F_P}^{(n)}) = G^n(P, E/F)$ if and only if $n \geq \nu_P(C)$. This means that $F^* \cdot J_S^D \subseteq F^* \cdot \mathrm{N}_{E/F}(J_E) = \ker(\theta_{E/F})$ if and only if $D \geq C = \mathrm{Cond}(E/F)$. This completes the proofs for both (i) and (ii). $\square$

Corollary 2.5.5 *Let S be a proper subset of $\mathbf{P}_F$ such that $\mathbf{P}_F \setminus S$ is finite and let D be a positive divisor of F with $\mathrm{supp}(D) \subseteq S$. Then $\mathrm{Cond}(F_S^D/F) = D$.*

For a positive divisor D of F and a proper subset S of $\mathbf{P}_F$ such that $\mathrm{supp}(D) \subseteq S$ and $\mathbf{P}_F \setminus S$ is finite, it follows from the above that the ray class field F_S^D has the property that F_S^D/F is a finite abelian extension with

$$\mathrm{Gal}(F_S^D/F) \simeq C_F/C_S^D.$$

By Proposition 2.4.3, C_F/C_S^D is isomorphic to $\mathrm{Cl}_D(O_S)$. Thus, we have

$$\mathrm{Gal}(F_S^D/F) \simeq \mathrm{Cl}_D(O_S).$$

Proposition 2.5.6 *Let S and D be as in Corollary 2.5.5. Then any place P of F not in $\mathrm{supp}(D)$ is unramified in F_S^D/F and the Artin symbol of P in F_S^D/F corresponds to the residue class of P in $\mathrm{Cl}_D(O_S)$ under the isomorphism $\mathrm{Gal}(F_S^D/F) \simeq \mathrm{Cl}_D(O_S)$.*

In the case $D = 0$, the ray class field F_S^0 has the property that F_S^0/F is an unramified abelian extension in which all places in $\mathbf{P}_F \setminus S$ split completely. Moreover, F_S^0 is maximal in the sense that if E/F is an unramified abelian extension with $E \subseteq F^{\mathrm{ab}}$ in which all places in $\mathbf{P}_F \setminus S$ split completely, then E is a subfield of F_S^0. We call F_S^0 the S-**Hilbert class field** of F and denote it by H_S. A detailed discussion of S-Hilbert class fields can be found in Rosen [131].

Proposition 2.5.7 *Let H_S be the S-Hilbert class field of $F/\mathbf{F}_q$. Then:*
(i) $\mathrm{Gal}(H_S/F) \simeq \mathrm{Cl}(O_S)$.
(ii) *The Artin symbol $\left[\frac{H_S/F}{P}\right]$ of a place P of F corresponds to the residue class of P in $\mathrm{Cl}(O_S)$ under the isomorphism $\mathrm{Gal}(H_S/F) \simeq \mathrm{Cl}(O_S)$.*
(iii) H_S *is a subfield of F_S^D for any positive divisor D of F with $\mathrm{supp}(D) \subseteq S$.*
(iv) *The full constant field of H_S is $\mathbf{F}_{q^d}$, where d is the greatest common divisor of the degrees of the places in $\mathbf{P}_F \setminus S$.*

2.6 Narrow Ray Class Fields

Let $F/\mathbf{F}_q$ be a global function field with $N(F) \geq 1$. We distinguish a rational place ∞ of F throughout this section. Let F_∞ be the ∞-adic completion of F. Then $\mathbf{F}_q$ is also the residue class field of F_∞.

Definition 2.6.1 A **sign function** sgn: $F_\infty^* \longrightarrow F_q^*$ is a multiplicative group homomorphism on F_∞^* such that:

(i) $\mathrm{sgn}(\alpha) = \alpha$ for any $\alpha \in F_q^*$;

(ii) $\mathrm{sgn}(U_{F_\infty}^{(1)}) = \{1\}$.

Lemma 2.6.2 *Let* sgn_1 *and* sgn_2 *be two sign functions on* F_∞^*. *Then there is an element* $\zeta \in F_q^*$ *such that*

$$\mathrm{sgn}_1(x) = \mathrm{sgn}_2(x)\zeta^{\nu_\infty(x)}$$

for all $x \in F_\infty^*$.

Proof. Consider the map

$$\kappa : F_\infty^* \longrightarrow F_q^*, \quad x \mapsto \frac{\mathrm{sgn}_1(x)}{\mathrm{sgn}_2(x)}.$$

For any $v \in U_{F_\infty}$, there are elements $\alpha \in F_q^*$ and $u \in U_{F_\infty}^{(1)}$ such that $v = \alpha u$. Thus,

$$\kappa(v) = \frac{\mathrm{sgn}_1(v)}{\mathrm{sgn}_2(v)} = \frac{\alpha \cdot \mathrm{sgn}_1(u)}{\alpha \cdot \mathrm{sgn}_2(u)} = 1.$$

Therefore κ is trivial on U_{F_∞}.

Let $t \in F$ be a fixed local parameter at ∞. Then for any $x \in F_\infty^*$ there exists an element $u_x \in U_{F_\infty}$ such that $x = u_x t^{\nu_\infty(x)}$. Hence

$$\kappa(x) = \frac{\mathrm{sgn}_1(x)}{\mathrm{sgn}_2(x)} = \frac{\mathrm{sgn}_1(u_x t^{\nu_\infty(x)})}{\mathrm{sgn}_2(u_x t^{\nu_\infty(x)})} = \left(\frac{\mathrm{sgn}_1(t)}{\mathrm{sgn}_2(t)}\right)^{\nu_\infty(x)}.$$

Define $\zeta := \mathrm{sgn}_1(t)/\mathrm{sgn}_2(t) \in F_q^*$. Then we obtain $\kappa(x) = \zeta^{\nu_\infty(x)}$, i.e., $\mathrm{sgn}_1(x) = \mathrm{sgn}_2(x)\zeta^{\nu_\infty(x)}$. $\quad\square$

Corollary 2.6.3 *There are exactly* $q - 1$ *sign functions on* F_∞^*.

Proof. Let $t \in F$ be a fixed local parameter at ∞. Define a multiplicative group homomorphism

$$s : F_\infty^* \longrightarrow F_q^*$$

such that $s(t) = 1$, $s(U_{F_\infty}^{(1)}) = \{1\}$, and $s(\alpha) = \alpha$ for all $\alpha \in F_q^*$. Then s is uniquely determined by these conditions. It is easy to see that s is actually a sign function. This shows the existence of sign functions. Now it follows from Lemma 2.6.2 that there exist exactly $q - 1$ sign functions on F_∞^*. $\quad\square$

From now on, we fix a sign function sgn on F_∞^* in this section. Define a subgroup of F_∞^* by

$$F_\infty^{\mathrm{sgn}} = \{x \in F_\infty^* : \mathrm{sgn}(x) = 1\}.$$

Then it is clear that $[F_\infty^* : F_\infty^{\mathrm{sgn}}] = q - 1$.

Let $D = \sum_P m_P P$ be a positive divisor of F with $\infty \notin \mathrm{supp}(D)$. Consider the subgroup of J_F given by

$$F_\infty^{\mathrm{sgn}} \times \prod_{P \neq \infty} U_{F_P}^{(m_P)}.$$

Then $F^*(F_\infty^{\mathrm{sgn}} \times \prod_{P \neq \infty} U_{F_P}^{(m_P)})/F^*$ is a subgroup of C_F of finite index. Thus, by global class field theory there exists an abelian extension $F^D(\infty)/F$ with $F^D(\infty) \subseteq F^{\mathrm{ab}}$ such that

$$\mathcal{N}_{F^D(\infty)/F} = F^*(F_\infty^{\mathrm{sgn}} \times \prod_{P \neq \infty} U_{F_P}^{(m_P)})/F^*.$$

The field $F^D(\infty)$ is called a **narrow ray class field** with modulus D and $F^D(\infty)/F$ the **narrow ray class extension** with modulus D.

Let $A := O_{\infty'}$ with $\infty' := \mathbf{P}_F \setminus \{\infty\}$ be the integral ring

$$A = \{x \in F : \nu_P(x) \geq 0 \text{ for all } P \in \mathbf{P}_F \text{ with } P \neq \infty\}.$$

For a positive divisor D of F with $\infty \notin \mathrm{supp}(D)$, let $\mathrm{Fr}_D(A)$ be the group of fractional ideals of A that are relatively prime to D, i.e.,

$$\mathrm{Fr}_D(A) = \{\mathsf{U} \in \mathrm{Fr}_{\infty'} : \nu_P(\mathsf{U}) = 0 \text{ for all } P \in \mathrm{supp}(D)\}.$$

Define

$$\mathrm{Princ}_D^+(A) = \{xA : x \in F^*,\ \mathrm{sgn}(x) = 1,\ x \equiv 1 \ (\mathrm{mod}\ D)\}.$$

The factor group

$$\mathrm{Cl}_D^+(A) := \mathrm{Fr}_D(A)/\mathrm{Princ}_D^+(A)$$

is called the **narrow ray class group** of A modulo D with respect to sgn. If we identify D with the nonzero ideal M of A (see e.g. (3.4) in Section 3.3 for the obvious connection), then we may speak also of the narrow ray class group $\mathrm{Cl}_\mathsf{M}^+(A)$ of A and the narrow ray class field $F^\mathsf{M}(\infty)$ with modulus M.

Proposition 2.6.4 (i) $\mathrm{Princ}_D^+(A)$ *is a subgroup of* $\mathrm{Princ}_D(A)$ *and* $\mathrm{Princ}_D(A)/\mathrm{Princ}_D^+(A) \simeq \mathbf{F}_q^*$.

(ii) *We have the isomorphisms*

$$\mathrm{Cl}_D^+(A)/\mathbf{F}_q^* \simeq \mathrm{Cl}_D^+(A)/(\mathrm{Princ}_D(A)/\mathrm{Princ}_D^+(A)) \simeq \mathrm{Cl}_D(A).$$

Proof. (i) It is clear from the definitions that $\mathrm{Princ}_D^+(A)$ is a subgroup of $\mathrm{Princ}_D(A)$. Let $\mathbf{F}_q^* = \{\alpha_1, \ldots, \alpha_{q-1}\}$. By the approximation theorem, there exist elements $x_1, \ldots, x_{q-1} \in F^*$ such that $x_i \equiv 1 \ (\mathrm{mod}\ D)$ and $\nu_\infty(x_i - \alpha_i) \geq 1$ for all $1 \leq i \leq q-1$. Thus $\alpha_i^{-1} x_i \in U_{F_\infty}^{(1)}$. This means that $1 = \alpha_i^{-1}\mathrm{sgn}(x_i)$, i.e., $\mathrm{sgn}(x_i) = \alpha_i$. We conclude that $(x_i A)(x_j A)^{-1} = (x_i x_j^{-1})A$ is not an element of $\mathrm{Princ}_D^+(A)$ for $i \neq j$ as $\mathrm{sgn}(x_i x_j^{-1}) = \mathrm{sgn}(x_i)\mathrm{sgn}(x_j^{-1}) = \alpha_i/\alpha_j \neq 1$. Therefore $\{(x_i A)\mathrm{Princ}_D^+(A)\}_{i=1}^{q-1}$ are $q-1$ distinct cosets of $\mathrm{Princ}_D(A)$ modulo $\mathrm{Princ}_D^+(A)$.

For any $xA \in \mathrm{Princ}_D(A)$ with $x \equiv 1 \ (\mathrm{mod}\ D)$, let $\mathrm{sgn}(x) = \alpha_k$ for some $1 \leq k \leq q-1$. Then $x x_k^{-1} A$ is an element of $\mathrm{Princ}_D^+(A)$, i.e., xA and $x_k A$ are in the same coset modulo $\mathrm{Princ}_D^+(A)$. This implies the desired result.

(ii) This follows directly from the definitions of $\mathrm{Cl}_D(A)$ and $\mathrm{Cl}_D^+(A)$ and the third isomorphism theorem for groups. □

In the following theorem we study, in particular, the relationship between the ray class field $F_{\infty'}^D$ with modulus D and the narrow ray class field $F^D(\infty)$ with modulus D.

Theorem 2.6.5 (i) *With the notation above we have the isomorphisms*

$$\mathrm{Gal}(F^D(\infty)/F) \;\simeq\; C_F/(F^*(F_\infty^{\mathrm{sgn}} \times \prod_{P \neq \infty} U_{F_P}^{(m_P)})/F^*) \simeq J_F/(F^*(F_\infty^{\mathrm{sgn}} \times \prod_{P \neq \infty} U_{F_P}^{(m_P)}))$$

$$\simeq\; \mathrm{Cl}_D^+(A).$$

(ii) $F_{\infty'}^D$ *is a subfield of* $F^D(\infty)/F$ *and* $\mathrm{Gal}(F^D(\infty)/F_{\infty'}^D) \simeq \mathbf{F}_q^*$.

(iii) *The ramification index of* ∞ *in* $F^D(\infty)/F$ *is equal to* $q-1$, *i.e.,* $e_\infty(F^D(\infty)/F) = e_\infty(F^D(\infty)/F_{\infty'}^D) = q-1$, *and* ∞ *splits into* $[F_{\infty'}^D : F]$ *places of* $F^D(\infty)$ *with residue class field* $\mathbf{F}_q$. *In particular,* $\mathbf{F}_q$ *is the full constant field of* $F^D(\infty)$.

Proof. (i) The first isomorphism follows from Theorem 2.5.1(i), the second isomorphism is trivial, and the third isomorphism follows from the approximation theorem.

(ii) Since $F^*(F_\infty^{\mathrm{sgn}} \times \prod_{P \neq \infty} U_{F_P}^{(m_P)})$ is a subgroup of $F^*(F_\infty^* \times \prod_{P \neq \infty} U_{F_P}^{(m_P)})$, it follows from Theorem 2.5.1(iii) that $F_{\infty'}^D$ is a subfield of $F^D(\infty)$. Furthermore,

$$\mathrm{Gal}(F^D(\infty)/F)/\big(\mathrm{Gal}(F^D(\infty)/F_{\infty'}^D)\big) \simeq \mathrm{Gal}(F_{\infty'}^D/F),$$

that is,

$$\mathrm{Cl}_D^+(A)/\mathrm{Gal}(F^D(\infty)/F_{\infty'}^D) \simeq \mathrm{Cl}_D(A).$$

It follows that $\mathrm{Gal}(F^D(\infty)/F_{\infty'}^D) \simeq \mathbf{F}_q^*$.

(iii) Considering the Artin map $\theta_{F^D(\infty)/F}$, we get

$$\theta_{F^D(\infty)/F}(F_\infty^{\mathrm{sgn}}) = \{\mathrm{id}\},$$

and therefore

$$\theta_{F^D(\infty)/F}(U_{F_\infty}^{(1)}) = \{\mathrm{id}\}$$

since $U_{F_\infty}^{(1)} \subseteq F_\infty^{\mathrm{sgn}}$. It is obvious that $\theta_{F^D(\infty)/F}(x) = 1$ for $x \in U_{F_\infty}$ if and only if $x \in U_{F_\infty} \cap F_\infty^{\mathrm{sgn}} = U_{F_\infty}^{(1)}$. Thus

$$\theta_{F^D(\infty)/F}(U_{F_\infty}) \simeq \mathbf{F}_q^*,$$

that is,

$$G_0(\infty, F^D(\infty)/F) \simeq \mathbf{F}_q^*.$$

So the ramification index of ∞ in $F^D(\infty)/F$ is equal to $q-1$.

It follows from Section 2.5 that ∞ splits completely in $F_{\infty'}^D/F$. This completes the proof. □

Proposition 2.6.6 *Let $P \neq \infty$ be a place of F and $F^D(\infty)/F$ the narrow ray class extension with modulus $D = nP$, $n \geq 1$. Then the conductor exponent of P in $F^D(\infty)/F$ is given by $c_P(F^D(\infty)/F) = n$.*

Proof. Compare with the proof of Theorem 2.5.4 and use the form of $\mathcal{N}_{F^D(\infty)/F}$.   □

Proposition 2.6.7 *Let $D = \sum_{j=1}^r n_j P_j$, where $P_1, \ldots, P_r$ are distinct places of F different from ∞ and $n_1, \ldots, n_r$ are positive integers. Then for any subfield K of the narrow ray class extension $F^D(\infty)/F$ with modulus D we have*

$$c_{P_j}(K/F) \leq n_j \qquad for \ 1 \leq j \leq r.$$

Proof. Note that $F^D(\infty)$ is the composite of the fields $F^{D_j}(\infty)$ with $D_j = n_j P_j$ for $1 \leq j \leq r$. Therefore, from Lemma 2.3.7 and Proposition 2.6.6 and the fact that the places of $F^D(\infty)$ lying over P_j are unramified in the extension $F^D(\infty)/F^{D_j}(\infty)$, we get

$$c_{P_j}(K/F) \leq c_{P_j}(F^D(\infty)/F) = n_j \qquad for \ 1 \leq j \leq r.$$

□

Theorem 2.6.8 *Let D be a positive divisor of F with $\infty \notin \mathrm{supp}(D)$. Then the conductor of $F^D(\infty)/F$ is equal to $D + \min(q - 2, 1)\infty$.*

Proof. Proceed as in the proof of Proposition 2.6.7 and use Theorem 2.6.5(iii) and Theorem 2.3.4.   □

Theorem 2.3.6 and the concept of the conductor can be combined to provide a genus formula for certain extension fields, in particular for a family of subfields of narrow ray class extensions.

Theorem 2.6.9 *Let K/F be an arbitrary finite abelian extension of global function fields with the same full constant field. Furthermore, let $P_1, \ldots, P_r$ be distinct places of F. If the conductor $\mathrm{Cond}(K/F) \leq \sum_{j=1}^r 2P_j$, then*

$$g(K) = 1 + [K : F](g(F) - 1) + [K : F] \sum_{j=1}^r \left(1 - \frac{b_j + 1}{2e_j}\right) \deg(P_j),$$

where $e_j = e_{P_j}(K/F)$ and b_j is the p-free part of e_j for $1 \leq j \leq r$, with p being the characteristic of F.

Proof. By the assumptions, the only possible ramified places in K/F are $P_1, \ldots, P_r$ and the corresponding different exponents are given by Theorem 2.3.6. The rest follows from the Hurwitz genus formula.   □

Corollary 2.6.10 *Let K be any subfield of the narrow ray class extension $F^D(\infty)/F$ with modulus $D \leq \sum_{j=1}^{r} 2P_j$, where $P_1, \ldots, P_r$ are distinct places of F different from ∞. Then*

$$g(K) = 1 + [K:F](g(F) - 1) + [K:F]\sum_{j=1}^{r}\left(1 - \frac{b_j + 1}{2e_j}\right)\deg(P_j)$$

$$+ [K:F]\left(\frac{1}{2} - \frac{1}{2e_\infty}\right),$$

where $e_j = e_{P_j}(K/F)$ and b_j is the p-free part of e_j for $1 \leq j \leq r$, with p being the characteristic of F, and where $e_\infty = e_\infty(K/F)$.

Proof. Proceed as in the proof of Theorem 2.6.9 and use the fact that ∞ is unramified or tamely ramified in $F^D(\infty)/F$, and so in K/F, by Theorem 2.6.5(iii). □

2.7 Class Field Towers

We start with a brief review of Tate cohomology (see [13, Chapter IV], [140, Part 3], [145] for more details). Let G be a finite group. By a G-**module** we mean a left Λ-module, with $\Lambda = \mathbb{Z}[G]$ being the integral group ring. If A is a G-module with action $(g, a) \mapsto ga$, then the set of elements of A invariant under the action of G is denoted by A^G, i.e.,

$$A^G = \{a \in A : ga = a \text{ for all } g \in G\}.$$

It is obvious that A^G is an abelian group.

Let I_G be the ideal of the group ring Λ generated by the set $\{g - 1 : g \in G\}$. Let $I_G A$ be the submodule of A generated by the set $\{ga - a : g \in G, a \in A\}$. The factor module $A/I_G A$ will be denoted by A_G.

Let n be the element $\sum_{g \in G} g$ of Λ. For any G-module A, multiplication by n defines an endomorphism $N : A \longrightarrow A$. Clearly,

$$I_G A \subseteq \ker(N), \qquad \operatorname{im}(N) \subseteq A^G.$$

Hence N induces a homomorphism

$$N^* : A_G \longrightarrow A^G.$$

For all integers r, the **Tate cohomology groups** $\hat{H}^r(G, A)$ can be defined. They satisfy the following conditions:

(i) $\hat{H}^0(G, A) = A^G/\operatorname{im}(N)$ and $\hat{H}^{-1}(G, A) = \ker(N^*)$;

(ii) for each exact sequence $0 \to A \to B \to C \to 0$ of G-modules, we have an exact sequence

$$\cdots \to \hat{H}^{r-1}(G, C) \to \hat{H}^r(G, A) \to \hat{H}^r(G, B) \to \hat{H}^r(G, C) \to \hat{H}^{r+1}(G, A) \to \cdots.$$

Proposition 2.7.1 *Let K be a complete valued field with finite residue class field and L/K be a finite unramified Galois extension with $G := \mathrm{Gal}(L/K)$. Then $\hat{H}^r(G, U_L) = 0$ and $\hat{H}^r(G, L^*) \simeq \hat{H}^r(G, \mathbf{Z})$ for any $r \in \mathbf{Z}$, where U_L is the group of units of the valuation ring of L.*

Proof. For the first part compare with the proof of [60, Proposition 2.64]. The second part follows from the first part and the exact sequence $1 \to U_L \to L^* \to \mathbf{Z} \to 0$. □

Proposition 2.7.2 *Let F be a global function field and E/F be a finite Galois extension with $G := \mathrm{Gal}(E/F)$. For a place P of F, let P' be a fixed place of E lying over P. Then we have*

$$\hat{H}^r\left(G, \prod_{Q|P} E_Q^*\right) \simeq \hat{H}^r(G^{P'}, E_{P'}^*) \quad and \quad \hat{H}^r\left(G, \prod_{Q|P} U_{E_Q}\right) \simeq \hat{H}^r(G^{P'}, U_{E_{P'}})$$

for all $r \in \mathbf{Z}$, where $G^{P'}$ is the decomposition group of P' over P.

Proof. See [13, Chapter VII, Proposition 7.2]. □

Proposition 2.7.3 *Let F be a global function field and E/F be a finite Galois extension with $G := \mathrm{Gal}(E/F)$. For every place P of F, let P' be a fixed place of E lying over P. Then we have*

$$\hat{H}^r(G, J_E) \simeq \prod_{P \in \mathbf{P}_F} \hat{H}^r(G^{P'}, E_{P'}^*) \qquad for \ all \ r \in \mathbf{Z}.$$

Proof. We observe that J_E is the limit of $J_{E,S}$ for finite sets $S \subset \mathbf{P}_F$ containing all ramified places in E/F, where

$$J_{E,S} = \prod_{P \in S}\left(\prod_{Q|P} E_Q^*\right) \times \prod_{P \notin S}\left(\prod_{Q|P} U_{F_Q}\right).$$

The limit is taken over an increasing sequence of sets S with $S \to \mathbf{P}_F$. The cohomology of finite groups commutes with direct limits and any cohomology theory commutes with products. So it is enough to look at the cohomology of the various parts. By Proposition 2.7.1,

$$\prod_{P \notin S}\left(\prod_{Q|P} U_{F_Q}\right)$$

has trivial cohomology since S contains all ramified places in E/F. By using also Proposition 2.7.2, we obtain

$$\hat{H}^r(G, J_E) = \lim_{S \to \mathbf{P}_F} \hat{H}^r(G, J_{E,S}) \simeq \lim_{S \to \mathbf{P}_F} \prod_{P \in S} \hat{H}^r(G^{P'}, E_{P'}^*) = \prod_{P \in \mathbf{P}_F} \hat{H}^r(G^{P'}, E_{P'}^*).$$

The proof is complete. □

Proposition 2.7.4 *Let E/F be a finite Galois extension of global function fields. Then*

$$\hat{H}^r(G, \mathbf{Z}) \simeq \hat{H}^{r+2}(G, C_E)$$

for all $r \in \mathbf{Z}$, where $G = \mathrm{Gal}(E/F)$ is the Galois group.

Proof. See [13, p. 197].   □

Let ℓ be a prime number and G be a finite ℓ-group. Then both $\hat{H}^{-2}(G, \mathbf{Z}/\ell\mathbf{Z})$ and $\hat{H}^{-3}(G, \mathbf{Z}/\ell\mathbf{Z})$ are $\mathbf{F}_\ell$-vector spaces.

Theorem 2.7.5 (Golod-Shafarevich Theorem) *Let G be a finite ℓ-group with $|G| > 1$. Put $d = \dim_{\mathbf{F}_\ell} \hat{H}^{-2}(G, \mathbf{Z}/\ell\mathbf{Z})$ and $c = \dim_{\mathbf{F}_\ell} \hat{H}^{-3}(G, \mathbf{Z}/\ell\mathbf{Z})$. Then*

$$c > \frac{1}{4}d^2.$$

Proof. See [13, Chapter IX] and observe the connection between homology and Tate cohomology stated on p. 102 of [13]. A proof in a more general context is given in [145, Chapter 1, Appendix 2].   □

The above theorem can also be expressed in terms of the minimal number of generators and relations of finite ℓ-groups. The version given here is a slightly refined form due independently to Gaschütz and Vinberg. The original version of Theorem 2.7.5 was used by Golod and Shafarevich in 1964 to construct infinite class field towers for algebraic number fields. In fact, their proof applies also to global function fields. In this section, we give a sufficient condition for an infinite Hilbert class field tower.

Let $F/\mathbf{F}_q$ be a global function field and S a proper subset of $\mathbf{P}_F$ such that $S' := \mathbf{P}_F \setminus S$ is finite. Let H_S denote the S-Hilbert class field of F, i.e., in the notation of Section 2.5 we have

$$N_{H_S/F}(J_{H_S}) = \prod_{P \in S'} F_P^* \times \prod_{P \in S} U_{F_P}.$$

All places of F are unramified in H_S/F and each place in S' splits completely in H_S/F. Moreover, H_S is the maximal abelian extension of F contained in F^{ab} satisfying these conditions. In order to define a tower, we denote F by F_0 and S by S_0 and we proceed recursively. For an integer $i \geq 1$, let F_i be the S_{i-1}-Hilbert class field of F_{i-1} and S_i be the set of places of F_i lying over the places in S_{i-1}. In this way we obtain a tower of fields

$$F = F_0 \subseteq F_1 \subseteq F_2 \subseteq \ldots \subseteq F_i \subseteq \ldots$$

and a sequence $\{S_i\}_{i=0}^{\infty}$ of sets of places such that $|S_i'| = [F_i : F_{i-1}]|S_{i-1}'|$ for all $i \geq 1$, where $S_i' = \mathbf{P}_{F_i} \setminus S_i$ for all $i \geq 0$. The above tower, together with the sequence $\{S_i\}_{i=0}^{\infty}$, is called the S-**Hilbert class field tower** of F. We say that this tower is finite if for some m we have $F_i = F_m$ for all $i \geq m$. Otherwise, the tower is said to be infinite. It is clear that the tower is infinite if and only if $F_i \neq F_{i-1}$ for all $i \geq 1$.

Instead of S-Hilbert class field towers we will consider (ℓ, S)-Hilbert class field towers since these are easier to handle. For a prime number ℓ we introduce the (ℓ, S)-Hilbert class field tower of a global function field F on the basis of the following definition: the (ℓ, S)-**Hilbert class field** $H_S(\ell)$ of F is the maximal unramified abelian ℓ-extension of F contained in F^{ab} in which all places in S' split completely. Put $F_0(\ell) = F$, $S_0(\ell) = S$, and for $i \geq 1$ let $F_i(\ell)$ be the $(\ell, S_{i-1}(\ell))$-Hilbert class field of $F_{i-1}(\ell)$ and $S_i(\ell)$ be the set of places of $F_i(\ell)$ lying over the places in $S_{i-1}(\ell)$. Thus, we obtain a tower

$$F = F_0(\ell) \subseteq F_1(\ell) \subseteq F_2(\ell) \subseteq \ldots \subseteq F_i(\ell) \subseteq \ldots,$$

which is called the (ℓ, S)-**Hilbert class field tower** of F. By induction on i it is easy to see that $F_i(\ell) \subseteq F_i$ for all $i \geq 0$. Hence F has an infinite S-Hilbert class field tower whenever F admits an infinite (ℓ, S)-Hilbert class field tower for some prime number ℓ.

Hilbert class field towers allow us to obtain some reasonable lower bounds on the number $A(q)$ which was introduced in Definition 1.6.20. The following result provides the connection.

Theorem 2.7.6 *Let $F/\mathbf{F}_q$ be a global function field of genus $g > 1$. Let S be a proper subset of $\mathbf{P}_F$ such that all places in $S' = \mathbf{P}_F \setminus S$ are rational. If F has an infinite S-Hilbert class field tower, then*

$$A(q) \geq \frac{|S'|}{g-1}.$$

Proof. Let $\{(F_i, S_i)\}_{i=0}^{\infty}$ with $F_0 = F$ and $S_0 = S$ be the infinite S-Hilbert class field tower of F. Then $\lim_{i \to \infty}[F_i : F] = \infty$. Moreover, the genus g_i of F_i satisfies

$$2g_i - 2 = [F_i : F](2g - 2)$$

since F_i/F is an unramified separable extension. The number N_i of rational places of F_i satisfies

$$N_i \geq |S_i'| = [F_i : F]\,|S'|$$

since all places in S' split completely in F_i/F.

By the definition of $A(q)$ we have

$$A(q) \geq \limsup_{i \to \infty} \frac{N_i}{g_i} \geq \lim_{i \to \infty} \frac{[F_i : F]\,|S'|}{[F_i : F](g-1)+1} = \frac{|S'|}{g-1},$$

which is the desired result. $\square$

From the above result, in order to get some lower bounds on $A(q)$, it is necessary to obtain global function fields F and infinite S-Hilbert class field towers for suitable $S \subseteq \mathbf{P}_F$. We are going to give a sufficient condition for global function fields with infinite (ℓ, S)-Hilbert class field towers.

For a prime number ℓ and a finitely generated (and, let us say, additively written) abelian group B, we let $d_\ell(B)$ denote the ℓ-**rank** of B, i.e., $d_\ell(B)$ is the dimension of the $\mathbf{F}_\ell$-vector space $B/\ell B$. If B is finite, then $d_\ell(B)$ is the number of summands that occur in a direct decomposition of the ℓ-Sylow subgroup of B into cyclic components.

Theorem 2.7.7 *Let $F/\mathbf{F}_q$ be a global function field and S a set of places of F such that $S' = \mathbf{P}_F \setminus S$ is a finite nonempty set. Suppose that for a prime number ℓ we have*

$$d_\ell(\mathrm{Cl}(O_S)) \geq 2 + 2(|S'| + \varepsilon_\ell(q))^{1/2},$$

where O_S is the S-integral ring of F and where $\varepsilon_\ell(q) = 1$ if $\ell|(q-1)$ and $\varepsilon_\ell(q) = 0$ otherwise. Then F admits an infinite (ℓ, S)-Hilbert class field tower.

Proof. Let $\{F_i(\ell)\}_{i=0}^\infty$ be the (ℓ, S)-Hilbert class field tower of F. We first show by induction on i that $F_i(\ell)/F$ is a Galois extension for all $i \geq 1$. This being trivial for $i = 1$, we can assume that it holds for some $i \geq 1$. Then $F_i(\ell)/F$ and $F_{i+1}(\ell)/F_i(\ell)$ are separable extensions, and so $F_{i+1}(\ell)/F$ is a separable extension. To prove that $F_{i+1}(\ell)/F$ is a normal extension, let F^{ac} be an algebraic closure of F containing $F_{i+1}(\ell)$ and let $\sigma : F_{i+1}(\ell) \hookrightarrow F^{\mathrm{ac}}$ be an embedding of $F_{i+1}(\ell)$ into F^{ac} over F. By induction hypothesis we have $\sigma(F_i(\ell)) = F_i(\ell)$. Then $\sigma(F_{i+1}(\ell))/F_i(\ell)$ is an unramified abelian ℓ-extension in which all places in $\mathbf{P}_{F_i(\ell)} \setminus S_i(\ell)$ split completely. By the maximality of $F_{i+1}(\ell)$ relative to these properties, we obtain $\sigma(F_{i+1}(\ell)) \subseteq F_{i+1}(\ell)$, and so the induction is complete.

Now suppose, by way of contradiction, that the (ℓ, S)-Hilbert class field tower $\{F_i(\ell)\}_{i=0}^\infty$ of F is finite. If $E = \cup_{i=0}^\infty F_i(\ell)$, then it follows from the above that E/F is a Galois extension for which $G := \mathrm{Gal}(E/F)$ is a finite ℓ-group. Let H be the maximal abelian factor group of G. Then there exists a subfield K of E/F such that $\mathrm{Gal}(K/F) \simeq H$. Thus, K/F is an unramified abelian ℓ-extension in which all places in S' split completely, and so K is the (ℓ, S)-Hilbert class field of F by the maximality of H. Hence H is isomorphic to the ℓ-Sylow subgroup of $\mathrm{Gal}(H_S/F) \simeq \mathrm{Cl}(O_S)$, which yields $d_\ell(H) = d_\ell(\mathrm{Cl}(O_S))$. On the other hand, $H \simeq \hat{H}^{-2}(G, \mathbf{Z})$ by Tate cohomology (see [13, Chapter IV, Proposition 1 and p. 102]), and so

$$d_\ell(\mathrm{Cl}(O_S)) = d_\ell(\hat{H}^{-2}(G, \mathbf{Z})).$$

The exact sequence $0 \to \mathbf{Z} \overset{\ell}{\to} \mathbf{Z} \to \mathbf{Z}/\ell\mathbf{Z} \to 0$, where $\overset{\ell}{\to}$ denotes "multiplication by ℓ", yields the exact cohomology sequence

$$\hat{H}^r(G, \mathbf{Z}) \overset{\ell}{\to} \hat{H}^r(G, \mathbf{Z}) \to \hat{H}^r(G, \mathbf{Z}/\ell\mathbf{Z}) \to \hat{H}^{r+1}(G, \mathbf{Z}) \overset{\ell}{\to} \hat{H}^{r+1}(G, \mathbf{Z}) \qquad (2.2)$$

for any integer r. From this we derive in turn the exact sequence

$$0 \to \hat{H}^r(G, \mathbf{Z})/\ell\hat{H}^r(G, \mathbf{Z}) \to \hat{H}^r(G, \mathbf{Z}/\ell\mathbf{Z}) \to \hat{H}^{r+1}(G, \mathbf{Z})_\ell \to 0,$$

where $\hat{H}^{r+1}(G, \mathbf{Z})_\ell$ is the kernel of the last map in (2.2). Comparing $\mathbf{F}_\ell$-dimensions we obtain

$$\dim_{\mathbf{F}_\ell}(\hat{H}^r(G, \mathbf{Z}/\ell\mathbf{Z})) = \dim_{\mathbf{F}_\ell}(\hat{H}^{r+1}(G, \mathbf{Z})_\ell) + d_\ell(\hat{H}^r(G, \mathbf{Z})). \qquad (2.3)$$

Put $r = -2$ and then $r = -3$ in (2.3), then in the notation of Theorem 2.7.5 we get

$$d = d_\ell(\hat{H}^{-2}(G, \mathbf{Z})) = d_\ell(\mathrm{Cl}(O_S)),$$

$$c = \dim_{\mathbf{F}_\ell}(\hat{H}^{-2}(G, \mathbf{Z})_\ell) + d_\ell(\hat{H}^{-3}(G, \mathbf{Z})) = d + d_\ell(\hat{H}^{-3}(G, \mathbf{Z})).$$

By Theorem 2.7.5 we have $c > \frac{1}{4}d^2$, which after simple algebraic manipulations yields

$$d_\ell(\mathrm{Cl}(O_S)) < 2 + 2(d_\ell(\hat{H}^{-3}(G, \mathbf{Z})) + 1)^{1/2}. \qquad (2.4)$$

Now we consider $\hat{H}^{-3}(G, \mathbf{Z})$. By Proposition 2.7.4 we have $\hat{H}^{-3}(G, \mathbf{Z}) \simeq \hat{H}^{-1}(G, C_E)$. Let $\mathcal{T}$ be the over-set of S with respect to the extension E/F. Then, by results and in the notation of Section 2.4, we have the exact sequence

$$1 \to C_\mathcal{T} \to C_E \to \mathrm{Cl}(O_\mathcal{T}) \to 1,$$

which leads to the exact cohomology sequence

$$\hat{H}^{-2}(G, \mathrm{Cl}(O_\mathcal{T})) \to \hat{H}^{-1}(G, C_\mathcal{T}) \to \hat{H}^{-1}(G, C_E) \to \hat{H}^{-1}(G, \mathrm{Cl}(O_\mathcal{T})).$$

Since the (ℓ, S)-Hilbert class field tower of F stabilizes at E, the group $\mathrm{Cl}(O_\mathcal{T})$ has order relatively prime to ℓ, and so it is cohomologically trivial as a G-module (note that G is an ℓ-group). Thus,

$$\hat{H}^{-1}(G, C_E) \simeq \hat{H}^{-1}(G, C_\mathcal{T}) = \hat{H}^{-1}(G, J_\mathcal{T}/E_\mathcal{T}^*).$$

Furthermore, from the trivial exact sequence

$$1 \to E_\mathcal{T}^* \to J_\mathcal{T} \to J_\mathcal{T}/E_\mathcal{T}^* \to 1$$

we get the exact cohomology sequence

$$\hat{H}^{-1}(G, J_\mathcal{T}) \to \hat{H}^{-1}(G, J_\mathcal{T}/E_\mathcal{T}^*) \to \hat{H}^0(G, E_\mathcal{T}^*) \to \hat{H}^0(G, J_\mathcal{T}).$$

Recall that

$$J_\mathcal{T} = \prod_{Q \notin \mathcal{T}} E_Q^* \times \prod_{Q \in \mathcal{T}} U_{E_Q}.$$

Since the extension E/F is unramified and the places in $\mathbf{P}_E \setminus \mathcal{T}$ are lying over places of F that split completely in E/F, it follows from Propositions 2.7.1 and 2.7.2 that $J_\mathcal{T}$ is cohomologically trivial as a G-module. Therefore

$$\hat{H}^{-1}(G, J_\mathcal{T}/E_\mathcal{T}^*) \simeq \hat{H}^0(G, E_\mathcal{T}^*).$$

Altogether, we have shown that

$$\hat{H}^{-3}(G, \mathbf{Z}) \simeq \hat{H}^0(G, E_\mathcal{T}^*).$$

Now $\hat{H}^0(G, E_\mathcal{T}^*)$ is a factor group of F_S^* (namely the norm factor group), and so

$$d_\ell(\hat{H}^{-3}(G, \mathbf{Z})) = d_\ell(\hat{H}^0(G, E_\mathcal{T}^*)) \le d_\ell(F_S^*),$$

where the inequality follows from [126, Proposition 6]. But F_S^* is also the group of units of the ring O_S, and so Dirichlet's unit theorem (see e.g. [173, Theorem 5-3-10]) shows that

$$F_S^* \simeq \mathbf{F}_q^* \times \mathbf{Z}^{|S'|-1}.$$

Hence $d_\ell(F_S^*) = |S'| - 1 + \varepsilon_\ell(q)$. Going back to the upper bound on $d_\ell(\mathrm{Cl}(O_S))$ shown in (2.4), we obtain

$$d_\ell(\mathrm{Cl}(O_S)) < 2 + 2(|S'| + \varepsilon_\ell(q))^{1/2},$$

which is a contradiction to the hypothesis. □

The condition on $d_\ell(\mathrm{Cl}(O_S))$ in Theorem 2.7.7 is called the **Golod-Shafarevich condition**. In many applications the following consequence of the above results turns out to be convenient.

Corollary 2.7.8 *Let $K/\mathbf{F}_q$ be a global function field of genus $g(K) > 1$ and let S be a nonempty set of rational places of K. Suppose that there exist an unramified abelian extension L/K in which all places in S split completely and a prime number ℓ such that*

$$d_\ell(\mathrm{Gal}(L/K)) \geq 2 + 2(|S| + \varepsilon_\ell(q))^{1/2},$$

where $\varepsilon_\ell(q) = 1$ if $\ell | (q-1)$ and $\varepsilon_\ell(q) = 0$ otherwise. Then

$$A(q) \geq \frac{|S|}{g(K) - 1}.$$

Proof. The properties of L imply that L is contained in the S'-Hilbert class field $H_{S'}$ of K with $S' = \mathbf{P}_K \setminus S$. Thus, $\mathrm{Gal}(L/K)$ is a factor group of $\mathrm{Gal}(H_{S'}/K) \simeq \mathrm{Cl}(O_{S'})$, and so

$$d_\ell(\mathrm{Cl}(O_{S'})) \geq d_\ell(\mathrm{Gal}(L/K)) \geq 2 + 2(|S| + \varepsilon_\ell(q))^{1/2}.$$

Theorem 2.7.7 shows then that K has an infinite (ℓ, S')-Hilbert class field tower, and so an infinite S'-Hilbert class field tower. The desired result follows now from Theorem 2.7.6.□

Chapter 3

Explicit Function Fields

In Chapter 2 we have seen how to obtain, in theory, all finite abelian extensions of a global function field $F/\mathbf{F}_q$. In this chapter we are going to construct some of these finite abelian extensions explicitly by considering Kummer and Artin-Schreier extensions, cyclotomic function fields, and Drinfeld modules of rank 1. This is particularly important for determining explicit equations which are satisfied by generators of global function fields.

3.1 Kummer and Artin-Schreier Extensions

In this section, two of the most common types of Galois extensions, namely Kummer extensions and Artin-Schreier extensions, are introduced. The advantage of these extensions is that their defining equations can be explicitly expressed. Another advantage is that genera of these two types of global function fields can be easily calculated. A detailed account of the theory of Kummer and Artin-Schreier extensions can be found in the book of Stichtenoth [152, Chapter III]. Thus, we state the results in this section without proof.

Definition 3.1.1 Let $n > 1$ be an integer. An element f of $F/\mathbf{F}_q$ is called nth **Kummer degenerate** if there exist an element u of F and a divisor d of n such that $d > 1$ and $f = u^d$. Otherwise, f is called nth **Kummer nondegenerate**.

If $\gcd(\nu_P(f), n) = 1$ for some $P \in \mathbf{P}_F$, then $f \in F$ is nth Kummer nondegenerate. An extension of the form considered in the next proposition is called a **Kummer extension**.

Proposition 3.1.2 Let $F/\mathbf{F}_q$ be a global function field and $n > 1$ a divisor of $q - 1$. Suppose that $f \in F$ is an nth Kummer nondegenerate element. Let y be a root of $T^n - f$. Then $E := F(y)$ is a cyclic extension field of F of degree n.

Proposition 3.1.3 Let $F/\mathbf{F}_q$ be a global function field and $n > 1$ a divisor of $q - 1$. Suppose that $f \in F$ is an nth Kummer nondegenerate element and y is a root of $T^n - f$. Then for any place Q of $E := F(y)$ and the place P of F lying under Q, the ramification index satisfies

$$e(Q|P) = \frac{n}{\gcd(\nu_P(f), n)}.$$

Theorem 3.1.4 *Let $F/\mathbf{F}_q$ be a global function field of genus g and $n > 1$ a divisor of $q - 1$. Suppose that $f \in F$ is an nth Kummer nondegenerate element and y is a root of $T^n - f$. Assume that there is a place R of F such that $\gcd(\nu_R(f), n) = 1$. Then $\mathbf{F}_q$ is still the full constant field of $E := F(y)$ and the genus of E satisfies*

$$g(E) = 1 + n(g - 1) + \frac{1}{2} \sum_{P \in \mathbf{P}_F} (n - \gcd(\nu_P(f), n)) \deg(P).$$

In a particular example of the above theorem, let F be the rational function field. In this case, f is just a rational function $g(x)/h(x)$, where $g(x)$ and $h(x) \neq 0$ are two coprime polynomials in $\mathbf{F}_q[x]$. If y is a root of $T^n - f$, then $h(x)y$ is a root of $T^n - g(x)h(x)^{n-1}$. Therefore we may assume that f is a polynomial in $\mathbf{F}_q[x]$.

Corollary 3.1.5 *Let $f(x) := \alpha \prod_{i=1}^{r} p_i(x)^{n_i} \in \mathbf{F}_q[x]$, where $p_1(x), \ldots, p_r(x)$ are $r \geq 1$ distinct monic irreducible polynomials in $\mathbf{F}_q[x]$ and α is a nonzero element of $\mathbf{F}_q$. Suppose that $n > 1$ is a divisor of $q - 1$ and $\gcd(n_i, n) = 1$ for some $1 \leq i \leq r$. Let y be a root of $T^n - f$. Then $E := \mathbf{F}_q(x, y)$ is a cyclic extension field of $\mathbf{F}_q(x)$ of degree n, the full constant field of E is $\mathbf{F}_q$, and the genus of E is given by*

$$g(E) = 1 - \frac{1}{2}(n + \gcd(\deg(f), n)) + \frac{1}{2} \sum_{i=1}^{r}(n - \gcd(n_i, n)) \deg(p_i).$$

Definition 3.1.6 An element f of $F/\mathbf{F}_q$ is called **Artin-Schreier degenerate** if there exists an element u of F such that $u^p - u = f$, where p is the characteristic of $\mathbf{F}_q$. Otherwise, f is called **Artin-Schreier nondegenerate**.

A sufficient condition for $f \in F$ to be Artin-Schreier nondegenerate is that for some $P \in \mathbf{P}_F$ and $z \in F$ we have $\nu_P(f - (z^p - z)) < 0$ and $\gcd(\nu_P(f - (z^p - z)), p) = 1$. An extension of the form considered in the next proposition is called an **Artin-Schreier extension**.

Proposition 3.1.7 *Let $F/\mathbf{F}_q$ be a global function field and f an Artin-Schreier nondegenerate element of F. Let y be a root of $T^p - T - f$. Then $E := F(y)$ is a cyclic extension field of F of degree p.*

Lemma 3.1.8 *Let $F/\mathbf{F}_q$ be a global function field and $f \in F$ an Artin-Schreier nondegenerate element. Then for any place P of F, there exists an element $z \in F$ such that:*
(i) $\nu_P(f - (z^p - z)) \geq 0$; *or*
(ii) $\nu_P(f - (z^p - z)) < 0$ *and* $\gcd(\nu_P(f - (z^p - z)), p) = 1$.
Furthermore, P is unramified in E/F in case (i) and totally ramified in E/F in case (ii), where $E := F(y)$ with y being a root of $T^p - T - f$.

Since the integer $\nu_P(f - (z^p - z))$ in case (ii) is uniquely determined, the following notation makes sense. Let $F/\mathbf{F}_q$ be a global function field and $f \in F$ an Artin-Schreier nondegenerate element: For a place P of F, define

$$m_P = \begin{cases} -1 & \text{if } \nu_P(f - (z^p - z)) \geq 0 \text{ for some } z \in F, \\ -\nu_P(f - (z^p - z)) & \text{if } \nu_P(f - (z^p - z)) < 0 \text{ is coprime to } p \text{ for some } z \in F. \end{cases}$$

Theorem 3.1.9 *Let $F/\mathbf{F}_q$ be a global function field of genus g. Suppose that $f \in F$ is an Artin-Schreier nondegenerate element and y is a root of $T^p - T - f$. Assume that there is a place R of F such that $m_R > 0$. Then $\mathbf{F}_q$ is still the full constant field of $E := F(y)$ and the genus of E satisfies*

$$g(E) = pg + \frac{p-1}{2}\left(-2 + \sum_{P \in \mathbf{P}_F} (m_P + 1)\deg(P)\right).$$

An Artin-Schreier extension is of degree p. In fact, we can consider extensions of degree p^r for some integer $r \geq 1$. The following proposition describes extensions of degree q for algebraic function fields with the full constant field $\mathbf{F}_{q^2}$.

Proposition 3.1.10 *Let q be an arbitrary prime power and $F/\mathbf{F}_{q^2}$ a global function field. Let $w \in F$ and assume that there exists a place $P \in \mathbf{P}_F$ such that*

$$\nu_P(w) = -m, \quad m > 0, \quad \text{and} \quad \gcd(m, q) = 1.$$

Let $E = F(y)$, where y satisfies the equation

$$y^q + y = w.$$

Then the following holds:

(i) *E/F is an abelian extension of degree q and $\mathbf{F}_{q^2}$ is algebraically closed in E.*

(ii) *The place P is totally ramified in E/F. For the unique place $P' \in \mathbf{P}_E$ lying over P, the different exponent of P' over P is given by*

$$d(P'|P) = (q-1)(m+1).$$

(iii) *Let $R \in \mathbf{P}_F$ and assume that*

$$\nu_R(w - u^q - u) \geq 0$$

for some element $u \in F$. Then the place R is unramified in E/F. In particular, this is the case whenever $\nu_R(w)$ is nonnegative.

(iv) *Suppose that the place $Q \in \mathbf{P}_F$ is a zero of $w - \gamma$, with $\gamma \in \mathbf{F}_q$. The equation $\alpha^q + \alpha = \gamma$ has q distinct roots $\alpha \in \mathbf{F}_{q^2}$, and for any such α there exists a unique place $Q_\alpha \in \mathbf{P}_E$ such that Q_α lies over Q and Q_α is a zero of $y - \alpha$. In particular, the place Q splits completely in E/F.*

Definition 3.1.11 A global function field $K/\mathbf{F}_q$ of genus g is called **maximal** if its number $N(K)$ of rational places meets the Hasse-Weil bound, i.e., if

$$N(K) = q + 1 + 2gq^{1/2}.$$

It is trivial that any maximal function field is optimal (compare with Definition 1.6.15). It is also obvious that a global function field $K/\mathbf{F}_q$ can be maximal only if either $g(K) = 0$ or q is a square.

Example 3.1.12 For an arbitrary prime power q, let $F = \mathbf{F}_{q^2}(x)$ be the rational function field over $\mathbf{F}_{q^2}$ and consider the extension field $E = F(y)$ determined by

$$y^q + y = x^{q+1}.$$

Then we can apply Proposition 3.1.10 with P being the infinite place ∞ of F and $m = q+1$. It follows from this proposition that $\mathbf{F}_{q^2}$ is the full constant field of E. Moreover, ∞ is the only ramified place in E/F and it is totally ramified and satisfies $d_\infty(E/F) = (q-1)(q+2)$. The Hurwitz genus formula then yields $g(E) = q(q-1)/2$. Furthermore, Proposition 3.1.10(iv) shows that all q^2 finite rational places of F split completely in E/F, and so $N(E) = q^3 + 1$. In particular, $E/\mathbf{F}_{q^2}$ is a maximal function field according to Definition 3.1.11. The global function field $E = \mathbf{F}_{q^2}(x,y)$ is called the **Hermitian function field** over $\mathbf{F}_{q^2}$.

It follows from Theorem 1.6.16 (with q replaced by q^2) that if $K/\mathbf{F}_{q^2}$ is a maximal function field of genus g, then necessarily $g \leq q(q-1)/2$. Thus, among all maximal function fields with full constant field $\mathbf{F}_{q^2}$, the Hermitian function field has the largest genus. Moreover, Rück and Stichtenoth [132] proved that any maximal function field $K/\mathbf{F}_{q^2}$ of genus $q(q-1)/2$ is isomorphic to the Hermitian function field over $\mathbf{F}_{q^2}$. The following is a useful property for the construction of maximal function fields: if $K/\mathbf{F}_{q^2}$ is a maximal function field, then any global function field $L/\mathbf{F}_{q^2}$ that is a subfield of K is also maximal.

A survey of early work on maximal function fields can be found in Garcia and Stichtenoth [29]. Considerable progress in the theory of maximal function fields has been achieved in recent years. Xing and Stichtenoth [186] conjectured and Fuhrmann and Torres [28] proved that if $K/\mathbf{F}_{q^2}$ is a maximal function field with $g(K) < q(q-1)/2$, then $g(K) \leq (q-1)^2/4$. A somewhat weaker result in this direction was already shown by Xing and Stichtenoth [186]. For odd q, Fuhrmann, Garcia, and Torres [27] characterized the maximal function fields $K/\mathbf{F}_{q^2}$ with $g(K) = (q-1)^2/4$. For even q, a partial characterization of the maximal function fields $K/\mathbf{F}_{q^2}$ with $g(K) = \lfloor (q-1)^2/4 \rfloor = q(q-2)/4$ was achieved by Abdón and Torres [1]. Garcia, Stichtenoth, and Xing [37] obtained many new constructions of maximal function fields over $\mathbf{F}_{q^2}$ by considering fixed fields under certain groups of automorphisms of the Hermitian function field over $\mathbf{F}_{q^2}$. A family of these maximal function fields was described explicitly in terms of generators and defining equations by Garcia and Stichtenoth [34]. Further constructions of maximal function fields are due to van der Geer and van der Vlugt [161], [162] who used, in particular, Artin-Schreier extensions obtained from methods of coding theory.

3.2 Cyclotomic Function Fields

The theory of cyclotomic function fields was initiated by Carlitz [12] and developed in the present form by Hayes [46]. Throughout this section, we fix the following notation:

F : the rational function field $\mathbf{F}_q(T)$;

R : the polynomial ring $\mathbf{F}_q[T]$;

F^{ac} : an algebraic closure of F.

Let $\varphi \in \mathrm{End}_{\mathbf{F}_q}(F^{\mathrm{ac}})$ be the endomorphism given by

$$\varphi(u) = u^q + Tu \qquad \text{for all } u \in F^{\mathrm{ac}}.$$

Define a ring homomorphism

$$R \longrightarrow \mathrm{End}_{\mathbf{F}_q}(F^{\mathrm{ac}}), \qquad f(T) \mapsto f(\varphi).$$

Then the $\mathbf{F}_q$-vector space F^{ac} is made into an R-module by introducing an action of R on F^{ac} by

$$u^{f(T)} = f(\varphi)(u) \qquad \text{for all } f(T) \in R \text{ and } u \in F^{\mathrm{ac}}.$$

In particular, for all $u \in F^{\mathrm{ac}}$ we have the following properties:

(i) $u^a = au$ for all $a \in \mathbf{F}_q$;

(ii) $u^T = u^q + Tu$;

(iii) $u^{f(T)+g(T)} = u^{f(T)} + u^{g(T)}$ for all $f(T), g(T) \in R$;

(iv) $(u^{f(T)})^{g(T)} = u^{f(T)g(T)}$ for all $f(T), g(T) \in R$.

Furthermore, we have

(v) $(u+v)^{f(T)} = u^{f(T)} + v^{f(T)}$ for all $u, v \in F^{\mathrm{ac}}$ and $f(T) \in R$.

For $i \in \mathbf{Z}$ and $M \in R$, we define a new polynomial $[M, i] \in R$ by:

(i) $[M, i] = 0$ for $i < 0$ or $i > \deg(M)$, where $\deg(0) = -1$;

(ii) $[M, 0] = M$;

(iii) $[T^{d+1}, i] = T[T^d, i] + [T^d, i-1]^q$;

(iv) $[aM + bN, i] = a[M, i] + b[N, i]$ for any $a, b \in \mathbf{F}_q$ and $M, N \in R$.

It is clear that $[M, i]$ is well defined, and the following lemma is easily verified.

Lemma 3.2.1 (i) $u^M = \sum_{i=0}^{d}[M, i]u^{q^i}$ for all nonzero $M \in R$, where $d = \deg(M)$.
(ii) $\deg([M, i]) = (d - i)q^i$ for all nonzero $M \in R$ and $0 \leq i \leq d$.

For a nonzero polynomial $M \in R$, define the set of M-torsion elements of F^{ac} by

$$\Lambda_M = \{z \in F^{\mathrm{ac}} : z^M = 0\}.$$

It is immediately seen that Λ_M is an R-submodule of F^{ac}.

By Lemma 3.2.1, the derivative of u^M with respect to u is equal to M. Hence u^M is a separable polynomial in u over F. The splitting field $F(\Lambda_M)$ of u^M over F is called the **cyclotomic function field** over F with modulus M. Clearly, $F(\Lambda_M)/F$ is a finite Galois extension. In order to investigate the degree and the Galois group of this extension, we first have to discuss the structure of Λ_M. We can always assume, without loss of generality, that M is monic.

Lemma 3.2.2 (i) *For a monic irreducible polynomial* $P \in R$, Λ_{P^n} *is a cyclic R-module for any $n \geq 0$. We have an R-module isomorphism*

$$\Lambda_{P^n} \simeq R/(P^n).$$

(ii) *Let $M \in R$ be a monic nonconstant polynomial. If M has the factorization $M = \prod_{j=1}^{s} P_j^{e_j}$ with distinct monic irreducible polynomials $P_1, \ldots, P_s \in R$, then*

$$\Lambda_M \simeq \bigoplus_{j=1}^{s} \Lambda_{P_j^{e_j}}.$$

(iii) *For any monic polynomial $M \in R$ we have an R-module isomorphism*

$$\Lambda_M \simeq R/(M).$$

Proof. (i) Let $d = \deg(P)$ be the degree of P. We proceed by induction on n. The case $n = 0$ is trivial. For $n = 1$, Λ_P can be viewed as an $R/(P)$-module, i.e., Λ_P is an $R/(P)$-vector space since $R/(P) \simeq \mathbf{F}_{q^d}$. Therefore $\Lambda_P \simeq R/(P)$ since Λ_P has exactly q^d elements.

Assume for some $n \geq 1$ that Λ_{P^n} is a cyclic R-module and $\Lambda_{P^n} \simeq R/(P^n)$. Consider the homomorphism

$$\Lambda_{P^{n+1}} \longrightarrow \Lambda_{P^n}, \quad u \mapsto u^P.$$

This homomorphism is surjective and its kernel is Λ_P. By the assumption that Λ_{P^n} is cyclic, there exists an element $\lambda \in \Lambda_{P^{n+1}} \backslash \Lambda_{P^n}$ such that λ^P is a generator of $\Lambda_{P^n} = \{\lambda^{Pf} : f \in R\}$. For any $u \in \Lambda_{P^{n+1}}$, there exists a polynomial $g \in R$ such that $u^P = \lambda^{Pg}$. Thus $u - \lambda^g \in \Lambda_P$. Since $\lambda^{P^n} \neq 0$ and $\lambda^{P^{n+1}} = 0$, we know that there exists a polynomial $h \in R$ such that $u - \lambda^g = \lambda^{P^n h}$, i.e., $u = \lambda^{g + P^n h}$. This shows that $\Lambda_{P^{n+1}}$ is a cyclic R-module and λ is a generator. Finally, consider the map

$$R/(P^{n+1}) \longrightarrow \Lambda_{P^{n+1}}, \quad f \pmod{P^{n+1}} \mapsto \lambda^f.$$

It is easy to verify that the above map is an R-module isomorphism.

(ii) This follows from the general theory of modules over principal ideal domains since $\Lambda_{P_j^{e_j}}$ is the P_j-primary submodule of Λ_M.

(iii) This follows immediately from (i) and (ii). $\qquad\square$

Denote by $(R/(M))^*$ the group of units of $R/(M)$, i.e., $(R/(M))^*$ consists of all residue classes $f + (M)$ with $\gcd(f, M) = 1$. We have $|(R/(M))^*| = |A(M)|$, where

$$A(M) := \{f \in R : \gcd(f, M) = 1 \text{ and } \deg(f) < \deg(M)\}.$$

Lemma 3.2.3 *Put $\Phi(M) := \Phi_q(M) := |(R/(M))^*|$. Then*

$$\Phi(M) = q^{\deg(M)} \prod_{P|M} (1 - q^{-\deg(P)}),$$

where the product is extended over all monic irreducible polynomials $P \in R$ dividing the monic polynomial $M \in R$.

Proof. We can assume that M is nonconstant. For a monic irreducible polynomial $P \in R$ and $n \geq 1$, it is easily seen that

$$\Phi(P^n) = |A(P^n)| = q^{(n-1)\deg(P)}(q^{\deg(P)} - 1).$$

The desired result follows from the isomorphism

$$(R/(M))^* \simeq \prod_{j=1}^{s} (R/(P_j^{e_j}))^*,$$

where M has the factorization $M = \prod_{j=1}^{s} P_j^{e_j}$ with distinct monic irreducible polynomials $P_1, \ldots, P_s \in R$. □

Proposition 3.2.4 *Let $P \in R$ be a monic irreducible polynomial and $n \geq 1$. Then:*

  (i) $[F(\Lambda_{P^n}) : F] = \Phi(P^n)$;

  (ii) $\mathrm{Gal}(F(\Lambda_{P^n})/F) \simeq (R/(P^n))^*$;

  (iii) P *is totally ramified in $F(\Lambda_{P^n})/F$ and all other finite places of F are unramified in $F(\Lambda_{P^n})/F$.*

Proof. (i) Let λ be a generator of the cyclic R-module Λ_{P^n}. Then $\lambda^f \in F(\lambda) = F(\Lambda_{P^n})$ for any $f \in R$ with $\gcd(f, P) = 1$. Let $u, v \in R$ satisfy $uf + vP^n = 1$. Thus $\lambda = \lambda^{uf + vP^n} = (\lambda^f)^u$, i.e., λ^f is also a generator of Λ_{P^n}. It is easy to see that λ^f is a root of the polynomial $g(z) := z^{P^n}/z^{P^{n-1}}$ for all $f \in R$ with $\gcd(f, P) = 1$. By Lemma 3.2.1, the degree of $g(z)$ as a polynomial in z is equal to $q^{\deg(P^n)} - q^{\deg(P^{n-1})} = \Phi(P^n)$. Therefore

$$g(z) = \prod_{f \in A(P^n)} (z - \lambda^f).$$

By comparing constant terms in this identity, we get

$$P = \pm \prod_{f \in A(P^n)} \lambda^f. \tag{3.1}$$

Consider the place of F which is the zero of P. We denote this place also by P. Let P' be an arbitrary place of $F(\Lambda_{P^n})$ lying over P. From $\lambda^{P^n} = 0$ we obtain $\nu_{P'}(\lambda) \geq 0$, hence also $\nu_{P'}(\lambda^f) \geq 0$ for all $f \in R$. Since $\nu_{P'}(P) > 0$, we must have $\nu_{P'}(\lambda^h) > 0$ for at least one $h \in A(P^n)$. If $f \in A(P^n)$ is arbitrary, then $f \equiv hk \pmod{P^n}$ for some $k \in A(P^n)$, and so $\lambda^f = (\lambda^h)^k$. It follows now from Lemma 3.2.1 and $\gcd(k, P) = 1$ that $\nu_{P'}(\lambda^f) = \nu_{P'}(\lambda^h)$. Thus, by going back to (3.1), we get

$$e(P'|P) = \Phi(P^n)\nu_{P'}(\lambda^h), \tag{3.2}$$

and so $e(P'|P) \geq \Phi(P^n)$. On the other hand,

$$e(P'|P) \leq [F(\Lambda_{P^n}) : F] \leq \deg(g(z)) = \Phi(P^n),$$

hence $e(P'|P) = [F(\Lambda_{P^n}) : F] = \Phi(P^n)$. Furthermore, $g(z)$ is the minimal polynomial of λ over F.

(ii) Any $\sigma \in \mathrm{Gal}(F(\Lambda_{P^n})/F)$ is uniquely determined by the element $\sigma(\lambda)$. It follows from the above that there exists a polynomial $f_\sigma \in A(P^n)$ such that $\sigma(\lambda) = \lambda^{f_\sigma}$. It is easy to verify that $\sigma \mapsto f_\sigma + (P^n) \in (R/(P^n))^*$ gives a canonical isomorphism between $\mathrm{Gal}(F(\Lambda_{P^n})/F)$ and $(R/(P^n))^*$.

(iii) We have already shown in the proof of (i) that P is totally ramified in $F(\Lambda_{P^n})/F$. Now let $Q \neq P$ be a finite place of F. By differentiating the identity $z^{P^n} = z^{P^{n-1}} g(z)$ with respect to z and using Lemma 3.2.1, we obtain

$$P^n = \lambda^{P^{n-1}} g'(\lambda). \tag{3.3}$$

Let Q' be an arbitrary place of $F(\Lambda_{P^n})$ lying over Q. From $\lambda^{P^n} = 0$ we get $\nu_{Q'}(\lambda) \geq 0$, hence also $\nu_{Q'}(\lambda^f) \geq 0$ for all $f \in R$, and so $\nu_{Q'}(g'(\lambda)) \geq 0$. But $\nu_{Q'}(P^n) = 0$, and so we must have $\nu_{Q'}(g'(\lambda)) = 0$. It follows then from a well-known sufficient condition for unramified places (see e.g. [152, Corollary III.5.11]) that Q is unramified in $F(\Lambda_{P^n})/F$. $\square$

Lemma 3.2.5 *Suppose that M and N are two monic polynomials in R with $\gcd(M, N) = 1$. Then $F(\Lambda_{MN})$ is the composite field of $F(\Lambda_M)$ and $F(\Lambda_N)$.*

Proof. For nonzero $K, L \in R$ with K dividing L it is clear that $\Lambda_K \subseteq \Lambda_L$. Therefore $F(\Lambda_M) \cdot F(\Lambda_N) \subseteq F(\Lambda_{MN})$. Conversely, we observe that Λ_{MN} is a cyclic R-module by Lemma 3.2.2(iii), and so we can choose a generator λ of Λ_{MN}. By Lemma 3.2.2(ii) we can write $\lambda = \lambda_1 + \lambda_2$ with $\lambda_1 \in \Lambda_M$, $\lambda_2 \in \Lambda_N$, and so $F(\Lambda_{MN}) = F(\lambda) \subseteq F(\Lambda_M) \cdot F(\Lambda_N)$. $\square$

Theorem 3.2.6 *Let M be a monic polynomial in R and put $E = F(\Lambda_M)$. Then:*

(i) $\mathrm{Gal}(E/F) \simeq (R/(M))^*$, *and therefore E/F is a finite extension of degree $\Phi(M)$.*

(ii) *For any generator λ of the cyclic R-module Λ_M,*

$$g_M(z) = \prod_{f \in A(M)} (z - \lambda^f)$$

is the minimal polynomial of λ over F.

(iii) *A monic irreducible polynomial $P \in R$ is unramified in E/F if P does not divide M.*

(iv) *For a monic irreducible polynomial $P \in R$ not dividing M, the Artin symbol $\left[\frac{E/F}{P}\right]$ satisfies*

$$\left[\frac{E/F}{P}\right]: \ \lambda \mapsto \lambda^P$$

for any generator λ of the cyclic R-module Λ_M.

(v) *For a monic irreducible polynomial $P \in R$, suppose that $P^m \| M$ for some $m \geq 0$. Then the ramification index $e_P(E/F)$ is equal to $\Phi(P^m)$ and the relative degree $f_P(E/F)$ is equal to the multiplicative order of P modulo M/P^m.*

Proof. (i) We can assume that M is nonconstant. Let M have the factorization $M = \prod_{j=1}^{s} P_j^{e_j}$ with distinct monic irreducible polynomials $P_1, \ldots, P_s \in R$. Lemma 3.2.5 shows that E is the composite field

$$F(\Lambda_{P_1^{e_1}}) \cdots F(\Lambda_{P_s^{e_s}}).$$

By looking at the ramification of P_j and using Proposition 3.2.4(iii), we see that each extension $F(\Lambda_{P_j^{e_j}})/F$ is linearly disjoint from the composite of the remaining ones. Then by Proposition 3.2.4(ii),

$$\mathrm{Gal}(E/F) \simeq \prod_{j=1}^{s} \mathrm{Gal}(F(\Lambda_{P_j^{e_j}})/F) \simeq \prod_{j=1}^{s} (R/(P_j^{e_j}))^{*}.$$

The desired result follows.

(ii) The degree of $g_M(z)$ is equal to $\Phi(M)$. Furthermore, we have $E = F(\lambda)$. If $\sigma \in \mathrm{Gal}(E/F)$, then $\sigma(\lambda)$ is again a generator of the cyclic R-module Λ_M, and so $\sigma(\lambda) = \lambda^f$ for some $f \in A(M)$. Moreover, it is easy to check that $\lambda^f \neq \lambda^h$ for any two distinct $f, h \in A(M)$. Thus, $g_M(z)$ is the minimal polynomial of λ over F.

(iii) With the notation as in the proof of (i), if P does not divide M, then by Proposition 3.2.4(iii), P is unramified in each component $F(\Lambda_{P_j^{e_j}})/F$ of $E = F(\Lambda_{P_1^{e_1}}) \cdots F(\Lambda_{P_s^{e_s}})$. Hence by Corollary 1.4.9(ii), P is unramified in E/F.

(iv) Let Q be a place of E lying over P and put $\sigma = \left[\frac{E/F}{P}\right]$. Then $\sigma(\lambda) \equiv \lambda^{q^d} \pmod{\mathsf{M}_Q}$, where $d = \deg(P)$. By considering the polynomial z^P in z, we obtain $\lambda^P \equiv \lambda^{q^d} \pmod{\mathsf{M}_Q}$. Furthermore, by differentiating

$$z^M = \prod_{\substack{f \in R \\ \deg(f) < \deg(M)}} (z - \lambda^f)$$

with respect to z, putting $z = \lambda^h$, and using $\nu_Q(M) = 0$, we see that $\lambda^f \not\equiv \lambda^h \pmod{\mathsf{M}_Q}$ for any two distinct $f, h \in R$ with $\deg(f) < \deg(M)$ and $\deg(h) < \deg(M)$. Thus, we must have $\sigma(\lambda) = \lambda^P$.

(v) Let $M = P^m N$ with $\gcd(P, N) = 1$. Then P is unramified in $F(\Lambda_N)/F$ as P is relatively prime to N, and P is totally ramified in $F(\Lambda_{P^m})/F$. Thus, we obtain

$$e_P(E/F) = e_P(F(\Lambda_{P^m})/F) = [F(\Lambda_{P^m}) : F] = \Phi(P^m).$$

Again, since P is totally ramified in $F(\Lambda_{P^m})/F$, the relative degree of P in E/F is the same as the relative degree of P in $F(\Lambda_N)/F$, which by Theorem 1.4.11(ii) is equal to the order r in the Galois group $\mathrm{Gal}(F(\Lambda_N)/F)$ of the Artin symbol of P in $F(\Lambda_N)/F$. By (iv), r is equal to the multiplicative order of P modulo $N = M/P^m$. $\square$

So far we have investigated the structure of the Galois group of cyclotomic function fields $F(\Lambda_M)/F$ and the decomposition of all places of F corresponding to monic irreducible polynomials. In order to obtain the genus of $F(\Lambda_M)$, we need to study also the behavior of the infinite place in the extension $F(\Lambda_M)/F$, i.e., of the place ∞ of $F = \mathbf{F}_q(T)$ which is the pole of T.

Lemma 3.2.7 *Let P be a monic irreducible polynomial in R. Then the infinite place ∞ of F splits into $\Phi(P)/(q-1)$ places of $F(\Lambda_P)$ and the ramification index $e_\infty(F(\Lambda_P)/F)$ is equal to $q - 1$.*

Proof. We have $E := F(\Lambda_P) = F(\lambda)$, where λ is a generator of the cyclic R-module Λ_P. Suppose that $d = \deg(P)$ and that the minimal polynomial of λ over F is

$$g_P(z) = g_0(T) + g_1(T)z^{q-1} + \cdots + g_d(T)z^{q^d-1},$$

where $g_i(T) \in \mathbf{F}_q[T]$ and $\deg(g_i) = (d-i)q^i$ for $0 \leq i \leq d$ by Lemma 3.2.1.

Let P_∞ be a place of E lying over ∞. Consider the P_∞-adic completion E_{P_∞} of E and the Newton polygon of $g_P(z)$ over E_{P_∞}:

$$A_0 = (0, -de), \quad A_1 = (q-1, -(d-1)qe), \ldots, \quad A_d = (q^d - 1, 0),$$

where e is the ramification index of ∞ in E/F. Then $\overline{A_0 A_1}$ is a segment of the Newton polygon. Hence there are exactly $q-1$ roots $\alpha_1, \ldots, \alpha_{q-1} \in E_{P_\infty}$ of $g_P(z)$ such that $\nu_{P_\infty}(\alpha_j) = e((d-1)q - d)/(q-1)$ for $1 \leq j \leq q-1$. As $(d-1)q - d$ is relatively prime to $q-1$, we have $e \geq q-1$. Furthermore, by considering the Newton polygon of $g_P(z)$ over the ∞-adic completion F_∞ of F, we see that $\prod_{j=1}^{q-1}(z - \alpha_j)$ is a polynomial belonging to $F_\infty[z]$. Therefore the number of places of E lying over ∞ is at least $\Phi(P)/(q-1)$. Thus, $e = q-1$ and there are exactly $\Phi(P)/(q-1)$ places of E lying over ∞. $\square$

Theorem 3.2.8 *Let $P \in R$ be a monic irreducible polynomial and $n \geq 1$. Then the infinite place ∞ of F splits into $\Phi(P^n)/(q-1)$ places of $F(\Lambda_{P^n})$ and the ramification index $e_\infty(F(\Lambda_{P^n})/F)$ is equal to $q-1$. In particular, the decomposition field and the inertia field of ∞ in $F(\Lambda_{P^n})/F$ coincide and $\mathbf{F}_q$ is still the full constant field of $F(\Lambda_{P^n})$.*

Proof. By looking at the Newton polygons of the minimal polynomial of a generator λ of the cyclic R-module Λ_{P^n} over F, as we did in the proof of the above lemma, we obtain the desired result. $\square$

Theorem 3.2.9 *Let $P \in R$ be a monic irreducible polynomial of degree d and $E := F(\Lambda_{P^n})$ for some $n \geq 1$. Then for the genus $g(E)$ of E we have*

$$2g(E) - 2 = q^{d(n-1)}\left((qdn - dn - q)\frac{q^d - 1}{q-1} - d\right).$$

Proof. Let Q be the unique place of E lying over P and $\lambda \in E$ a root of $g(z) := z^{P^n}/z^{P^{n-1}}$. Then it follows from (3.2) and the identity $e(P'|P) = \Phi(P^n)$ in the proof of Proposition 3.2.4(i) that λ is a local parameter at Q. Furthermore, the identity (3.3) in the proof of Proposition 3.2.4(iii) shows that $g'(\lambda) = P^n/\lambda^{P^{n-1}}$. By Proposition 1.3.13, the different exponent $d(Q|P)$ satisfies

$$\begin{aligned} d(Q|P) &= \nu_Q(g'(\lambda)) = \nu_Q(P^n) - \nu_Q(\lambda^{P^{n-1}}) \\ &= n\Phi(P^n) - q^{\deg(P^{n-1})} = n(q^d - 1)q^{d(n-1)} - q^{d(n-1)}. \end{aligned}$$

Here we used the fact that $z^{P^{n-1}} \in R[z]$ is an Eisenstein polynomial at P for $n \geq 2$. Since according to Theorem 3.2.8, ∞ is unramified (if $q = 2$) or tamely ramified in E/F with ramification index $q-1$, our result follows from the Hurwitz genus formula. $\square$

3.3   Drinfeld Modules of Rank 1

The theory of cyclotomic function fields presented in the previous section is a special case of the theory of Drinfeld modules of rank 1. Instead of the rational function field over $\mathbf{F}_q$, any global function field can serve as a base field in the more general theory. In this section we summarize those results on Drinfeld modules of rank 1 that are relevant for our purposes. The development of the theory of Drinfeld modules in the way in which it is presented here is due to Hayes [47], [48]. The proofs proceed in analogy with those in Section 3.2. Detailed accounts of the theory of Drinfeld modules can be found in Goss [42] and Hayes [49].

Throughout this section, we work over a global function field $F/\mathbf{F}_q$ with characteristic p and $q = p^m$. We assume that $N(F) \geq 1$ and distinguish a rational place ∞ of F. Denote by A the ∞'-integral ring of F with $\infty' := \mathbf{P}_F \setminus \{\infty\}$, that is,

$$A = \{x \in F : \nu_P(x) \geq 0 \text{ for all } P \in \mathbf{P}_F \text{ with } P \neq \infty\}.$$

Then A is a Dedekind ring with class number $h(A) = |\mathrm{Cl}(O_{\infty'})| = h(F)$, where $h(F)$ is the divisor class number of the global function field F (compare with Proposition 1.2.5).

Let H_A be the ∞'-Hilbert class field of F and let $\pi : c \mapsto c^p$ be the Frobenius endomorphism of H_A. Consider the left twisted polynomial ring $H_A[\pi]$ whose elements are polynomials in π with coefficients from H_A written on the left; but multiplication in $H_A[\pi]$ is twisted by the rule

$$\pi u = u^p \pi \qquad \text{for all } u \in H_A.$$

Let $D : H_A[\pi] \longrightarrow H_A$ be the map which assigns to each polynomial in $H_A[\pi]$ its constant term.

Definition 3.3.1 A **Drinfeld A-module** of rank 1 over H_A is a ring homomorphism $\phi : A \longrightarrow H_A[\pi]$, $a \mapsto \phi_a$, such that:

(i) not all elements of $H_A[\pi]$ in the image of ϕ are constant polynomials;

(ii) $D \circ \phi$ is the identity on A;

(iii) $\deg(\phi_a) = -m\nu_\infty(a)$ for all nonzero $a \in A$, where $\deg(\phi_a)$ is the degree of ϕ_a as a polynomial in π.

Remark 3.3.2 For a Drinfeld A-module ϕ as above, we have $\phi_\alpha = \alpha$ for any $\alpha \in \mathbf{F}_q$. Note that ϕ actually takes its values in the subring $H_A[\pi^m]$ of $H_A[\pi]$ and that ϕ_a is an $\mathbf{F}_q$-linear map on H_A for any $a \in A$. Furthermore, the map ϕ is always injective.

Example 3.3.3 Consider the rational function field $F = \mathbf{F}_q(T)$ with ∞ being the infinite place of F. Then we have $A = \mathbf{F}_q[T]$ and $H_A = F = \mathbf{F}_q(T)$. A Drinfeld A-module ϕ of rank 1 over F is uniquely determined by the image ϕ_T of T. By the definition we must have

$$D(\phi_T) = (D \circ \phi)(T) = T,$$

i.e., ϕ_T is a nonconstant polynomial in π with the constant term T. Since $\deg(\phi_T) = -m\nu_\infty(T) = m$, we know that ϕ_T is of the form $T + f(\pi)\pi + x\pi^m$ for an element $x \in F^*$

and $f(\pi) \in F[\pi]$ with $\deg(f(\pi)) \leq m - 2$. Taking $x = 1$ and $f(\pi) = 0$ gives the so-called **Carlitz module** in the previous section which formed the starting point of the theory of cyclotomic function fields.

Definition 3.3.4 We fix a sign function sgn as in Definition 2.6.1. We say that a Drinfeld A-module ϕ of rank 1 over H_A is sgn-**normalized** if $\text{sgn}(a)$ is equal to the leading coefficient of ϕ_a for all $a \in A$.

It is shown in [49] that sgn-normalized Drinfeld A-modules of rank 1 over H_A always exist. The following result is a special case of [49, Proposition 4.1].

Lemma 3.3.5 *The twisted polynomial ring $H_A[\pi]$ is a left principal ideal ring.*

Given a Drinfeld A-module ϕ of rank 1 over H_A and a nonzero ideal M of A, let $I_{M,\phi}$ be the left ideal generated in $H_A[\pi]$ by the twisted polynomials ϕ_a, $a \in M$. As left ideals are principal, $I_{M,\phi} = H_A[\pi]\phi_M$ for a unique monic twisted polynomial $\phi_M \in H_A[\pi]$.

Let L be any H_A-algebra. Then for a polynomial $f(\pi) = \sum_{i=0}^{k} b_i \pi^i \in H_A[\pi]$ the action of $f(\pi)$ on L is defined by

$$f(\pi)(t) = \sum_{i=0}^{k} b_i t^{p^i} \quad \text{for all } t \in L.$$

Let $\overline{H_A}$ denote a fixed algebraic closure of H_A whose additive group $(\overline{H_A}, +)$ is equipped with an A-module structure under the action of ϕ.

Definition 3.3.6 Let ϕ be a sgn-normalized Drinfeld A-module of rank 1 over H_A and M be a nonzero ideal of A. The M-**torsion module** $\Lambda_\phi(M)$ associated with ϕ is defined by

$$\Lambda_\phi(M) = \{t \in (\overline{H_A}, +) : \phi_M(t) = 0\}.$$

The following are a few basic facts on $\Lambda_\phi(M)$:

(i) $\Lambda_\phi(M)$ is a finite set of cardinality $|\Lambda_\phi(M)| = p^{\deg(\phi_M)}$;

(ii) $\Lambda_\phi(M)$ is an A-submodule of $(\overline{H_A}, +)$ and a cyclic A-module isomorphic to A/M;

(iii) $\Lambda_\phi(M)$ has $\Phi(M) := |(A/M)^*|$ generators as a cyclic A-module, where $(A/M)^*$ is the group of units of the ring A/M.

The elements of $\Lambda_\phi(M)$ are also called the M-**torsion elements** in $(\overline{H_A}, +)$. By adjoining these M-torsion elements to H_A, we obtain the field $K_M := H_A(\Lambda_\phi(M))$, which turns out to be a narrow ray class field (compare with Section 2.6). For a nonzero ideal M of A, we define the corresponding positive divisor

$$D := \sum_{P \in \infty'} \nu_P(M)P \in \text{Div}(F). \tag{3.4}$$

Thus, we also denote $\Lambda_\phi(M)$ by $\Lambda_\phi(D)$. In this way we can also identify prime ideals P of the ring A with places P of the field F, and this will be done in the remainder of the section.

In the case where F is the rational function field and ϕ is the Carlitz module in Example 3.3.3, the field K_M is the cyclotomic function field over F with modulus M in Section 3.2.

Theorem 3.3.7 *Let ϕ be a sgn-normalized Drinfeld A-module of rank 1 over H_A. Then the field $H_A(\Lambda_\phi(D))$ is F-isomorphic to the narrow ray class field $F^D(\infty)$ with modulus D for any positive divisor D of F with $\infty \notin \mathrm{supp}(D)$.*

Proposition 3.3.8 *Let $K_\mathsf{M} = H_A(\Lambda_\phi(\mathsf{M})) = H_A(\Lambda_\phi(D))$. Then:*

(i) *K_M is independent of the specific choice of the sgn-normalized Drinfeld A-module ϕ of rank 1 over H_A.*

(ii) *K_M/F is unramified away from ∞ and the places $P \in \infty'$ with $\nu_P(\mathsf{M}) > 0$.*

(iii) *The extension K_M/F is abelian and there is an isomorphism*

$$\sigma : \mathrm{Cl}_D^+(A) \longrightarrow \mathrm{Gal}(K_\mathsf{M}/F),$$

determined by $\sigma_\mathsf{J}\phi = \mathsf{J} * \phi$ (see [49, Section 4] for the notation) for any nonzero ideal J of A coprime to M, and $\lambda^{\sigma_\mathsf{J}} = \phi_\mathsf{J}(\lambda)$ for any generator λ of the cyclic A-module $\Lambda_\phi(\mathsf{M})$. Moreover, for any prime ideal P of A that is coprime to M, the corresponding Artin symbol in K_M/F is exactly σ_P. Furthermore, if $\mathsf{M} = P^n$ for some prime ideal P of A and $n \geq 1$, then both the decomposition group and the inertia group of ∞ in K_M/F are isomorphic to $\mathbb{F}_q^*$.

(iv) *The multiplicative group $(A/\mathsf{M})^*$ is isomorphic to $\mathrm{Gal}(K_\mathsf{M}/H_A)$ by means of*

$$b \mapsto \sigma_{bA},$$

where $b \in A$ satisfies $\mathrm{sgn}(b) = 1$ and is coprime to M.

(v) *Suppose that $\mathsf{M} = P^n$ for some prime ideal P of A and $n \geq 1$. Let λ be a generator of the cyclic A-module $\Lambda_\phi(\mathsf{M})$. Then $K_\mathsf{M} = H_A(\lambda)$ and the minimal polynomial of λ over H_A is*

$$f(z) := \frac{\phi_{P^n}(z)}{\phi_{P^{n-1}}(z)}.$$

Moreover, $f(z)$ is an Eisenstein polynomial at any place Q of H_A lying over P. Thus, Q is totally ramified in K_M/H_A and $\nu_R(\lambda) = 1$ for the place R of K_M lying over Q.

(vi) *Let $\mathsf{M} = P_1^{n_1} P_2^{n_2}$ be a product of two coprime primary ideals of A. Then K_M is the composite field of $K_{P_1^{n_1}}$ and $K_{P_2^{n_2}}$ and furthermore $K_{P_1^{n_1}} \cap K_{P_2^{n_2}} = H_A$.*

The genus formula for K_{P^n} can be obtained as in the case of cyclotomic function fields. The following result was established by Xing and Niederreiter [180], [181].

Proposition 3.3.9 *Let $P \neq \infty$ be a place of $F/\mathbb{F}_q$ of degree d and put $E = K_{P^n}$ for an integer $n \geq 1$. Then for the genus $g(E)$ of E we have*

$$2g(E) - 2 = h(F)q^{d(n-1)} \left((2g(F) - 2)(q^d - 1) + dn(q^d - 1) - d + \frac{(q^d - 1)(q - 2)}{q - 1} \right).$$

Proof. Only the places P and ∞ can be ramified in the extension E/F, so it suffices to calculate the different exponents of P and ∞. By Proposition 3.3.8(iii), we have $e_\infty(E/F) = q - 1$, and so $d_\infty(E/F) = q - 2$.

Now let Q be a place of H_A lying over P and R the unique place of E lying over Q. Let λ be a generator of the cyclic A-module $\Lambda_\phi(\mathsf{P}^n)$ with the minimal polynomial $f(z)$ over H_A. Then by Proposition 1.3.13, the different exponent $d(R|Q)$ of R over Q is

$$d(R|Q) = \nu_R(f'(\lambda)) = \nu_Q(\mathrm{N}_{E/H_A}(f'(\lambda))) = \nu_Q(\mathrm{N}_{E/H_A}(\frac{\phi'_{\mathsf{P}^n}(\lambda)}{\phi_{\mathsf{P}^{n-1}}(\lambda)})).$$

Since $\phi_{\mathsf{P}^n}(z)$ is a q-polynomial, $\phi'_{\mathsf{P}^n}(\lambda)$ is equal to the constant term $D(\phi_{\mathsf{P}^n})$ of ϕ_{P^n} as a polynomial in the Frobenius endomorphism π. From the proof of [49, Proposition 11.4] it follows that $\nu_Q(D(\phi_{\mathsf{P}})) = 1$.

Let ψ be the Drinfeld A-module $\mathsf{P}^{n-1} * \phi$ (see [49, Section 4] for the notation). Then $g(z) := \psi_{\mathsf{P}}(z)/z$ is a polynomial in $H_A[z]$ which is Eisenstein at Q by Proposition 3.3.8(v) and satisfies

$$\phi_{\mathsf{P}^n}(z) = g(\phi_{\mathsf{P}^{n-1}}(z)) \cdot \phi_{\mathsf{P}^{n-1}}(z). \tag{3.5}$$

Note also that $D(\phi_{\mathsf{P}^n}) = g(0)D(\phi_{\mathsf{P}^{n-1}})$. Then by induction one finds $\nu_Q(D(\phi_{\mathsf{P}^n})) = n$.

Applying $\phi_{\mathsf{P}^n}(\lambda) = 0$, we have $g(\phi_{\mathsf{P}^{n-1}}(\lambda)) = 0$ by (3.5). Let G be the field generated over H_A by the elements $\alpha \in \overline{H_A}$ with $\psi_{\mathsf{P}}(\alpha) = 0$. Then $H_A \subseteq G \subseteq E$ and $[G : H_A] = \Phi(\mathsf{P})$. Thus,

$$\mathrm{N}_{E/H_A}(\phi_{\mathsf{P}^{n-1}}(\lambda)) = \mathrm{N}_{G/H_A}(\mathrm{N}_{E/G}(\phi_{\mathsf{P}^{n-1}}(\lambda))) = (\pm D(\psi_{\mathsf{P}}))^{\Phi(\mathsf{P}^n)/\Phi(\mathsf{P})}.$$

Therefore

$$d(R|Q) = n\Phi(\mathsf{P}^n) - \frac{\Phi(\mathsf{P}^n)}{\Phi(\mathsf{P})}.$$

Since the extension H_A/F is unramified, the tower formula for different exponents (see Proposition 1.3.11) shows that

$$d_P(E/F) = n\Phi(\mathsf{P}^n) - \frac{\Phi(\mathsf{P}^n)}{\Phi(\mathsf{P})} = nq^{d(n-1)}(q^d - 1) - q^{d(n-1)}.$$

Thus, the Hurwitz genus formula and the fact that $[H_A : F] = h(F)$ yield the desired result.   □

Chapter 4

Function Fields with Many Rational Places

We say informally that a global function field $K/\mathbf{F}_q$ has many rational places if its number $N(K)$ of rational places is reasonably close to $N_q(g(K))$ or to a known upper bound on $N_q(g(K))$. Recall from Definition 1.6.14 that $N_q(g)$ is the maximum number of rational places that a global function field with full constant field $\mathbf{F}_q$ and genus g can have. We refer to Section 1.6 for upper bounds on $N_q(g)$.

In this chapter, we employ class field theory as well as explicit function fields such as Kummer and Artin-Schreier extensions to construct global function fields with many rational places. Ray class fields (including Hilbert class fields) and narrow ray class fields are also good candidates as their Galois groups and Artin symbols are known.

4.1   Function Fields from Hilbert Class Fields

For a global function field $F/\mathbf{F}_q$ with a rational place ∞, the ∞'-Hilbert class field H of F with $\infty' := \mathbf{P}_F \setminus \{\infty\}$ is a finite unramified abelian extension of F in which ∞ splits completely. Moreover, $[H : F] = h(F)$, the divisor class number of F, and the Galois group $\mathrm{Gal}(H/F)$ is isomorphic to the group $\mathrm{Cl}(F)$ of divisor classes of degree zero of F. With this canonical isomorphism, the Artin symbol in H/F of a place P of F corresponds to the divisor class $[P - \deg(P)\infty]$. As in Section 1.2, we write $[D]$ for the divisor class of $D \in \mathrm{Div}(F)$. According to Proposition 1.4.12, in order to construct a subfield K of H/F with rational places $P_1, \ldots, P_r$ of F splitting completely in K/F, one has to choose a field K such that $\mathrm{Gal}(H/K)$ contains the divisor classes $[P_i - \infty]$ for all $i = 1, \ldots, r$. Based on this idea, we establish some general principles for the construction of global function fields with many rational places in this section. We first present a simple example to illustrate the power of this method.

Example 4.1.1 Let $F = \mathbf{F}_2(x, y)$ be the Artin-Schreier extension of the rational function field $\mathbf{F}_2(x)$ defined by

$$y^2 + y = \frac{(x+1)(x^2+x+1)(x^3+x+1)}{x^3}.$$

Then $g(F) = 3$ by Theorem 3.1.9 and $h(F) = 24$. Let ∞ be the unique place of F lying over the infinite place of $\mathbf{F}_2(x)$ and let H be the ∞'-Hilbert class field of F. Let P be the unique zero of x in F and C the cyclic subgroup of $\mathrm{Cl}(F)$ generated by the divisor class $[P - \infty]$. From $\mathrm{div}(x) = 2P - 2\infty$ and the Weierstrass gap theorem (see Corollary 1.1.5) it follows that $|C| = 2$. Thus, if K is the subfield of H/K fixed by C, then K/F is an unramified extension of degree 12 in which ∞ and P split completely. Therefore $g(K) = 25$ and $N(K) \geq 24$. But $N_2(25) \leq 24$ by the method in Theorem 1.6.18, hence $N(K) = 24$ and K is an optimal function field. This example was found by Niederreiter and Xing [108] (see also Xing and Niederreiter [181]) and it was the first optimal function field for the case $q = 2$ and $g = 25$ at the time of its discovery.

The following result is due to Niederreiter and Xing [111].

Theorem 4.1.2 *Let q be odd, let R be a subset of $\mathbf{F}_q$, and put $n = |R|$. Choose a polynomial $f \in \mathbf{F}_q[x]$ such that $\deg(f)$ is odd, f has no multiple roots, and $f(c) = 0$ for all $c \in R$. For the global function field $F = \mathbf{F}_q(x, y)$ with $y^2 = f(x)$, assume that its divisor class number $h(F)$ is divisible by $2^n m$ for some positive integer m. Then there exists a global function field $K/\mathbf{F}_q$ such that*

$$g(K) = \frac{h(F)}{2^{n+1}m}(\deg(f) - 3) + 1 \quad and \quad N(K) \geq \frac{(n+1)h(F)}{2^n m},$$

with equality if $n = q$.

Proof. Note that F is a Kummer extension of the rational function field $\mathbf{F}_q(x)$ with

$$g(F) = \frac{1}{2}(\deg(f) - 1)$$

by Corollary 3.1.5. For each $c \in R$ the place $x - c$ of $\mathbf{F}_q(x)$ is totally ramified in $F/\mathbf{F}_q(x)$, and so is the pole of x in $\mathbf{F}_q(x)$. Let ∞ denote the unique place of F lying over the pole of x in $\mathbf{F}_q(x)$. For the principal divisor $\mathrm{div}(x - c)$ of F we thus have

$$\mathrm{div}(x - c) = 2P_c - 2\infty,$$

where all $P_c, c \in R$, are rational places of F. Consequently, the divisor class $[P_c - \infty]$ has order 1 or 2 in the group $\mathrm{Cl}(F)$, and so the subgroup J of $\mathrm{Cl}(F)$ generated by all divisor classes $[P_c - \infty], c \in R$, has order dividing 2^n. It follows that there exists a subgroup G of $\mathrm{Cl}(F)$ with $|G| = 2^n m$ and $G \supseteq J$. Let H be the ∞'-Hilbert class field of F and let K be the subfield of the extension H/F fixed by G. Then

$$[K : F] = \frac{h(F)}{2^n m}.$$

By construction, the places ∞ and $P_c, c \in R$, split completely in the extension K/F, and this yields the desired lower bound for $N(K)$. Furthermore, K/F is an unramified extension, and so the formula for $g(K)$ follows immediately from the Hurwitz genus formula. $\square$

In the following tables we list a few examples of global function fields $K/\mathbf{F}_q$ ($q = 3, 5$) with many rational places that are obtained from Theorem 4.1.2. The tables contain the following data: the value of the genus $g(K)$, the value or a lower bound for the number $N(K)$ of rational places, the values of n and m, the polynomial $f(x)$, and the value of the divisor class number $h(F)$ of $F = \mathbf{F}_q(x, y)$ with $y^2 = f(x)$. In the cases where the exact value of $N(K)$ is indicated, it can be obtained using Theorem 4.1.2 or by other simple arguments. The divisor class numbers $h(F)$ have been calculated by the standard method based on the results in Section 1.6 and with the help of the software package Mathematica. The first three examples in Table 4.1.1 are from Niederreiter and Xing [105] and the remaining two examples are from Niederreiter and Xing [110]. All examples in Table 4.1.2 are from Niederreiter and Xing [111], where also many more examples for $q = 5$ can be found.

Table 4.1.1 $(q = 3)$

$g(K)$	$N(K)$	n	m	$f(x)$	$h(F)$
11	$= 20$	1	1	$x(x^2 + 1)(x^2 - x - 1)$	20
12	$= 22$	1	1	$x(x^4 + x - 1)$	22
13	$= 24$	1	1	$x(x^4 - x^3 + x^2 - x + 1)$	24
27	$= 39$	2	1	$2x(x + 2)(x^5 + x^4 + x^2 + x + 1)$	52
29	$= 42$	2	1	$2x(x + 2)(x^5 + x^3 + x + 1)$	56

Table 4.1.2 $(q = 5)$

$g(K)$	$N(K)$	n	m	$f(x)$	$h(F)$
15	$= 35$	4	1	$x(x + 1)(x + 2)(x - 1)(x^3 + x^2 + x - 2)$	112
19	≥ 45	4	1	$x(x + 1)(x + 2)(x - 2)(x^3 - 2x^2 + 2x - 2)$	144
21	$= 50$	4	1	$(x^5 - x)(x^2 - x + 1)$	160
23	$= 55$	4	1	$x(x + 1)(x + 2)(x - 1)(x^3 + x^2 - 2x + 1)$	176
24	$= 46$	1	1	$x(x^4 + x^3 + 2x^2 + x - 2)$	46
27	$= 52$	1	1	$x(x - 1)(x^3 - x + 2)$	52
28	$= 54$	5	2	$(x^5 - x)(x^2 - 2x - 2)(x^2 - 2x - 1)$	576
29	≥ 56	3	1	$x(x + 1)(x + 2)(x - 1)(x^3 + x^2 + x - 2)$	112
30	$= 58$	1	1	$x(x^4 + x^2 + 2)$	58
32	$= 62$	1	1	$x(x^4 + 2x^3 - 2x^2 - 2x + 2)$	62
35	≥ 68	3	1	$x(x + 1)(x + 2)(x^4 + x^2 - 2x - 2)$	136
37	$= 72$	3	1	$x(x + 1)(x + 2)(x^4 - 2x - 1)$	144
39	$= 76$	3	1	$x(x + 1)(x + 2)(x^4 + x^3 - 2x^2 + 2x + 1)$	152
40	$= 65$	4	3	$x(x + 1)(x + 2)(x - 2)(x^5 + 2x^2 - 2x + 1)$	624
41	$= 80$	3	1	$x(x + 1)(x + 2)(x^4 + x - 1)$	160
43	$= 84$	3	1	$x(x + 1)(x + 2)(x^4 - 2x^2 - 2)$	168
45	$= 88$	3	1	$x(x + 1)(x + 2)(x^4 + 2x^2 + 2x + 1)$	176
46	≥ 75	4	4	$x(x + 1)(x + 2)(x - 2)(x^3 - x^2 - x + 2)(x^2 + x + 2)$	960
47	$= 92$	3	1	$x(x + 1)(x + 2)(x^4 - 2x^2 - x - 2)$	184
49	$= 96$	3	1	$x(x + 1)(x + 2)(x^4 + x^3 + 2x^2 + 2)$	192
52	$= 102$	5	1	$(x^5 - x)(x^4 + x^2 + 2x + 2)$	544
53	$= 104$	3	1	$x(x + 1)(x + 2)(x^2 + x + 2)(x^2 - x + 1)$	208
55	$= 108$	3	1	$x(x + 1)(x + 2)(x^4 + x^2 + 2x + 2)$	216
57	$= 112$	3	1	$x(x + 1)(x + 2)(x^4 - 2x^2 + x + 1)$	224

Theorem 4.1.2 gives a construction based on quadratic extensions of rational function fields. This construction can be generalized as follows (see Niederreiter and Xing [107]).

Theorem 4.1.3 *Let $F/\mathbf{F}_q$ be a global function field and $L/\mathbf{F}_q$ a finite separable extension of F. Let $S = \mathbf{P}_F \setminus \{P, P_1, \ldots, P_m\}$ with P a rational place of F and $P_1, \ldots, P_m$ arbitrary places of F different from P. Suppose that S satisfies the condition that some place in S is totally ramified in L/F. Let $\mathcal{T}'$ be the over-set of $S' = \{P, P_1, \ldots, P_m\}$ with respect to the extension L/F and assume that the number n of rational places in $\mathcal{T}'$ is positive. Then there exists a global function field $K/\mathbf{F}_q$ with*

$$g(K) = \frac{h(F)}{|G|}(g(L) - 1) + 1 \quad and \quad N(K) \geq \frac{h(F)n}{|G|},$$

where G is the subgroup of $\mathrm{Cl}(F)$ generated by the divisor classes $[P_1 - \deg(P_1)P], \ldots, [P_m - \deg(P_m)P]$.

Proof. Let $B := \mathrm{Div}(F)/\mathrm{Princ}(F)$ be the divisor class group of F and let J be the subgroup of B generated by the divisor classes $[P], [P_1], \ldots, [P_m]$. Since S' contains the rational place P, the group B is generated by $\mathrm{Cl}(F)$ and J. Thus, from the exact sequence

$$0 \rightarrow \mathrm{Cl}(F)/(J \cap \mathrm{Cl}(F)) \rightarrow \mathrm{Cl}(O_S) \rightarrow B/\mathrm{Cl}(F)J \rightarrow 0$$

we obtain

$$\mathrm{Cl}(O_S) \simeq \mathrm{Cl}(F)/(J \cap \mathrm{Cl}(F)),$$

where $\mathrm{Cl}(O_S)$ is the fractional S-ideal class group of F. It follows that

$$r := |\mathrm{Cl}(O_S)| = \frac{h(F)}{|G|}.$$

From the condition on S and [131, Proposition 2.2] we deduce that r divides $|\mathrm{Cl}(O_T)|$, where $\mathrm{Cl}(O_T)$ is the fractional $\mathcal{T}$-ideal class group of L with $\mathcal{T} = \mathbf{P}_L \setminus \mathcal{T}'$. Let H_T be the $\mathcal{T}$-Hilbert class field of L. Then $\mathrm{Gal}(H_T/L) \simeq \mathrm{Cl}(O_T)$ and $\mathbf{F}_q$ is the full constant field of H_T since $n \geq 1$. Let $K/\mathbf{F}_q$ be a subfield of the extension H_T/L which is obtained as the fixed field of a subgroup of $\mathrm{Cl}(O_T)$ of order $\frac{1}{r}|\mathrm{Cl}(O_T)|$. Then $[K : L] = r$. Since H_T/L is an unramified extension, the Hurwitz genus formula yields

$$g(K) - 1 = r(g(L) - 1) = \frac{h(F)}{|G|}(g(L) - 1).$$

Furthermore, all places in $\mathcal{T}'$ split completely in K/L, hence $N(K) \geq rn$. $\square$

In the following we give several examples from Niederreiter and Xing [110] to illustrate Theorem 4.1.3.

Example 4.1.4 $g(K/\mathbf{F}_3) = 37$, $N(K/\mathbf{F}_3) = 48$. Consider the function field $F = \mathbf{F}_3(x, y_1)$ with

$$y_1^2 = x(x^4 - x^3 + x^2 - x + 1).$$

Then $g(F) = 2$, $N(F) = 6$, and F has six places of degree 2, hence $h(F) = 24$. In F we have $\text{div}(x) = 2P_1 - 2P_\infty$, and this identity uniquely determines the rational places P_1 and P_∞ of F. Furthermore, let $L = F(y_2)$ with

$$y_2^2 = x + 1.$$

Then $g(L) = 4$ since the only places of F ramifying in the Kummer extension L/F are those lying over $x + 1$. Now K is obtained using Theorem 4.1.3 with $S = \mathbf{P}_F \setminus \{P_\infty, P_1\}$. Note that the condition on S in Theorem 4.1.3 is satisfied since the places of F lying over $x + 1$ are totally ramified in L/F. Furthermore, we have $n = 4$ since both places in S' split completely in L/F, and also $|G| = 2$. Theorem 4.1.3 yields $g(K) = 37$ and $N(K) \geq 48$, but since $N_3(37) \leq 54$ by the method in Theorem 1.6.18, we get $N(K) = 48$.

Example 4.1.5 $g(K/\mathbf{F}_3) = 51$, $N(K/\mathbf{F}_3) = 60$. Consider the function field $F = \mathbf{F}_3(x, y_1)$ with

$$y_1^2 = (x + 1)(x - 1)(x^2 + x - 1)(x^3 - x + 1).$$

Then $g(F) = 3$, $N(F) = 5$, F has three places of degree 2 and five places of degree 3, hence $h(F) = 40$. In F we have $\text{div}(x + 1) = 2P_1 - 2P_\infty$ and $\text{div}(x - 1) = 2P_2 - 2P_\infty$, and these identities uniquely determine the rational places P_1, P_2, and P_∞ of F. Furthermore, let $L = F(y_2)$ with

$$y_2^2 = x(x^2 + x - 1).$$

Then $g(L) = 6$ since the only places of F ramifying in the Kummer extension L/F are those lying over x. Now K is obtained using Theorem 4.1.3 with $S = \mathbf{P}_F \setminus \{P_\infty, P_1, P_2\}$. Note that the condition on S in Theorem 4.1.3 is satisfied since the places of F lying over x are totally ramified in L/F. Furthermore, we have $n = 6$ since all places in S' split completely in L/F, and also $|G| = 4$. Theorem 4.1.3 yields $g(K) = 51$ and $N(K) \geq 60$, but since $N_3(51) \leq 69$ by the method in Theorem 1.6.18, we get $N(K) = 60$.

Example 4.1.6 $g(K/\mathbf{F}_3) = 71$, $N(K/\mathbf{F}_3) = 84$. Consider the function field $F = \mathbf{F}_3(x, y_1)$ with

$$y_1^2 = (x + 1)(x - 1)(x^2 + x - 1)(x^3 - x^2 + 1).$$

Then $g(F) = 3$, $N(F) = 5$, F has five places of degree 2 and eleven places of degree 3, hence $h(F) = 56$. In F we have $\text{div}(x + 1) = 2P_1 - 2P_\infty$ and $\text{div}(x - 1) = 2P_2 - 2P_\infty$. Furthermore, let $L = F(y_2)$ with

$$y_2^2 = x(x^2 + x - 1).$$

Now K is obtained using Theorem 4.1.3 by proceeding in a similar way to the previous example. Initially we obtain $g(K) = 71$ and $N(K) \geq 84$, but since $N_3(71) \leq 90$ by the method in Theorem 1.6.18, we have $N(K) = 84$.

4.2   Function Fields from Narrow Ray Class Fields

In the previous section, for a given global function field $F/\mathbf{F}_q$ we considered an unramified abelian extension K/F. This restricts the search for more global function fields with many rational places. In order to construct a ramified abelian extension K/F, we can employ some theoretical results from Chapters 2 and 3. We know that the conductor of an abelian extension K/F controls, in a sense, the genus of K. Hence if we first fix a conductor D and let a rational place ∞ of F split completely in K/F, then by class field theory, K is a subfield of the narrow ray class extension $H(\Lambda_\phi(D))/F$, where H is the ∞'-Hilbert class field of F with $\infty' := \mathbf{P}_F \setminus \{\infty\}$. From now on in this section, we always assume that ∞ is a rational place of F and A is the ∞'-integral ring of F, i.e., A consists of all elements of F whose only possible pole is ∞. Since $H(\Lambda_\phi(D))$ is independent of the specific choice of ϕ by Proposition 3.3.8(i), we can use the simpler notation $H(\Lambda(D))$.

We recall from Section 3.3 that there is a canonical way of identifying nonzero ideals M of A with positive divisors D of F satisfying $\operatorname{supp}(D) \subseteq \infty'$; see (3.4) for the concrete connection. If D is the divisor above and M is the corresponding nonzero ideal of A, then we write also $H(\Lambda(\mathsf{M}))$ for $H(\Lambda(D))$, in line with a similar convention in Section 3.3. Note that by Proposition 3.3.8(iii) we have

$$\operatorname{Gal}(H(\Lambda(\mathsf{M}))/F) = \operatorname{Gal}(H(\Lambda(D))/F) \simeq \operatorname{Cl}_D^+(A) = \operatorname{Cl}_{\mathsf{M}}^+(A),$$

where $\operatorname{Cl}_D^+(A)$, respectively $\operatorname{Cl}_{\mathsf{M}}^+(A)$, is the narrow ray class group of A modulo D, respectively with modulus M, defined in Section 2.6. In view of Theorem 3.3.7, we may also call $H(\Lambda(\mathsf{M}))$ the narrow ray class field with modulus M.

4.2.1   The First Construction

The material in this subsection is based on the papers of Xing and Niederreiter [180], [181]. We start with some auxiliary results. We refer to Section 2.7 for the definition of the p-rank of a finite abelian group.

Lemma 4.2.1 *Let $F/\mathbf{F}_q$ be a global function field with a fixed rational place ∞ and let A be the ∞'-integral ring of F. Let $P_1, \ldots, P_m \in \infty'$ be distinct places with $\deg(P_i) = d_i$ for $1 \le i \le m$. If $q = p^r$ for a prime p, then the p-rank of $(A/(\mathsf{P}_1^2 \cdots \mathsf{P}_m^2))^*$ is $r \sum_{i=1}^m d_i$, where P_i is the prime ideal $P_i \cap A$ of A corresponding to P_i.*

Proof. For each $i = 1, \ldots, m$, let $t_i \in A$ be a local parameter at P_i. Then it is easy to see that $(A/\mathsf{P}_i^2)^* \simeq (\tilde{F}_{P_i}[t_i]/(t_i^2))^* \simeq \mathbf{F}_{q^{d_i}}^* \times \mathbf{F}_{q^{d_i}}$, where $\tilde{F}_{P_i}$ is the residue class field of P_i. Therefore we have

$$(A/(\mathsf{P}_1^2 \cdots \mathsf{P}_m^2))^* \simeq \prod_{i=1}^m (A/\mathsf{P}_i^2)^* \simeq \prod_{i=1}^m \left(\mathbf{F}_{q^{d_i}}^* \times \mathbf{F}_{q^{d_i}} \right).$$

The desired result follows from the above isomorphisms.   □

For a nonempty subset $\mathcal{P}$ of $\infty' = \mathbf{P}_F \setminus \{\infty\}$ we write $\langle \mathcal{P} \rangle$ for the subgroup of $\mathrm{Cl}(A)$ generated by the ideal classes in $\mathrm{Cl}(A)$ determined by the prime ideals Q of A for all $Q \in \mathcal{P}$. Here, with a similar notation as in Lemma 4.2.1, we write $\mathsf{Q} = Q \cap A$ for the prime ideal of $A = O_{\infty'}$ corresponding to $Q \in \infty'$. Since $\mathrm{Cl}(A) \simeq \mathrm{Cl}(F)$ by Proposition 1.2.5, we can describe $\langle \mathcal{P} \rangle$ also as the subgroup of $\mathrm{Cl}(F)$ generated by the divisor classes $[Q - \deg(Q)\infty]$ for all $Q \in \mathcal{P}$. For a fixed $P \in \infty'$ and a nonempty subset $\mathcal{P}$ of $P_F \setminus \{\infty, P\}$, we let $\langle \mathcal{P} \rangle_n$ be the subgroup of the narrow ray class group $\mathrm{Cl}_{\mathsf{P}^n}^+(A)$ generated by the ideal classes in $\mathrm{Cl}_{\mathsf{P}^n}^+(A)$ determined by the prime ideals Q of A for all $Q \in \mathcal{P}$. In the case of a finite set $\mathcal{P} = \{P_1, \ldots, P_m\}$, we also write $\langle P_1, \ldots, P_m \rangle$ and $\langle P_1, \ldots, P_m \rangle_n$ for $\langle \mathcal{P} \rangle$ and $\langle \mathcal{P} \rangle_n$, respectively.

Lemma 4.2.2 *For $m \geq 1$ let $P_1, \ldots, P_m, P, \infty$ be $m + 2$ distinct rational places of $F/\mathbf{F}_q$. Let G be a subgroup of $\mathrm{Cl}_{\mathsf{P}^n}^+(A)$ containing $\langle P_1, \ldots, P_m \rangle_n$ and K the subfield of $H(\Lambda(\mathsf{P}^n))/F$ fixed by G. Suppose that $\langle P_1, \ldots, P_m \rangle = \mathrm{Cl}(A)$ (or, equivalently, that $[P_1 - \infty], \ldots, [P_m - \infty]$ generate the divisor class group of F) and that K/F is an extension of p-power degree. Then P is totally ramified in K/F.*

Proof. Let E be the field $H(\Lambda(\mathsf{P}^n))$ and put $\mathcal{P} = \{P_1, \ldots, P_m\}$. Since $\langle \mathcal{P} \rangle = \mathrm{Cl}(A)$ and $\mathrm{Cl}_{\mathsf{P}^n}^+(A)/\mathrm{Gal}(E/H)$ is isomorphic to $\mathrm{Cl}(A)$, we have $\langle \mathcal{P} \cup \mathrm{Gal}(E/H) \rangle_n = \mathrm{Cl}_{\mathsf{P}^n}^+(A)$, hence $\langle G \cup \mathrm{Gal}(E/H) \rangle_n = \mathrm{Cl}_{\mathsf{P}^n}^+(A)$. It follows from the Galois correspondence that the field $K \cap H$ is F. The only possible ramified places in K/F are ∞ and P. The ramification index of ∞ in E/F is $q - 1$ which is coprime to the degree of the extension K/F, so ∞ splits completely in K/F. Since H/F is the maximal unramified abelian extension in F^{ab} in which ∞ splits completely, P is totally ramified in K/F. $\square$

In Lemma 4.2.2 it is an important condition that the ideal classes determined by $P_1, \ldots, P_m$ generate the group $\mathrm{Cl}(A)$. The following lemma gives a sufficient condition to guarantee that the ideal classes determined by any m rational places of F generate $\mathrm{Cl}(A)$.

Lemma 4.2.3 *Let a, b be two real numbers satisfying $N_q(g) \leq ag + b$ for all $g \geq 0$. Let $F/\mathbf{F}_q$ be a global function field with at least $m + 1$ distinct rational places $P_1, \ldots, P_m, \infty$. Suppose that $m > ag(F) + \frac{1}{2}(b - a) - 1$. Then $\langle P_1, \ldots, P_m \rangle = \mathrm{Cl}(A)$.*

Proof. Suppose that $\langle P_1, \ldots, P_m \rangle$ is a proper subgroup of $\mathrm{Cl}(A)$ of the order h_1. Let L be the subfield of H/F fixed by $\langle P_1, \ldots, P_m \rangle$. Then all places $P_1, \ldots, P_m, \infty$ split completely in L/F, hence

$$N(L) \geq \frac{h(F)}{h_1}(m + 1).$$

Moreover,

$$g(L) = 1 + \frac{h(F)}{h_1}(g(F) - 1)$$

since L/F is unramified. Thus,

$$\frac{h(F)}{h_1}(m + 1) \leq N_q(g(L)) \leq ag(L) + b = \frac{h(F)}{h_1}a(g(F) - 1) + a + b,$$

from which we get

$$m \leq ag(F) - a - 1 + \frac{h_1}{h(F)}(a+b) \leq ag(F) - a - 1 + \frac{1}{2}(a+b) = ag(F) + \frac{1}{2}(b-a) - 1,$$

which is a contradiction.   □

For instance, let $F/\mathbf{F}_2$ be a global function field of genus 9 with at least 10 rational places (see e.g. [103] for an explicit example). By Example 1.6.19, we have $N_2(g) \leq (0.83)g + 5.35$ for all $g \geq 0$. Hence the ideal classes determined by any 9 rational places of F different from ∞ generate $\mathrm{Cl}(A)$.

From now on in this subsection, let $q = p$ be a prime. Let $P_1, \ldots, P_m, P, \infty$ be $m+2$ distinct rational places of $F/\mathbf{F}_p$. We define the multiplicative semigroups

$$S = \{f \in A : f(P) = 1, \mathrm{sgn}(f) = 1, \text{ and } \nu_Q(f) = 0 \text{ for all places } Q \neq \infty, P_1, \ldots, P_m \text{ of } F\},$$

$$S(n) = \{\overline{f} \in (A/\mathbf{P}^n)^* : f \in S\},$$

where n is a positive integer and $\overline{f}$ is the residue class of f modulo $\mathbf{P}^n$. In fact, $S(n)$ is a subgroup of $(A/\mathbf{P}^n)^*$, hence it can be naturally identified with a subgroup of $\mathrm{Cl}_{\mathbf{P}^n}^+(A)$. For $r \geq 1$ we put

$$S_r = \{f \in S : \nu_P(f-1) = r\},$$

$$S_r(n) = \{\overline{f} \in S(n) : f \in S_r\}.$$

We note that S_r is possibly empty. By starting from the smallest index, we successively list all nonempty sets $S_{i_1}, S_{i_2}, S_{i_3}, \ldots$. If $t \in A$ is a fixed local parameter at P, then a complete system of representatives of $(A/\mathbf{P}^n)^*$ is given by

$$\mathbf{F}_p^* \times \{1 + \sum_{v=1}^{n-1} a_v t^v : a_v \in \mathbf{F}_p\}.$$

Lemma 4.2.4 *With the above notation, suppose that the positive integers l and n satisfy $i_l < n \leq i_{l+1}$. Then:*

    (i) $|S_{i_j}(n)| = (p-1)p^{l-j}$ for all $j = 1, \ldots, l$;

    (ii) $|S(n)| = p^l$.

Proof. (i) We first settle the case $j = l$. Choose an element $f \in S_{i_l}$. Suppose that at P it has the local expansion

$$f = 1 + t^{i_l} \sum_{v \geq 0} a_v t^v$$

with all $a_v \in \mathbf{F}_p$ and $a_0 \neq 0$. Then $f^r \equiv 1 + r a_0 t^{i_l} \pmod{\mathbf{P}^{i_l+1}}$ for all $r \geq 1$. Hence $\overline{f}, \overline{f}^2, \ldots, \overline{f}^{p-1}$ are $p-1$ distinct elements in $S_{i_l}(n)$. Now let g be an arbitrary element in S_{i_l}. It has a local expansion at P of the form

$$g = 1 + t^{i_l} \sum_{v \geq 0} b_v t^v$$

with all $b_v \in \mathbf{F}_p$ and $b_0 \neq 0$. There exists an s with $1 \leq s \leq p - 1$ such that $b_0 = s a_0$ in $\mathbf{F}_p$. Then we have $\nu_P(f^{p-s}g - 1) \geq i_l + 1$ and $\nu_P(f^{p-s}f^s - 1) \geq i_l + 1$. Since all S_r are empty for $i_l + 1 \leq r \leq n - 1$, we conclude that $\nu_P(f^{p-s}g - 1) \geq n$ and $\nu_P(f^{p-s}f^s - 1) \geq n$. This implies that $\bar{g} = (\overline{f^{p-s}})^{-1} = \overline{f}^s$. Thus, (i) is shown for $j = l$.

Now suppose that (i) is correct for $S_{i_{j+1}}(n), \ldots, S_{i_l}(n)$. Let e be an element in S_{i_j} and consider the set

$$\bar{e} \cdot (\cup_{r=j+1}^{l} S_{i_r}(n)) \cup \{\bar{e}\}.$$

The above set is a subset of $S_{i_j}(n)$, and it has p^{l-j} elements by induction hypothesis. Let $e_1 \in S$ satisfy $\bar{e}_1 = \bar{e}^{-1}$ in $S(n)$, then $e_1 \in S_{i_j}$. Choose an arbitrary element g in S_{i_j}. Then there exists an s with $1 \leq s \leq p - 1$ such that $\nu_P(e_1^s g - 1) > i_j$. Thus $\overline{e_1^s g} \in (\cup_{r=j+1}^{l} S_{i_r}(n)) \cup \{\bar{1}\}$ and $\bar{g} = \bar{e}^s \overline{e_1^s g} \in \bar{e}^s \cdot (\cup_{r=j+1}^{l} S_{i_r}(n)) \cup \{\bar{e}^s\}$.

(ii) The result follows directly from the fact that $S(n) = (\cup_{j=1}^{l} S_{i_j}(n)) \cup \{\bar{1}\}$. $\square$

If m, n are two positive integers, then we denote by $T_p(m)$ the set of the first m positive integers which are not divisible by p, that is,

$$T_p(m) = \{1 \leq i \leq m + \left\lfloor \frac{m}{p-1} \right\rfloor : p \nmid i\},$$

and by $s_p(m, n)$ the number

$$s_p(m, n) = \sum_{i \in T_p(m)} \left\lceil \log_p \frac{n}{i} \right\rceil, \tag{4.1}$$

where $\lceil u \rceil$ is the least integer $\geq u$ and $\log_p$ denotes the logarithm to the base p.

Lemma 4.2.5 *Let P be a rational place of $F/\mathbf{F}_p$ and $n \geq 2$. Then:*
  (i) *The p-rank of $(A/P^n)^*$ is equal to $n - \lfloor \frac{n-1}{p} \rfloor - 1$.*
  (ii) *If G is the p-Sylow subgroup of $(A/P^n)^*$ and $1 \leq m \leq n - \lfloor \frac{n-1}{p} \rfloor - 1$, then any m elements of G generate a subgroup of G of order at most $p^{s_p(m,n)}$, and equality can be achieved.*

Proof. (i) Let $t \in A$ be a local parameter at P. Then $(A/P^n)^* \simeq (\mathbf{F}_p[t]/(t^n))^*$. It is easy to verify that $(\mathbf{F}_p[t]/(t^n))^*$ has the decomposition as a direct product

$$(\mathbf{F}_p[t]/(t^n))^* = \mathbf{F}_p^* \times \left(\prod_{\substack{i=1 \\ \gcd(p,i)=1}}^{n-1} \langle 1 + t^i \rangle \right),$$

where $\langle 1 + t^i \rangle$ denotes the subgroup of $(\mathbf{F}_p[t]/(t^n))^*$ generated by the residue class of $1 + t^i$ modulo t^n. Our result follows.

(ii) Let J be a subgroup of G generated by m elements. We identify G with

$$\prod_{\substack{i=1 \\ \gcd(p,i)=1}}^{n-1} \langle 1 + t^i \rangle.$$

The order of $\langle 1 + t^i \rangle$ is the smallest power p^a satisfying $ip^a \geq n$, that is, a is equal to $\lceil \log_p \frac{n}{i} \rceil$. Therefore $|\langle 1 + t^i \rangle| \geq |\langle 1 + t^j \rangle|$ if $i \leq j$. Hence

$$|J| \leq | \prod_{i \in T_p(m)} \langle 1 + t^i \rangle | = p^{s_p(m,n)},$$

and it is obvious that equality can be achieved. $\square$

Lemma 4.2.6 Let $E = H(\Lambda(\mathsf{P}^n))$ be the narrow ray class field with modulus P^n defined by the sgn-normalized Drinfeld A-module ϕ of rank 1 over H. Let $t \in A$ be a local parameter at P and λ a generator of the cyclic A-module $\Lambda(\mathsf{P}^n)$. If Q is a place of E lying over P, then for $1 \leq i \leq n-1$ we have

$$\nu_Q(\phi_{t^i}(\lambda)) = p^i.$$

Proof. Write $t^i A = \mathsf{P}^i J$ with an ideal J of A which is relatively prime to P. Then by [49, Lemma 4.5],

$$\phi_{t^i A} = \phi_{\mathsf{P}^i J} = (\mathsf{P}^i * \phi)_J \cdot \phi_{\mathsf{P}^i}.$$

Thus,

$$\nu_Q(\phi_{t^i}(\lambda)) = \nu_Q(\phi_{t^i A}(\lambda)) = \nu_Q \left((\mathsf{P}^i * \phi)_J(\phi_{\mathsf{P}^i}(\lambda)) \right).$$

Hence it suffices to prove the following two formulas:

$$\nu_Q(\phi_{\mathsf{P}^i}(\lambda)) = p^i \quad \text{for} \quad 0 \leq i \leq n-1, \tag{4.2}$$

$$\nu_Q(D((\mathsf{P}^i * \phi)_J)) = 0 \quad \text{for} \quad i \geq 1. \tag{4.3}$$

We prove (4.2) by induction on i, with $i = 0$ being trivial. Suppose that we have shown (4.2) for some i and consider $i + 1 < n$. By Proposition 11.4 and its proof in [49], we have

$$\phi_{\mathsf{P}^{i+1}}(u) = f(\phi_{\mathsf{P}^i}(u))\phi_{\mathsf{P}^i}(u), \tag{4.4}$$

where the polynomial $f(u)$ is Eisenstein at the place R of H lying under Q. Moreover, $\deg(f) = p - 1$ by formula (5.8) in [49]. Thus, if

$$f(u) = \sum_{j=0}^{p-1} c_j u^j \quad \text{with all } c_j \in H,$$

then

$$\nu_Q(c_j \phi_{\mathsf{P}^i}(\lambda)^j) \geq p^{n-1}(p-1)$$

for $0 \leq j \leq p-2$, whereas

$$\nu_Q(c_{p-1} \phi_{\mathsf{P}^i}(\lambda)^{p-1}) = p^i(p-1).$$

Hence (4.2) for $i + 1$ follows from (4.4).

To prove (4.3), put $\psi = \mathsf{P}^{i-1} * \phi$, then by the proof of Proposition 11.4 in [49] we obtain

$$\nu_R(D((\mathsf{P} * \psi)_J)) = 0.$$

Hence $\nu_Q(D((\mathsf{P}^i * \phi)_J)) = 0$ since $\mathsf{P}^i * \phi = \mathsf{P} * \psi$. ☐

We are now ready to state our first theorem in this section which covers a construction of global function fields with many rational places using narrow ray class extensions.

Theorem 4.2.7 *Let $F/\mathbf{F}_p$ be a global function field with at least $m+2$ distinct rational places $P_1, \ldots, P_m, P, \infty$, where $m \geq 1$. Let $E = H(\Lambda(\mathsf{P}^n))$ be the narrow ray class field with modulus P^n determined by the sgn-normalized Drinfeld A-module ϕ of rank 1 over H. Suppose that $n - \lfloor \frac{n-1}{p} \rfloor - 1 \geq m$ and $\langle P_1, \ldots, P_m \rangle = \mathrm{Cl}(A)$. Then there exists a subfield $K/\mathbf{F}_p$ of the extension E/F such that:*

(i)
$$g(K) = 1 + \frac{1}{2}p^{n-1-s}(2g(F) - 2 + n) - \frac{1}{2}\left(\sum_{r=1}^{s} p^{j_r - r} + \frac{p^{n-1-s} - 1}{p-1} + 1\right),$$

where $s = s_p(m,n)$ is given by (4.1) and $j_1, j_2, \ldots, j_s$ are s integers satisfying $1 \leq j_1 < j_2 < \cdots < j_s < n$. Moreover, in the case where $i_s < n \leq i_{s+1}$, we have $j_r = i_r$ for $r = 1, 2, \ldots, s$, where the i_r are as in Lemma 4.2.4.

(ii) *$N(K) \geq p^{n-1-s}(m+1) + 1$. Furthermore, if $P_1, \ldots, P_m, P, \infty$ are all rational places of F, then equality holds.*

Proof. (i) Let G_1 be the p-Sylow subgroup of $(A/\mathsf{P}^n)^*$. Then there exists a subgroup G_2 of $\mathrm{Cl}_{\mathsf{P}^n}^+(A)$ such that $\mathrm{Cl}_{\mathsf{P}^n}^+(A)/G_2 \simeq G_1$. By Lemma 4.2.5, we can find a subgroup G of $\mathrm{Cl}_{\mathsf{P}^n}^+(A)$ with $G_2 \subseteq G$, $\langle P_1, \ldots, P_m \rangle_n \subseteq G$, and $|G/G_2| = p^s$. A simple calculation shows that $|\mathrm{Cl}_{\mathsf{P}^n}^+(A)/G| = p^{n-1-s}$.

Let K be the subfield of E/F fixed by G. Let Q be a place of E lying over P. By Lemma 4.2.2, P is totally ramified in K/F. Hence the ramification index of Q in E/K is
$$\frac{(p-1)p^{n-1}}{p^{n-1-s}} = (p-1)p^s.$$

Then the inertia group of Q in E/K is $G_3 := \mathrm{Gal}(E/H) \cap \mathrm{Gal}(E/K)$, whose order is $(p-1)p^s$.

Let B be the p-Sylow subgroup of G_3, which by Proposition 3.3.8(iv) can be identified with a subgroup of $(A/\mathsf{P}^n)^*$ that we also call B. It is obvious that $f(P) = 1$ and $\mathrm{sgn}(f) = 1$ for any $f \in B \subseteq (A/\mathsf{P}^n)^*$. For any $1 \leq r \leq n-1$ we define
$$B_r = \{f \in B : \nu_P(f - 1) = r\}.$$

By starting from the smallest index, we list all nonempty sets $B_{j_1}, B_{j_2}, \ldots, B_{j_u}$ with $1 \leq j_1 < j_2 < \cdots < j_u < n$. Similar arguments to those used in the proof of Lemma 4.2.4 show that $|B_{j_r}| = (p-1)p^{u-r}$ for $r = 1, 2, \ldots, u$ and $|B| = p^u$. Hence $u = s$.

Let L be the subfield of E/F fixed by G_3. Let λ be a generator of the cyclic A-module $\Lambda(\mathsf{P}^n)$. Then by Proposition 1.3.13 the different exponent $d_Q(E/L)$ of Q in E/L is
$$d_Q(E/L) = \sum_{\gamma \in G_3 \setminus \{\mathrm{id}\}} \nu_Q(\lambda - \lambda^\gamma).$$

Let $t \in A$ be a local parameter at P. For $\gamma \in G_3$ we get in accordance with Proposition 3.3.8(iv) that $\gamma = \sigma_{gA}$ for some $g \in A$ with $\mathrm{sgn}(g) = 1$ and $g = \sum_{i=0}^{n-1} a_i t^i$, where all $a_i \in \mathbf{F}_p$ and $a_0 \neq 0$. Among the $(p-1)p^s$ elements g that arise in this way, those corresponding to some $\gamma \in B$ are characterized by $a_0 = 1$. By Proposition 3.3.8(iii), [49, Lemma 4.4], and Lemma 4.2.6, we have

$$\nu_Q(\lambda - \lambda^\gamma) = \nu_Q(\lambda - \phi_g(\lambda)) = \nu_Q((1 - a_0)\lambda - \sum_{i=1}^{n-1} a_i \phi_{t^i}(\lambda)) = p^{j_r}$$

for $\gamma \in B_{j_r}$, and $\nu_Q(\lambda - \lambda^\gamma) = 1$ for $\gamma \in G_3 \setminus B$. Therefore

$$d_Q(E/L) = \sum_{r=1}^{s}(p-1)p^{s-r}p^{j_r} + (p-2)p^s = (p-2)p^s + (p-1)p^s \sum_{r=1}^{s} p^{j_r-r}.$$

Thus, the global different divisor of E/L is

$$\mathrm{Diff}(E/L) = \sum_{Q|P} d_Q(E/L)Q + \sum_{R|\infty}(p-2)R.$$

Applying the Hurwitz genus formula, we obtain

$$
\begin{aligned}
2g(E) - 2 &= \deg(\mathrm{Diff}(E/L)) + (p-1)p^s(2g(L) - 2) \\
&= h(F)d_Q(E/L) + (p-2)h(F)p^{n-1} + (p-1)p^s(2g(L) - 2),
\end{aligned}
$$

and combining this with the result of Proposition 3.3.9, we get

$$2g(L) - 2 = h(F)p^{n-1-s}(2g(F) + n - 2) - h(F)\left(\sum_{r=1}^{s} p^{j_r-r} + \frac{p^{n-1-s} - 1}{p-1} + 1\right).$$

Our result on the genus of K follows from $2g(L) - 2 = h(F)(2g(K) - 2)$, which holds since L/K is an unramified extension of degree $h(F)$.

From the definitions of $S(n)$ and G_3 it is clear that $S(n) \subseteq G_3$ (with the proper identification), and so $S(n) \subseteq B$ by Lemma 4.2.4(ii). Hence $\{i_1, i_2, \ldots, i_l\} \subseteq \{j_1, j_2, \ldots, j_s\}$ with l satisfying $i_l < n \leq i_{l+1}$. So if $i_s < n \leq i_{s+1}$, then $l = s$ and $\{i_1, i_2, \ldots, i_s\} = \{j_1, j_2, \ldots, j_s\}$. This implies that $i_r = j_r$ for $r = 1, 2, \ldots, s$.

(ii) We know from Proposition 3.3.8(iii) that $\mathbf{F}_p^*$ is the decomposition group of ∞ in E/F, so ∞ splits completely in K/F. For the places $P_1, \ldots, P_m$, their Artin symbols in E/F are contained in $\mathrm{Gal}(E/K)$. Thus, the rational places $P_1, \ldots, P_m$ split completely in K/F. The extension K/F has degree p^{n-1-s}. By Lemma 4.2.2, P is totally ramified in K/F, hence our result follows. $\square$

Before discussing some examples, we give explicit generators of $S(n)$ in the case where $\gcd(h(F), p) = 1$. For $g \in A$ we again write $\bar{g}$ for the residue class of g modulo P^n.

Lemma 4.2.8 *Let* $f_i \in A$ *satisfy* $\mathrm{div}(f_i) = h(F)P_i - h(F)\infty$ *and* $f_i \equiv 1 \pmod{P}$ *for* $i = 1, \ldots, m$. *Suppose* $\gcd(h(F), p) = 1$. *Then* $S(n)$ *is generated by the set* $\{\bar{f}_i : 1 \leq i \leq m\}$.

Proof. We first show that each $\overline{f_i}$ is contained in $S(n)$. We know that $\mathrm{sgn}(f_i^{p^{n-1}-1}) = 1$, thus $\overline{f_i^{p^{n-1}-1}} \in S(n)$. It follows from $f_i \equiv 1 \pmod{P}$ that $\overline{f_i}$ belongs to the p-Sylow subgroup of $(A/P^n)^*$, which has order p^{n-1}. Therefore $\overline{f_i} = \left(\overline{f_i^{p^{n-1}-1}}\right)^{-1} \in S(n)$.

Let $g \in S$ with $\mathrm{div}(g) = \sum_{i=1}^m n_i P_i - (\sum_{i=1}^m n_i)\infty$. Then

$$\mathrm{div}(g^{h(F)}) = h(F)\sum_{i=1}^m n_i P_i - h(F)(\sum_{i=1}^m n_i)\infty = \mathrm{div}(\prod_{i=1}^m f_i^{n_i}).$$

Hence $g^{h(F)} = \prod_{i=1}^m f_i^{n_i}$ since both are congruent to 1 modulo P. By the given condition $\gcd(h(F), p) = 1$, there exists an integer a satisfying $ah(F) \equiv 1 \pmod{p^{n-1}}$. Therefore $\overline{g} = \prod_{i=1}^m \overline{f_i}^{an_i}$. $\qquad\square$

From Example 4.2.9 to Example 4.2.13 below, we always have $\langle P_1, \ldots, P_m \rangle = \mathrm{Cl}(A)$ with $m = N(F) - 2$ and $P_1, \ldots, P_m, P, \infty$ are all rational places of F. Moreover, the condition $i_{s_p(m,n)} < n \le i_{s_p(m,n)+1}$ is also satisfied for any n with $n - \lfloor \frac{n-1}{p} \rfloor - 1 \ge m$.

Example 4.2.9 Let F be the rational function field $\mathbf{F}_2(x)$. Let P be the place of F which is the zero of x and P_1 be the place of F which is the zero of $x + 1$. For $r \ge 1$ let r be equal to $2^u r_0$ with odd r_0, then

$$(1 + x)^r = 1 + x^{2^u} + \cdots.$$

By Lemma 4.2.8, $S_r \neq \emptyset$ if and only if r is a 2-power. Hence the jth nonempty set S_{i_j} is $S_{2^{j-1}}$. For a given $n \ge 2$, the largest integer l satisfying $2^{l-1} < n$ is $l = \lfloor \log_2(n-1) \rfloor + 1$. Hence by Theorem 4.2.7, for each $n \ge 2$ we get a field K_n with

$$g(K_n) = 1 + (n-3)2^{n-l-2} - \sum_{j=0}^{l-1} 2^{2^j - j - 2}$$

and

$$N(K_n) = 1 + 2^{n-l}.$$

Taking some small values of n into consideration, we get the examples of global function fields $K_n/\mathbf{F}_2$ in the following table.

n	4	6	7	8	10	11
$g(K_n)$	1	5	15	39	103	247
$N(K_n)$	5	9	17	33	65	129

By the method in Theorem 1.6.18, we know that K_4, K_6, K_7, K_8 are optimal, and $65 = N(K_{10}) \le N_2(103) \le 72$. This example was first presented in Niederreiter and Xing [103, Theorem 2].

Example 4.2.10 Consider the elliptic function field $F = \mathbf{F}_2(x, y)$ defined by

$$y^2 + y = x^3 + x.$$

This field has five rational places, one is the place ∞ which is the pole of x, and the other four places are $P = (0, 0)$, $P_1 = (0, 1)$, $P_2 = (1, 0)$, and $P_3 = (1, 1)$, where $Q = (a, b)$ is the rational place of F determined by $(x, y) \equiv (a, b)$ (mod Q). Consider the following three elements of A:

$$f_1 := xy + y + 1, \quad f_2 := xy + x^2 + 1, \quad f_3 := xy + x^2 + x + 1.$$

The principal divisors associated with them are

$$\operatorname{div}(f_i) = 5P_i - 5\infty \qquad \text{for } i = 1, 2, 3.$$

Choose x as the local parameter at P. Then f_1, f_2, f_3 have the following local expansions at P:

$$f_1 = 1 + x + x^5 + x^6 + x^7 + x^8 + x^9 + x^{12} + x^{13} + x^{16} + x^{17} + \cdots,$$
$$f_2 = 1 + x^3 + x^4 + x^5 + x^7 + x^9 + x^{13} + x^{17} + \cdots,$$
$$f_3 = 1 + x + x^3 + x^4 + x^5 + x^7 + x^9 + x^{13} + x^{17} + \cdots.$$

For $n \geq 6$ let E be the field $H(\Lambda(\mathsf{P}^n))$. The class number $h(F)$ of F is 5. By Lemma 4.2.8, $S(n)$ is generated by $\overline{f}_1$, $\overline{f}_2$, and $\overline{f}_3$. Obviously, S_1, S_2, S_3, S_4, S_6, S_8, S_{12}, and S_{16} are nonempty. Furthermore, $f_1^3 f_2 f_3$ has the local expansion

$$f_1^3 f_2 f_3 = 1 + x^5 + x^7 + x^8 + \cdots$$

at P. Hence S_5 and S_{10} are also nonempty.

Now we consider the group $S(16)$. Note that $\overline{f}_1$ generates a cyclic group of order 16 in $S(16)$ and $\overline{f}_2$ generates a cyclic group of order 8. At the same time we have

$$f_3^4 \equiv (f_1 f_2)^4 \pmod{\mathsf{P}^{16}}.$$

Hence the set $\{\overline{f}_1, \overline{f}_2, \overline{f}_3\}$ generates a subgroup of order at most 2^9 in $S(16)$. By Lemma 4.2.4, there are at most nine S_r's with $r < 16$ satisfying $S_r \neq \emptyset$. Therefore $S_7 = S_9 = S_{11} = S_{13} = S_{14} = S_{15} = \emptyset$.

Taking some special values of n and combining with Theorem 4.2.7 gives the fields $K_n/\mathbf{F}_2$ in the following table.

n	8	10	12	14	15	16
$g(K_n)$	5	15	39	95	215	471
$N(K_n)$	9	17	33	65	129	257

Hence $65 \leq N_2(95) \leq 68$, $129 \leq N_2(215) \leq 135$, and $257 \leq N_2(471) \leq 272$, where the upper bounds for $N_2(95)$, $N_2(215)$, and $N_2(471)$ are obtained by using the method in Theorem 1.6.18. This example was first presented in Niederreiter and Xing [108] (see also Xing and Niederreiter [181]).

Example 4.2.11 Let $F = \mathbf{F}_2(x,y)$ be the global function field over $\mathbf{F}_2$ defined by

$$y^2 + y = \frac{x(x+1)}{x^3 + x + 1}.$$

Then $g(F) = 2$ and F has six rational places and no places of degree 2. So we get the class number $h(F) = 19$. Choose ∞ as the place of F which is the common zero of $1/x$ and y. Let $P \in \mathbf{P}_F$ be the common zero of $1/x$ and $y+1$. For $n \geq 8$ consider the field $E = H(\Lambda(P^n))$. After a computation, we find that S_1, S_2, S_3, S_4, S_5, S_6, S_7, S_8, S_{10}, S_{12}, S_{14}, and S_{16} are nonempty. But $S_9 = S_{11} = S_{13} = S_{15} = \emptyset$. Hence from Theorem 4.2.7 we get the fields $K_n/\mathbf{F}_2$ in the following table.

n	10	12	14	16
$g(K_n)$	8	22	54	126
$N(K_n)$	11	21	41	81

Here K_{10} is optimal, and $21 \leq N_2(22) \leq 22$, $41 \leq N_2(54) \leq 43$, and $81 \leq N_2(126) \leq 86$.

Example 4.2.12 Let F be the rational function field $\mathbf{F}_3(x)$. The pole of x is viewed as ∞. Consider the cyclotomic function field $E = F(\Lambda_M)$ with $M = x^n$ for $n \geq 3$. Then $\overline{1+x}$ and $\overline{1+2x}$ generate $S(n)$ by Lemma 4.2.8. Since $(1+x)(1+2x) = 1+2x^2$, S_2 is nonempty. Hence S_1, S_2, S_3, S_6, S_9, and S_{18} are nonempty. Furthermore, $\overline{1+x}$ and $\overline{1+2x}$ generate a subgroup of $S(18)$ of order at most $3^3 \times 3^2 = 3^5$. By Lemma 4.2.4, $S_r = \emptyset$ for all $r \leq 17$ with $r \neq 1,2,3,6,9$. Hence from Theorem 4.2.7 we get the fields $K_n/\mathbf{F}_3$ in the following table.

n	5	6	8	9
$g(K_n)$	3	15	69	258
$N(K_n)$	10	28	82	244

Here K_5 and K_6 are optimal, and $82 \leq N_3(69) \leq 88$, and $244 \leq N_3(258) \leq 275$.

Example 4.2.13 Let $F = \mathbf{F}_3(x,y)$ be the global function field over $\mathbf{F}_3$ defined by

$$y^2 = x^3 - x + 1.$$

Then $g(F) = 1$ and F has seven rational places. One is the pole of x which is viewed as ∞. The other rational places are $P = (0,1)$, $P_1 = (0,2)$, $P_2 = (1,1)$, $P_3 = (1,2)$, $P_4 = (2,1)$, and $P_5 = (2,2)$. Let E be the field $H(\Lambda(P^n))$. Consider the following five elements of A:

$$f_1 := 2(x+1)^2 y + 2x^3 + 2x^2 + 2,$$

$$f_2 := 2x^2 y + (x+2)^3 + (x+2)^2 + 1, \quad f_3 := x^2 y + (x+2)^3 + (x+2)^2 + 1,$$

$$f_4 := (x+2)^2 y + 2(x+1)^3 + 2(x+1)^2 + 2, \quad f_5 := (x+2)^2 y + (x+1)^3 + (x+1)^2 + 1.$$

The corresponding principal divisors are

$$\mathrm{div}(f_i) = 7P_i - 7\infty \qquad \text{for } 1 \le i \le 5.$$

Choose x as the local parameter at P. Then the f_i have the following local expansions at P:

$$f_1 = 1 + x^2 + x^3 + x^7 + \cdots,$$

$$f_2 = 1 + x + 2x^4 + 2x^5 + x^7 + \cdots,$$

$$f_3 = 1 + x + 2x^2 + 2x^3 + x^4 + x^5 + 2x^7 + \cdots,$$

$$f_4 = 1 + 2x^2 + 2x^3 + 2x^4 + x^6 + x^8 + \cdots,$$

$$f_5 = 1 + x + x^2 + x^3 + 2x^4 + x^6 + x^8 + \cdots.$$

A further computation yields the following local expansions at P:

$$f_1 f_4 = 1 + x^4 + \cdots,$$

$$f_1 f_2^5 f_3 = 1 + x^4 + \cdots,$$

$$f_2^2 f_4 f_5 = 1 + x^4 + \cdots,$$

$$f_1^2 f_2^2 f_4^3 f_5 = 1 + 2x^5 + \cdots,$$

$$f_1^5 f_2^7 f_3 f_4^5 f_5 = 1 + 2x^7 + \cdots.$$

Hence S_1, S_2, S_3, S_4, S_5, S_6, S_7, S_9, and S_{12} are nonempty. In view of the above expansions we find that the set $\{\overline{f}_i : 1 \le i \le 5\}$ generates a subgroup of $S(12)$ of order at most $3^2 \times 3^3 \times 3 \times 3 \times 3 = 3^8$, hence $|S(12)| \le 3^8$ by Lemma 4.2.8. It follows from Lemma 4.2.4 that $S_8 = S_{10} = S_{11} = \emptyset$. Therefore from Theorem 4.2.7 we get the fields $K_n/\mathbf{F}_3$ in the following table.

n	9	11	12
$g(K_n)$	10	43	151
$N(K_n)$	19	55	163

Hence we obtain $19 \le N_3(10) \le 21$, $55 \le N_3(43) \le 60$, and $163 \le N_3(151) \le 171$.

4.2.2 The Second Construction

Now we give the second construction based on narrow ray class fields which is due to Niederreiter and Xing [106]. Other variants of the following theorem can be found in Niederreiter and Xing [116].

Theorem 4.2.14 *Let $q = p^r$ with a prime p and $r \geq 1$, and for a given integer $m \geq 1$ let $F/\mathbf{F}_q$ be a global function field of genus $g(F)$ with $N(F) \geq m + 1$. Suppose that F has at least one place of degree $d > 1$ with $rd > m$. Assume also that $N_q(1 + p(g(F) - 1)) < (m+1)p$ when $g(F) \geq 1$. Then for every integer l with $1 \leq l \leq rd - m$ there exists a global function field $K_l/\mathbf{F}_q$ such that:*

(i) The number of rational places of $K_l/\mathbf{F}_q$ satisfies $N(K_l) \geq (m+1)p^l$ and $p^l | N(K_l)$. Furthermore, $N(K_l) = (m+1)p^l$ if $N(F) = m + 1$.

(ii) The genus of $K_l/\mathbf{F}_q$ is given by

$$g(K_l) = p^l(g(F) + d - 1) + 1 - d.$$

Proof. (i) Let $\infty, P_1, \ldots, P_m$ be $m + 1$ distinct rational places of F and let A be the ∞'-integral ring of F. Let Q be a place of F of degree d. Consider the $\mathbf{F}_p$-vector space

$$V := \mathrm{Cl}_{Q^2}^+(A)/\mathrm{Cl}_{Q^2}^+(A)^p.$$

Then $\dim_{\mathbf{F}_p}(V)$ is equal to the p-rank of $\mathrm{Cl}_{Q^2}^+(A)$, which is at least the p-rank of $(A/Q^2)^*$ (compare with Section 2.7). By Lemma 4.2.1 we get $\dim_{\mathbf{F}_p}(V) \geq rd$. If we view $P_1, \ldots, P_m$ as elements of the vector space V in an obvious sense, then they generate a subspace of V of dimension at most m. For a given l with $1 \leq l \leq rd - m$, let W_l be a subspace of V of dimension $\dim_{\mathbf{F}_p}(V) - l$ containing all the P_i. Let G_l be the subgroup of $\mathrm{Cl}_{Q^2}^+(A)$ that contains $\mathrm{Cl}_{Q^2}^+(A)^p$ and satisfies $G_l/\mathrm{Cl}_{Q^2}^+(A)^p = W_l$. Then G_l contains all the P_i and we have $[\mathrm{Cl}_{Q^2}^+(A) : G_l] = p^l$. Let

$$E = H(\Lambda(Q^2)) = F^{Q^2}(\infty)$$

be the narrow ray class field with modulus Q^2 determined by a sgn-normalized Drinfeld A-module of rank 1 over H. Let K_l be the subfield of the extension E/F fixed by G_l. Then K_l/F is an extension of degree p^l and $\infty, P_1, \ldots, P_m$ split completely in K_l/F, hence $N(K_l) \geq (m+1)p^l$. The remaining assertions in part (i) of the theorem follow from the fact that Q is the only possible ramified place in K_l/F.

(ii) We first show that Q is totally ramified in K_l/F. Otherwise, one could find a subfield J of K_l/F such that J/F is an unramified extension of degree p. This is impossible if $g(F) = 0$. If $g(F) \geq 1$, then the genus of J is $1 + p(g(F) - 1)$ and the number of rational places of J is at least $(m+1)p$. This yields the contradiction $(m+1)p \leq N(J) \leq N_q(g(J)) < (m+1)p$.

By Proposition 2.6.6, the conductor exponent $c_Q(E/F)$ of Q in E/F is 2, and hence the conductor exponent $c_Q(K_l/F)$ is at most 2 by Lemma 2.3.7. Thus by Theorem 2.3.6, the different exponent of Q in K_l/F is

$$d_Q(K_l/F) = 2p^l - 2.$$

By the Hurwitz genus formula, we have

$$2g(K_l) - 2 = [K_l : F](2g(F) - 2) + d_Q(K_l/F)\deg(Q),$$

i.e., $g(K_l) = 1 + p^l(g(F) + d - 1) - d$. $\qquad\square$

Remark 4.2.15 The formula $c_Q(E/F) = 2$ used in the above proof can also be demonstrated by employing the theory of Drinfeld modules of rank 1; see Niederreiter and Xing [117, Lemma 2] for a more general result proved in this way.

Remark 4.2.16 If $l = 1$ and we drop the condition on $N_q(1 + p(g(F) - 1))$ in Theorem 4.2.14, then in Theorem 4.2.14(ii) we either have the stated formula for $g(K_1)$ or $g(K_1) = p(g(F) - 1) + 1$. This holds since then $[K_1 : F] = p$, so that either Q is totally ramified in K_1/F or the extension K_1/F is unramified.

Example 4.2.17 Let $F/\mathbf{F}_2$ be the elliptic function field $F = \mathbf{F}_2(x, y)$ defined by $y^2 + y = x^3 + x$ with five rational places. Then there exist places of F of degree d for any $d \geq 5$. Let $m = 4$ in Theorem 4.2.14. Then it is obvious that $N_2(1 + 2(g(F) - 1)) = N_2(1) = 5 < 10 = 2(m + 1)$. Hence for each $d \geq 5$ there exists a global function field $K_d/\mathbf{F}_2$ such that

$$g(K_d) = 1 + (2^{d-4} - 1)d, \qquad N(K_d) = 5 \cdot 2^{d-4}.$$

The fields K_5, K_6, K_7 are optimal. This example was obtained via a different method by Serre [141].

Example 4.2.18 Let $F/\mathbf{F}_2$ be the global function field of genus 2 with six rational places as in Example 4.2.11. Then there exist places of F of degree d for any $d \geq 6$ (this is easy for $d = 6$ and follows from Lemma 1.6.13 for $d \geq 7$). Let $m = 5$ in Theorem 4.2.14. Then all conditions in Theorem 4.2.14 are satisfied. Therefore for each $d \geq 6$ there exists a global function field $K_d/\mathbf{F}_2$ satisfying

$$g(K_d) = 1 + 2^{d-5}(d + 1) - d, \qquad N(K_d) = 6 \cdot 2^{d-5}.$$

The field K_6 is optimal, and $24 = N(K_7) \leq N_2(26) \leq 25$ and $48 = N(K_8) \leq N_2(65) \leq 50$.

4.2.3   The Third Construction

For the above constructions, we always let the distinguished rational place ∞ split completely. In order to construct more global function fields with many rational places, we consider also extensions K/F where ∞ could ramify. The constructions in this subsection apply specifically to nonprime constant fields.

Let $F/\mathbf{F}_q$ again be a global function field, let ∞ be a rational place of F and A the ∞'-integral ring of F. For $r \geq 2$ we consider the constant field extension $F_r = F \cdot \mathbf{F}_{q^r}$. Then ∞ can be viewed as a rational place of $F_r/\mathbf{F}_{q^r}$ with a corresponding $(\mathbf{P}_{F_r} \setminus \{\infty\})$-integral ring A_r of F_r. Let $P \neq \infty$ be a place of F of degree d with $\gcd(d, r) = 1$. Then similarly, P is a place of $F_r/\mathbf{F}_{q^r}$ of the same degree d by Theorem 1.5.2. We consider now the narrow ray class group $\mathrm{Cl}_{\mathbf{P}^n}^+(A_r)$ for a given $n \geq 1$. Note that $(A_r/\mathbf{P}^n)^*$ can be viewed as a subgroup of $\mathrm{Cl}_{\mathbf{P}^n}^+(A_r)$ in the following way: for every $a \in A_r$ there is a $b \in F_r$ satisfying $\mathrm{sgn}(b) = 1$ and $b \equiv a \pmod{\mathbf{P}^n}$; then we have the embedding $\bar{a} \in (A_r/\mathbf{P}^n)^* \mapsto \overline{bA} \in \mathrm{Cl}_{\mathbf{P}^n}^+(A_r)$, with the bars denoting appropriate residue classes.

Next we observe that $\mathrm{Cl}_{\mathbf{P}^n}^+(A)$ can also be viewed as a subgroup of $\mathrm{Cl}_{\mathbf{P}^n}^+(A_r)$, as can be shown in analogy with Theorem 1.5.4.

Let I_∞ be the subgroup of $(A_r/\mathbf{P}^n)^*$ formed by the residue classes modulo $\mathbf{P}^n$ of the elements of $\mathbf{F}_{q^r}^*$, so that in particular $|I_\infty| = q^r - 1$. According to Proposition 3.3.8(iii), I_∞ is both the decomposition group and the inertia group of ∞ in the extension $H_r(\Lambda(\mathbf{P}^n))/F_r$, where H_r is the $(\mathbf{P}_{F_r} \setminus \{\infty\})$-Hilbert class field of F_r. The following lemma and Theorem 4.2.20 below were established in Niederreiter and Xing [106] (see also [114]).

Lemma 4.2.19 *Let B be a subgroup of I_∞. Then:*
  (i) $(A_r/\mathbf{P}^n)^* \cap (B \cdot \mathrm{Cl}_{\mathbf{p}^n}^+(A)) = B \cdot (A/\mathbf{P}^n)^*$;
  (ii) $|B \cap \mathrm{Cl}_{\mathbf{p}^n}^+(A)| = \gcd(|B|, q - 1)$.

Proof. (i) It is trivial that $B \cdot (A/\mathbf{P}^n)^* \subseteq (A_r/\mathbf{P}^n)^* \cap (B \cdot \mathrm{Cl}_{\mathbf{p}^n}^+(A))$. Conversely, consider an element of $(A_r/\mathbf{P}^n)^* \cap (B \cdot \mathrm{Cl}_{\mathbf{p}^n}^+(A))$. This element is a coset modulo $\mathrm{Princ}_{nP}^+(A_r)$ determined by an $\mathbf{F}_q$-rational divisor D relatively prime to P and ∞. Since D represents an element of $(A_r/\mathbf{P}^n)^*$, we can write $D = (a)_\pi$ with $a \in F_r$, $\mathrm{sgn}(a) = 1$, and $a \not\equiv 0 \pmod{\mathbf{P}}$, where $(a)_\pi$ is the divisor corresponding to the principal ideal aA_r. Now D also represents an element of $B \cdot \mathrm{Cl}_{\mathbf{p}^n}^+(A)$, hence modulo $\mathrm{Princ}_{nP}^+(A_r)$ we can write $D = (b)_\pi + D_1$, where $b \in F_r$, $\mathrm{sgn}(b) = 1$, $b \equiv \beta \pmod{\mathbf{P}^n}$ for some $\beta \in \mathbf{F}_{q^r}^*$ with $\beta + \mathbf{P}^n \in B$, and D_1 is an $\mathbf{F}_q$-rational divisor relatively prime to P and ∞. Thus, modulo $\mathrm{Princ}_{nP}^+(A_r)$ we have

$$(a)_\pi = (b)_\pi + D_1,$$

and so

$$(ab^{-1})_\pi - D_1 = (c)_\pi$$

for some $c \in F_r$ with $\mathrm{sgn}(c) = 1$ and $c \equiv 1 \pmod{\mathbf{P}^n}$. This means that $D_1 = (ab^{-1}c^{-1})_\pi$. Since D_1 and ∞ are $\mathbf{F}_q$-rational, it follows that $ab^{-1}c^{-1} \in F$, hence D_1 represents an element of $(A/\mathbf{P}^n)^*$. In view of the fact that $D = (b)_\pi + D_1$, we obtain that D represents an element of $B \cdot (A/\mathbf{P}^n)^*$.

(ii) This follows from the fact that I_∞ is formed by the residue classes modulo $\mathbf{P}^n$ of the elements of $\mathbf{F}_{q^r}^*$, so that $I_\infty \cap \mathrm{Cl}_{\mathbf{p}^n}^+(A)$ is a cyclic subgroup of $(A/\mathbf{P}^n)^*$ of order $q - 1$. $\square$

Theorem 4.2.20 *Let $F/\mathbf{F}_q$ be a global function field of genus $g(F)$ with $N(F) \geq 2$. Then for all integers $n \geq 1$ and $r \geq 2$ there exists a global function field $K_{n,r}/\mathbf{F}_{q^r}$ such that:*
  (i) *The number of rational places of $K_{n,r}/\mathbf{F}_{q^r}$ satisfies*

$$N(K_{n,r}) \geq \frac{h(F_r)}{h(F)}\left(1 + q^{(r-1)(n-1)}(N(F) - 1)\right).$$

  (ii) *The genus of $K_{n,r}/\mathbf{F}_{q^r}$ satisfies*

$$\frac{h(F)}{h(F_r)}(2g(K_{n,r}) - 2) = q^{(r-1)(n-1)}(2g(F) + n - 2) - \frac{q^{(r-1)(n-1)} - 1}{q^{r-1} - 1} - 1.$$

Proof. (i) Let P and ∞ be two different rational places of $F/\mathbf{F}_q$. For a given $r \geq 2$ consider the constant field extension $F_r = F \cdot \mathbf{F}_{q^r}$, and let A and A_r be as above. For fixed $n \geq 1$ let

$$E = H_r(\Lambda(\mathbf{P}^n))$$

be the narrow ray class field with modulus P^n determined by a sgn-normalized Drinfeld A_r-module ϕ of rank 1 over H_r. Let $K = K_{n,r}$ be the subfield of the extension E/F_r fixed by the subgroup $I_\infty \cdot \mathrm{Cl}_{P^n}^+(A)$ of $\mathrm{Cl}_{P^n}^+(A_r) = \mathrm{Gal}(E/F_r)$. Since $|I_\infty \cap \mathrm{Cl}_{P^n}^+(A)| = q - 1$ by Lemma 4.2.19(ii), we have

$$[K : F_r] = \frac{[E : F_r]}{|I_\infty \cdot \mathrm{Cl}_{P^n}^+(A)|} = \frac{h(F_r)}{h(F)} q^{(r-1)(n-1)}. \tag{4.5}$$

By the construction of K, the place ∞ of F_r splits completely in the extension K/F_r. A rational place of $F_r/\mathbf{F}_{q^r}$ different from P and ∞ splits completely in K/F_r if its Artin symbol in E/F_r is contained in $\mathrm{Cl}_{P^n}^+(A)$, and this holds if the restriction of this rational place to $F/\mathbf{F}_q$ is rational. In this way we get

$$[K : F_r](N(F) - 1) = \frac{h(F_r)}{h(F)} q^{(r-1)(n-1)}(N(F) - 1) \tag{4.6}$$

rational places of $K/\mathbf{F}_{q^r}$. In order to prove (i), it remains to study the decomposition of P in the extension K/F_r.

Let Q be a place of K lying over P and R a place of E lying over Q. Then the inertia group $G(R|Q)$ of R over Q is

$$G(R|Q) = \mathrm{Gal}(E/K) \cap G(R|P),$$

where $G(R|P)$ is the inertia group of R over P, which is equal to $(A_r/P^n)^*$ (recall that the extension H_r/F_r is unramified). By Lemma 4.2.19(i) we conclude that $G(R|Q) = I_\infty \cdot (A/P^n)^*$, and so for the ramification indices we get

$$e(Q|P) = \frac{e(R|P)}{e(R|Q)} = \frac{(q^r - 1)q^{r(n-1)}}{|I_\infty \cdot (A/P^n)^*|} = q^{(r-1)(n-1)}, \tag{4.7}$$

where we used also that $|I_\infty \cap (A/P^n)^*| = q - 1$.

Let T be the inertia field of Q in the extension K/F_r. We have already noted that ∞ splits completely in K/F_r, and so by Proposition 3.3.8(ii) the only ramified place in K/F_r can be P. Consequently, T/F_r is an unramified abelian extension in which ∞ splits completely, and so it follows from the definition of the Hilbert class field that $T \subseteq H_r$. We observe also that

$$[T : F_r] = \frac{[K : F_r]}{e(Q|P)} = \frac{h(F_r)}{h(F)} \tag{4.8}$$

in view of (4.5) and (4.7). Let $J = H_r \cap K$, then $F_r \subseteq T \subseteq J$. On the one hand, the extension J/T is unramified, and on the other hand, any place of T lying over P is totally ramified in J/T. Thus, we must have $J = H_r \cap K = T$. It follows that $\mathrm{Gal}(E/T)$ is the subgroup of $\mathrm{Cl}_{P^n}^+(A_r)$ generated by $(A_r/P^n)^*$ and $I_\infty \cdot \mathrm{Cl}_{P^n}^+(A)$. By applying Lemma 4.2.19(i), we get

$$\begin{aligned}
\mathrm{Gal}(H_r/T) &= \mathrm{Gal}(E/T)/(A_r/P^n)^* \\
&\simeq \left(I_\infty \cdot \mathrm{Cl}_{P^n}^+(A)\right)/\left(I_\infty \cdot (A/P^n)^*\right) \simeq \mathrm{Cl}_{P^n}^+(A)/(A/P^n)^*.
\end{aligned}$$

Let $t \in A$ be a local parameter at P. Then for the divisor $(t)_\pi$ corresponding to the principal ideal tA_r we have

$$(t)_\pi = P + D,$$

where $P \notin \operatorname{supp}(D)$ and D is a positive $\mathbf{F}_q$-rational divisor relatively prime to ∞. For the corresponding fractional ideals we have $\mathsf{P} \equiv \mathsf{D}^{-1}$ modulo principal ideals, and so for the corresponding Galois automorphisms in $\operatorname{Gal}(H_r/F_r) = \operatorname{Cl}(A_r)$ we get $\tau_\mathsf{P} = \tau_{\mathsf{D}^{-1}}$. Since D is $\mathbf{F}_q$-rational and relatively prime to P and ∞, it follows from the formula for $\operatorname{Gal}(H_r/T)$ above that $\tau_\mathsf{P} = \tau_{\mathsf{D}^{-1}} \in \operatorname{Gal}(H_r/T)$, and so the theory of Hilbert class fields shows that P splits completely in T/F_r. By taking into account (4.8), we see that P splits into $h(F_r)/h(F)$ rational places of K. Together with (4.6) this yields the bound for $N(K) = N(K_{n,r})$ in the theorem.

(ii) Let L be the inertia field of R in E/K. Then $\operatorname{Gal}(E/L) = G(R|Q) = I_\infty \cdot (A/\mathsf{P}^n)^*$ by part (i) of the proof, and $|\operatorname{Gal}(E/L)| = (q^r - 1)q^{n-1}$. Furthermore,

$$\operatorname{Gal}(E/L) \subseteq G(R|P) = \operatorname{Gal}(E/H_r),$$

hence $H_r \subseteq L$. Thus, the place S of L lying under R is totally ramified in E/L. Then by Proposition 1.3.13 the different exponent $d(R|S)$ of R over S is given by

$$d(R|S) = \sum_{\gamma \in \operatorname{Gal}(E/L)\setminus\{\mathrm{id}\}} \nu_R(\lambda - \lambda^\gamma),$$

where λ is a generator of the cyclic A_r-module $\Lambda(\mathsf{P}^n)$. In accordance with Proposition 3.3.8(iv), for $\gamma \in \operatorname{Gal}(E/L)$ we have $\gamma = \sigma_{gA_r}$ for some $g \in A_r$ with $\operatorname{sgn}(g) = 1$ and $g = \sum_{i=0}^{n-1} \alpha_i u^i$, where all $\alpha_i \in \mathbf{F}_{q^r}$ and $u \in A_r$ is a local parameter at P. Using the special form of $\operatorname{Gal}(E/L)$, the n-tuple $(\alpha_0, \ldots, \alpha_{n-1})$ of coefficients can be written in the form $\beta(1, b_1, \ldots, b_{n-1})$ with $\beta \in \mathbf{F}_{q^r}^*$ and $b_1, \ldots, b_{n-1} \in \mathbf{F}_q$. By Proposition 3.3.8(iii) and [49, Lemma 4.4] we have

$$\nu_R(\lambda - \lambda^\gamma) = \nu_R(\lambda - \phi_g(\lambda)) = \nu_R\left((1 - \beta)\lambda - \sum_{i=1}^{n-1} \beta b_i \phi_{u^i}(\lambda)\right).$$

As in Lemma 4.2.6 we see that

$$\nu_R(\phi_{u^i}(\lambda)) = q^{ri} \qquad \text{for } 0 \le i \le n - 1.$$

Thus, if $\beta \ne 1$, then $\nu_R(\lambda - \lambda^\gamma) = 1$, and if $\beta = 1$ and $g \ne 1$, then

$$\nu_R(\lambda - \lambda^\gamma) = q^{rj},$$

where j is the least positive integer with $b_j \ne 0$. This yields

$$d(R|S) = (q^r - 2)q^{n-1} + \sum_{j=1}^{n-1}(q - 1)q^{n-1-j}q^{rj} = (q^r - 2)q^{n-1} + (q - 1)q^{n-1}\frac{q^{(r-1)n} - q^{r-1}}{q^{r-1} - 1}.$$

$$(4.9)$$

Since $\text{Gal}(E/L)$ contains I_∞, the place ∞ splits completely in L/F_r. By the definition of L, the place Q is unramified in L/K, and this holds for any place of K lying over P. Thus, L/K is an unramified extension. Furthermore, we have

$$[L:K] = \frac{[E:F_r]}{[E:L][K:F_r]} = h(F).$$

Hence the Hurwitz genus formula yields

$$2g(L) - 2 = h(F)(2g(K) - 2). \tag{4.10}$$

For the extension E/L the Hurwitz genus formula shows that

$$2g(E) - 2 = (q^r - 1)q^{n-1}(2g(L) - 2) + \deg(\text{Diff}(E/L)). \tag{4.11}$$

Only places of E lying over P or ∞ can contribute to $\deg(\text{Diff}(E/L))$. In part (i) of the proof we have shown that there are exactly $h(F_r)/h(F)$ rational places of K lying over P. If we also use the facts that the extension L/K of degree $h(F)$ is unramified and that the places of L lying over P are totally ramified in E/L, then we can conclude that the sum of the degrees of the places of E lying over P is equal to $h(F_r)$. Recall that I_∞ is both the decomposition group and the inertia group of ∞ in E/F_r. Therefore we get

$$\deg(\text{Diff}(E/L)) = d(R|S)h(F_r) + (q^r - 2)h(F_r)q^{r(n-1)}.$$

If we now combine this formula with Proposition 3.3.9 (of course with q replaced by q^r), (4.9), (4.10), and (4.11), and if we note that $g(F_r) = g(F)$, then we arrive at the desired formula for $g(K) = g(K_{n,r})$. $\square$

Corollary 4.2.21 *Let $F/\mathbf{F}_q$ be a global function field of genus $g(F)$ with $N(F) \geq 2$. Then for every integer $r \geq 2$ there exists a global function field $K_r/\mathbf{F}_{q^r}$ with*

$$g(K_r) = \frac{h(F_r)}{h(F)}(g(F) - 1) + 1 \quad and \quad N(K_r) \geq \frac{h(F_r)N(F)}{h(F)}.$$

Proof. Apply Theorem 4.2.20 with $n = 1$. $\square$

It is useful to note that if $L(F,t) = L(t) = (1-t)(1-qt)Z(t) = \prod_{i=1}^{2g}(1 - \omega_i t)$ is the L-polynomial of $F/\mathbf{F}_q$ (see Definition 1.6.7), then

$$\frac{h(F_2)}{h(F)} = \frac{\prod_{i=1}^{2g}(1 - \omega_i^2)}{\prod_{i=1}^{2g}(1 - \omega_i)} = L(-1). \tag{4.12}$$

In the theory of algebraic curves over $\mathbf{F}_q$ of genus 2 (see [141], [144]), the prime power $q = p^e$, p prime, $e \geq 1$, is called **nonspecial** if either (i) e is even and $q \neq 4, 9$; or (ii) e is odd, p does not divide $\lfloor 2q^{1/2} \rfloor$, and q is not of the form $k^2 + 1, k^2 + k + 1$, or $k^2 + k + 2$ for some integer k.

Corollary 4.2.22 *If the prime power q is nonspecial, then there exists a global function field $K/\mathbf{F}_{q^2}$ with*

$$g(K) = (q - m + 1)^2 + 1 \quad and \quad N(K) \geq (q + 2m + 1)(q - m + 1)^2,$$

where $m = \lfloor 2q^{1/2} \rfloor$.

Proof. Since q is nonspecial, there is a global function field $F/\mathbf{F}_q$ with $g(F) = 2$ and $N(F) = q + 2m + 1$ (see [141], [144]). By [144] we can have $g(F) = 2$ and $N(F) = q + 2m + 1$ only if the eigenvalues of the Frobenius are α and $\overline{\alpha}$ (each with multiplicity 2) with $\alpha + \overline{\alpha} = -m$ and $\alpha\overline{\alpha} = q$. Therefore

$$L(F, t) = (1 - \alpha t)^2 (1 - \overline{\alpha} t)^2 = (qt^2 + mt + 1)^2.$$

By Corollary 4.2.21 and (4.12) we get a global function field $K/\mathbf{F}_{q^2}$ with the desired properties of $g(K)$ and $N(K)$. □

Corollary 4.2.23 *Let q be a nonsquare and let the characteristic p of $\mathbf{F}_q$ satisfy $p \equiv 1 \pmod 4$. Then there exists a global function field $K/\mathbf{F}_{q^2}$ with*

$$g(K) = q^2 + 2q + 2 \quad and \quad N(K) \geq (q + 1)^3.$$

Proof. It is well known that under our conditions on q there exists an elliptic curve E over $\mathbf{F}_q$ with $q+1$ $\mathbf{F}_q$-rational points (see e.g. [137] and [171]). Then E is a supersingular elliptic curve with a cyclic group of $\mathbf{F}_q$-rational points (see [137, Lemma 4.8]). Furthermore, the order of the Frobenius acting on the group of 2-division points of E is at most 2. Thus according to [144], E can be glued to itself if the j-invariant of E is not equal to 1728. By [146, p. 144, Example 4.5] an elliptic curve with the j-invariant 1728 is not supersingular if $p \equiv 1 \pmod 4$. Hence under our assumptions, E can be glued to itself. If C is the projective curve over $\mathbf{F}_q$ with Jacobian isogenous to $E \times E$, then for its function field $F/\mathbf{F}_q$ we have $g(F) = 2$, $N(F) = q + 1$, and $h(F) = (q+1)^2$. This yields $L(F, t) = (qt^2 + 1)^2$, and so the desired result follows from Corollary 4.2.21 and (4.12). □

Example 4.2.24 Let F be the rational function field $\mathbf{F}_2(x)$. Then with $n = 4$ and $r = 2$ in Theorem 4.2.20 we get a global function field $K/\mathbf{F}_4$ with $g(K) = 5$ and $N(K) \geq 17$. Note that $N_4(5) \leq 18$.

Example 4.2.25 Let $F = \mathbf{F}_2(x, y)$ be the global function field defined by

$$y^2 + y = \frac{x}{x^2 + x + 1}.$$

Then $g(F) = 1$, $N(F) = 4$, and $L(F, t) = 2t^2 + t + 1$. Thus, by using (4.12) and Theorem 4.2.20 with $n = 3$ and $r = 2$, we get a global function field $K/\mathbf{F}_4$ with $g(K) = 9$ and $N(K) = 26$. The function field K is optimal.

Example 4.2.26 Let $F = \mathbf{F}_2(x, y)$ be the global function field defined by

$$y^2 + y = x^3 + x.$$

Then $g(F) = 1$, $N(F) = 5$, and $L(F, t) = 2t^2 + 2t + 1$. Thus, by using (4.12) and Theorem 4.2.20 with $n = 3, 4, 5$ and $r = 2$, we get three global function fields $K_n/\mathbf{F}_4, n = 3, 4, 5$, with

$$g(K_3) = 5, \quad N(K_3) = 17;$$
$$g(K_4) = 13, \quad N(K_4) = 33;$$
$$g(K_5) = 33, \quad N(K_5) = 65.$$

The function field K_4 is optimal.

Example 4.2.27 Let $F = \mathbf{F}_2(x, y)$ be the global function field defined by

$$y^2 + y = \frac{x}{x^3 + x + 1}.$$

Then $g(F) = 2$ and $N(F) = 4$. Since F has exactly three places of degree 2, we obtain

$$L(F, t) = 4t^4 + 2t^3 + 3t^2 + t + 1.$$

Thus, by using (4.12) and Theorem 4.2.20 with $n = 1$ and $r = 2$, we get a global function field $K/\mathbf{F}_4$ with $g(K) = 6$ and $N(K) = 20$. The function field K is optimal.

Example 4.2.28 Let F be the rational function field $\mathbf{F}_q(x)$, where q is an arbitrary prime power. Then with $n = 3$ and $r = 2$ in Theorem 4.2.20 we get a global function field $K/\mathbf{F}_{q^2}$ with $g(K) = q(q-1)/2$ and $N(K) = q^3 + 1$. The field K is the well-known Hermitian function field (see Example 3.1.12), it is optimal and even maximal.

The following construction is due to Niederreiter and Xing [114].

Theorem 4.2.29 Let $F/\mathbf{F}_q$ be a global function field of genus $g(F)$. For an integer $r \geq 2$ let b be a positive divisor of $q^r - 1$ and put $s = \gcd(q - 1, b)$. Suppose that F has a place of degree d with $\gcd(d, r) = 1$ and that $N(F) \geq 1 + \varepsilon_d$, where $\varepsilon_d = 1$ if $d = 1$ and $\varepsilon_d = 0$ if $d \geq 2$. Then for every integer $n \geq 1$ there exists a global function field $K_n/\mathbf{F}_{q^r}$ with

$$2g(K_n) - 2 = \frac{s(q^r - 1)h(F_r)}{b(q-1)h(F)} \cdot$$
$$\cdot \left(\frac{(q-1)(q^{dr}-1)}{(q^d-1)(q^r-1)} q^{d(r-1)(n-1)} (2g(F) + dn - 2) - \frac{d(q-1)(q^{dr}-1)(q^{d(r-1)(n-1)}-1)}{(q^d-1)(q^r-1)(q^{d(r-1)}-1)} - d \right)$$
$$+ \left(\frac{s(q^r - 1)}{b(q-1)} - 1 \right) \frac{h(F_r)}{h(F)} d + \left(\frac{s(q^r - 1)}{b(q-1)} - 1 \right) \frac{(q-1)(q^{dr}-1)h(F_r)}{(q^d-1)(q^r-1)h(F)} q^{d(r-1)(n-1)}$$

and

$$N(K_n) \geq \frac{s(q^{dr}-1)h(F_r)}{b(q^d-1)h(F)} q^{d(r-1)(n-1)} (N(F) - 1 - \varepsilon_d)$$
$$+ \frac{(q-1)(q^{dr}-1)h(F_r)}{(q^d-1)(q^r-1)h(F)} q^{d(r-1)(n-1)} + \frac{h(F_r)}{h(F)} \varepsilon_d.$$

Proof. Let ∞ be a distinguished rational place of $F/\mathbf{F}_q$ and let $Q \neq \infty$ be a place of $F/\mathbf{F}_q$ of degree d. Then Q is still a place of degree d of $F_r/\mathbf{F}_{q^r}$. For a given $n \geq 1$ let $E_n = H_r(\Lambda(\mathbf{M}))$ be the narrow ray class field with the modulus being the ideal $\mathbf{M} = Q^n$ of A_r. Then we can identify $\mathrm{Gal}(E_n/F_r)$ with $\mathrm{Cl}_{\mathbf{M}}^+(A_r)$. Let B be a subgroup of I_∞ with $|B| = b$. Now let K_n be the subfield of the extension E_n/F_r fixed by the subgroup $G = B \cdot \mathrm{Cl}_{\mathbf{M}}^+(A)$ of $\mathrm{Cl}_{\mathbf{M}}^+(A_r)$. We have

$$|G| = \frac{|B| \cdot |\mathrm{Cl}_{\mathbf{M}}^+(A)|}{|B \cap \mathrm{Cl}_{\mathbf{M}}^+(A)|} = \frac{b}{s} h(F)(q^d - 1)q^{d(n-1)}$$

by Lemma 4.2.19(ii), and so

$$[K_n : F_r] = \frac{|\mathrm{Cl}_{\mathbf{M}}^+(A_r)|}{|G|} = \frac{s(q^{dr} - 1)h(F_r)}{b(q^d - 1)h(F)} q^{d(r-1)(n-1)}. \tag{4.13}$$

The rational place of $F_r/\mathbf{F}_{q^r}$ lying over ∞ is again denoted by ∞. Let P_∞ be a place of K_n lying over ∞. Then the inertia group of P_∞ in the extension E_n/K_n is $I_\infty \cap G$, and so the ramification index $e(P_\infty|\infty)$ of P_∞ over ∞ is given by

$$e(P_\infty|\infty) = \frac{|I_\infty|}{|I_\infty \cap G|} = \frac{|I_\infty \cdot G|}{|G|} = \frac{|I_\infty \cdot \mathrm{Cl}_{\mathbf{M}}^+(A)|}{|G|} = \frac{|I_\infty| \cdot |\mathrm{Cl}_{\mathbf{M}}^+(A)|}{|I_\infty \cap \mathrm{Cl}_{\mathbf{M}}^+(A)| \cdot |G|} = \frac{s(q^r - 1)}{b(q - 1)},$$

where we used Lemma 4.2.19(ii) in the last step.

Let R be a place of K_n lying over Q. Since the inertia group of Q in E_n/F_r is $(A_r/\mathbf{M})^*$ by the theory of narrow ray class fields, the inertia group of R in E_n/K_n is $(A_r/\mathbf{M})^* \cap G = B \cdot (A/Q^n)^*$ in view of Lemma 4.2.19(i). Thus, the ramification index $e(R|Q)$ of R over Q is given by

$$e(R|Q) = \frac{|(A_r/\mathbf{M})^*|}{|B \cdot (A/Q^n)^*|} = \frac{|(A_r/\mathbf{M})^*| \cdot |B \cap (A/Q^n)^*|}{|B| \cdot |(A/Q^n)^*|} = \frac{s(q^{dr} - 1)}{b(q^d - 1)} q^{d(r-1)(n-1)}. \tag{4.14}$$

Let L_n be the subfield of E_n/F_r fixed by $I_\infty \cdot \mathrm{Cl}_{\mathbf{M}}^+(A)$. Then ∞ is unramified in L_n/F_r, and from the special case $b = q^r - 1$ in (4.14) we get that Q is ramified in L_n/F_r with ramification index

$$\frac{(q-1)(q^{dr} - 1)}{(q^r - 1)(q^d - 1)} q^{d(r-1)(n-1)}.$$

Furthermore, from (4.13) applied to K_n and L_n we obtain

$$[K_n : L_n] = \frac{s(q^r - 1)}{b(q - 1)}.$$

It follows that the places of L_n lying over ∞ or Q are all totally ramified and thus tamely ramified or unramified in the extension K_n/L_n, and by the theory of narrow ray class fields these are the only possible ramified places in this extension. By the proof of Theorem 4.2.20, the sum of the degrees of the places of L_n lying over Q is $dh(F_r)/h(F)$. Now we apply the Hurwitz genus formula to the extension K_n/L_n and note that ∞ splits completely in L_n/F_r. Then we obtain

$$
\begin{aligned}
2g(K_n) - 2 &= \frac{s(q^r - 1)}{b(q - 1)}(2g(L_n) - 2) + \left(\frac{s(q^r - 1)}{b(q - 1)} - 1 \right) \frac{h(F_r)}{h(F)} d \\
&\quad + \left(\frac{s(q^r - 1)}{b(q - 1)} - 1 \right) \frac{(q-1)(q^{dr} - 1)h(F_r)}{(q^d - 1)(q^r - 1)h(F)} q^{d(r-1)(n-1)}.
\end{aligned}
$$

The desired formula for $g(K_n)$ follows now from a formula for $g(L_n)$ that is proved like that for $g(K_{n,r})$ in Theorem 4.2.20.

By construction, all rational places of F_r counted by $N(F)$, with the possible exception of ∞ and Q, split completely in K_n/F_r. By what we have shown above, ∞ splits into

$$\frac{(q-1)(q^{dr}-1)h(F_r)}{(q^d-1)(q^r-1)h(F)}q^{d(r-1)(n-1)}$$

rational places in K_n/F_r. If $d = 1$, then Q splits into $h(F_r)/h(F)$ rational places of L_n, as shown in the proof of Theorem 4.2.20, and we have noted above that these are totally ramified in K_n/L_n. Putting these facts together and using (4.13), we get the desired lower bound on $N(K_n)$. $\square$

Lots of global function fields with many rational places can be obtained from the above theorem. The base function fields F in Theorem 4.2.29 play an important role. We first list some base fields in Table 4.2.1 and then these base fields are used to obtain many global function fields from Theorem 4.2.29. The quotient $h(F_r)/h(F)$ of divisor class numbers is simply denoted by h_r/h in Table 4.2.1. In the second column of Tables 4.2.2–6, the first number is a lower bound for $N_{q^r}(g)$ obtained from Theorem 4.2.29 and the second is an upper bound for $N_{q^r}(g)$ obtained by the method in Theorem 1.6.18. If only one number is given, then this is the exact value of $N_{q^r}(g)$. Tables 4.2.1–6 are taken from Niederreiter and Xing [114].

Table 4.2.1. Base fields $F/\mathbf{F}_q$

	q	$g(F)$	$N(F)$	equation or reference	$L(F,t)$	h_2/h	h_3/h
F.1	2	1	5	$y^2 + y = x^3 + x$	$2t^2 + 2t + 1$	1	1
F.2	2	1	4	$y^2 + y = (x+1)^2/x$	$2t^2 + t + 1$	2	1
F.3	2	1	3	$y^2 + y = x(x^2 + x + 1)$	$2t^2 + 1$	3	
F.4	2	2	6	$y^2 + y =$ $x(x+1)/(x^3 + x + 1)$	$4t^4 + 6t^3 + 5t^2 +$ $3t + 1$	1	4
F.5	2	2	5	$y^2 + y =$ $x^2(x+1)(x^2 + x + 1)$	$4t^4 + 4t^3 + 4t^2 +$ $2t + 1$	3	
F.6	2	2	4	$y^2 + y = x/(x^3 + x + 1)$	$4t^4 + 2t^3 + 3t^2 +$ $t + 1$	5	
F.7	2	2	3	$y^2 + y = x(x^2 + x + 1)^2$	$4t^4 + 2t^2 + 1$	7	
F.8	2	3	7	[103, Example 3A]	$8t^6 + 16t^5 + 18t^4 +$ $15t^3 + 9t^2 + 4t + 1$	1	7
F.9	3	1	7	$y^2 = x^3 - x + 1$	$3t^2 + 3t + 1$	1	4
F.10	3	1	6	$y^2 = x(x^2 + x - 1)$	$3t^2 + 2t + 1$	2	3
F.11	3	1	5	$y^2 = x^3 - x^2 + 1$	$3t^2 + t + 1$	3	
F.12	3	1	4	$y^2 = -x(x^2 + 1)$	$3t^2 + 1$	4	7
F.13	3	2	8	$y^2 = x^6 - x^2 + 1$	$9t^4 + 12t^3 + 9t^2 +$ $4t + 1$	3	16
F.14	3	2	7	$y^2 = x^5 - x + 1$	$9t^4 + 9t^3 + 7t^2 +$ $3t + 1$	5	19
F.15	4	1	9	$y^2 + y = x^3$	$4t^2 + 4t + 1$	1	
F.16	4	1	8	$y^2 + y = (x^2 + x + 1)/x$	$4t^2 + 3t + 1$	2	
F.17	4	1	7	$y^2 + y = \alpha x(x+1)(x+\alpha)$ with $\alpha^2 + \alpha = 1$	$4t^2 + 2t + 1$	3	
F.18	4	1	6	$y^2 + y = \alpha(x^2 + x + 1)/x$ with $\alpha^2 + \alpha = 1$	$4t^2 + t + 1$	4	
F.19	4	1	5	$y^2 + y = x^2(x+1)$	$4t^2 + 1$	5	
F.20	4	2	10	$y^2 + y = x/(x^3 + x + 1)$	$16t^4 + 20t^3 + 13t^2 +$ $5t + 1$	5	

Table 4.2.2. Constructions over $\mathbf{F}_4$

g	$N_4(g)$	$g(F)$	F	d	b	n	g	$N_4(g)$	$g(F)$	F	d	b	n
3	14	1	F.2	1	3	2	94	129-155	3	F.8	1	1	4
4	15	1	F.3	1	3	2	97	99-159	1	F.3	1	3	5
5	17-18	0		1	3	4	101	125-165	2	F.6	1	3	4
6	20	2	F.6	1	3	1	105	129-170	0		7	3	1
7	21-22	0		3	1	1	109	165-176	2	F.5	5	3	1
8	21-24	2	F.7	1	3	1	113	161-181	2	F.4	1	3	6
9	26	1	F.2	1	3	3	114	161-183	1	F.1	1	1	5
10	27-28	1	F.3	3	3	1	115	168-184	3	F.8	3	3	2
13	33	1	F.1	1	3	4	121	150-192	2	F.6	3	1	1
15	33-37	0		5	3	1	125	176-198	2	F.4	5	1	1
18	41-42	1	F.1	1	1	3	145	195-225	2	F.5	1	3	5
21	41-47	2	F.4	1	3	4	148	215-229	1	F.1	7	3	1
25	51-53	2	F.5	1	3	3	154	168-237	0		3	1	2
26	55	1	F.1	5	3	1	158	209-243	3	F.8	5	1	1
29	49-60	3	F.8	1	3	4	161	194-247	1	F.2	1	3	6
30	53-61	2	F.4	1	1	3	162	209-248	2	F.4	1	1	5
31	60-63	2	F.6	3	3	1	181	220-274	2	F.6	5	3	1
33	65-66	1	F.1	1	3	5	183	220-276	1	F.2	5	1	1
34	57-68	3	F.8	3	1	1	191	258-287	2	F.4	7	3	1
35	58-69	1	F.2	1	1	3	193	257-290	1	F.1	1	3	7
37	66-72	2	F.4	5	3	1	199	216-298	1	F.3	3	3	2
41	65-78	2	F.6	1	3	3	208	243-309	2	F.5	1	1	4
43	72-81	0		3	3	2	210	257-312	3	F.8	1	1	5
46	81-86	1	F.1	1	1	4	234	301-343	3	F.8	7	3	1
48	77-89	3	F.8	5	3	1	241	245-353	2	F.6	1	3	5
49	81-90	2	F.4	1	3	5	257	321-373	2	F.4	1	3	7
51	88-93	1	F.2	5	3	1	274	321-396	1	F.1	1	1	6
57	63-102	2	F.7	1	3	3	295	344-423	1	F.2	7	3	1
59	77-105	0		5	1	1	298	384-427	2	F.4	3	1	2
61	99-108	2	F.5	1	3	4	321	385-456	3	F.8	1	3	7
65	98-114	1	F.2	1	3	5	337	387-477	2	F.5	1	3	6
70	105-121	2	F.4	1	1	4	370	417-520	2	F.4	1	1	6
76	99-130	1	F.3	5	3	1	373	429-523	2	F.5	5	1	1
81	129-137	1	F.1	1	3	6	379	456-531	3	F.8	3	1	2
88	123-147	2	F.5	1	1	3	449	513-619	1	F.1	1	3	8
91	144-151	2	F.4	3	3	2	451	480-622	2	F.6	3	3	2
92	143-152	1	F.1	5	1	1	466	513-641	3	F.8	1	1	6

Table 4.2.2. Constructions over $\mathbf{F}_4$ (cont.)

g	$N_4(g)$	$g(F)$	F	d	b	n
492	559-673	1	F.1	7	1	1
571	645-772	2	F.5	7	3	1
577	641-779	2	F.4	1	3	8
621	688-834	2	F.4	7	1	1
705	769-939	3	F.8	1	3	8
750	817-994	3	F.8	7	1	1
766	855-1014	1	F.1	9	3	1
769	771-1018	2	F.5	1	3	7
937	1026-1223	2	F.4	9	3	1
1015	1152-1318	2	F.4	3	3	3
1108	1197-1430	3	F.8	9	3	1
1207	1344-1550	3	F.8	3	3	3
1731	1760-2179	1	F.1	5	3	2
2083	2112-2596	2	F.4	5	3	2
2435	2464-3011	3	F.8	5	3	2

Table 4.2.3. Constructions over $\mathbf{F}_8$

g	$N_8(g)$	$g(F)$	F	d	b	n
6	33-36	0		1	7	3
9	45-47	0		2	1	1
45	144-156	0		2	7	2
53	120-179	2	F.4	1	1	1
54	129-181	0		1	7	4
77	195-242	1	F.1	4	7	1
78	175-245	3	F.8	1	7	2
93	192-284	1	F.2	2	7	2
118	257-348	1	F.1	1	7	4
141	259-407	3	F.8	1	1	1
149	324-428	2	F.4	1	7	3
225	453-616	0		5	7	1
376	755-977	1	F.1	5	7	1
461	936-1178	2	F.4	4	7	1

Table 4.2.4. Constructions over $\mathbf{F}_9$

g	$N_9(g)$	$g(F)$	F	d	b	n	g	$N_9(g)$	$g(F)$	F	d	b	n
1	16	0		1	4	2	95	272-318	1	F.10	1	8	4
3	28	0		1	1	2	101	275-335	2	F.14	1	8	3
5	32-36	1	F.10	1	8	2	102	244-338	0		1	8	5
6	35-40	2	F.14	1	8	1	109	298-358	1	F.9	1	4	4
7	39-43	1	F.11	1	8	2	112	315-366	2	F.13	3	4	1
9	40-51	1	F.12	1	8	2	119	308-386	1	F.10	1	1	3
12	55-63	1	F.9	1	8	3	131	320-419	2	F.14	1	1	2
15	64-74	1	F.9	1	1	2	136	354-433	2	F.13	1	4	3
19	84-88	1	F.10	3	8	1	142	327-449	1	F.11	1	8	4
21	82-95	0		1	8	4	151	427-474	1	F.9	5	8	1
22	78-98	2	F.13	1	1	1	154	357-483	1	F.11	3	1	1
23	92-101	1	F.10	1	8	3	183	487-563	1	F.9	1	8	5
24	91-104	0		3	1	1	186	455-571	2	F.14	3	4	1
25	64-108	1	F.12	1	4	2	212	427-642	0		5	4	1
28	105-117	1	F.11	3	8	1	217	488-656	1	F.10	1	4	4
29	104-120	1	F.10	1	1	2	223	570-672	2	F.13	1	8	4
34	111-136	1	F.11	1	8	3	226	500-681	2	F.14	1	4	3
36	110-142	2	F.14	1	1	1	231	568-694	1	F.9	1	1	4
37	120-145	2	F.13	1	4	2	238	609-713	2	F.13	3	1	1
43	120-164	1	F.11	1	1	2	286	678-840	2	F.13	1	1	3
45	112-170	1	F.12	1	8	3	301	732-879	1	F.10	5	8	1
47	154-177	1	F.10	3	4	1	334	793-965	1	F.9	5	4	1
48	163-180	1	F.9	1	8	4	365	812-1045	1	F.10	1	8	5
49	168-183	2	F.13	3	8	1	367	756-1050	0		3	8	2
52	175-192	1	F.9	3	1	1	371	815-1061	2	F.14	1	8	4
55	164-201	1	F.10	1	4	3	396	875-1125	2	F.14	3	1	1
60	190-217	1	F.9	1	1	3	406	892-1151	1	F.9	1	4	5
61	192-220	2	F.13	1	8	3	451	915-1267	1	F.11	5	8	1
70	189-247	1	F.11	3	4	1	487	1056-1359	2	F.13	1	4	4
79	228-273	2	F.13	1	1	2	556	1323-1536	1	F.9	3	8	2
81	245-279	2	F.14	3	8	1	634	1464-1735	2	F.13	5	8	1
82	192-282	1	F.11	1	4	3	667	1342-1819	1	F.10	5	4	1
90	244-304	0		5	8	1	669	1459-1824	1	F.9	1	8	6
93	196-313	1	F.12	3	4	1	700	1525-1903	1	F.9	5	1	1

Table 4.2.5. Constructions over $\mathbf{F}_{16}$

g	$N_{16}(g)$	$g(F)$	F	d	b	n	g	$N_{16}(g)$	$g(F)$	F	d	b	n
6	65	0		1	1	2	140	577-685	1	F.15	1	1	3
36	185-223	2	F.20	1	5	2	156	650-754	2	F.20	3	5	1
37	208-228	1	F.16	3	5	1	186	725-884	2	F.20	1	5	3
43	226-259	1	F.16	1	5	3	226	825-1054	2	F.20	1	1	2
51	250-295	1	F.16	1	1	2	235	898-1090	1	F.16	1	5	4
54	257-309	0		1	5	4	245	936-1131	1	F.16	3	1	1
55	273-313	1	F.17	3	5	1	279	994-1267	1	F.16	1	1	3
58	273-327	0		3	1	1	306	1025-1374	0		5	5	1
60	257-336	0		1	1	3	352	1155-1557	1	F.17	1	5	4
64	291-354	1	F.17	1	5	3	367	1209-1616	1	F.17	3	1	1
73	312-393	1	F.18	3	5	1	511	1845-2181	1	F.15	5	5	1
76	315-407	1	F.17	1	1	2	598	2049-2521	1	F.15	1	5	5
85	324-446	1	F.18	1	5	3	716	2305-2980	1	F.15	1	1	4
91	325-472	1	F.19	3	5	1	906	2885-3719	2	F.20	1	5	4
101	340-516	1	F.18	1	1	2	936	2990-3835	2	F.20	3	1	1
106	325-538	1	F.19	1	5	3	1021	3280-4163	1	F.16	5	5	1
118	513-590	1	F.15	1	5	4	1195	3586-4812	1	F.16	1	5	5
123	533-611	1	F.15	3	1	1	2476	7488-9525	1	F.15	3	5	2

Table 4.2.6. Constructions over $\mathbf{F}_{27}$

g	$N_{27}(g)$	$g(F)$	F	d	b	n
17	128-185	2	F.13	1	13	1
19	126-199	1	F.10	2	13	1
20	133-207	2	F.14	1	13	1
25	196-242	1	F.9	2	13	1
33	220-298	1	F.9	1	13	2
36	244-319	0		1	13	3
37	162-326	1	F.10	1	1	1
42	280-360	0		2	1	1
43	196-367	1	F.12	2	13	1
48	244-402	0		1	1	2
49	268-409	1	F.9	1	1	1
209	896-1404	2	F.13	2	13	1

4.3   Function Fields from Cyclotomic Fields

Cyclotomic function fields are a special kind of narrow ray class fields where the base fields
are rational function fields. Because everything in rational function fields is explicit, this
allows us to compute the genus and the number of rational places easily. In this section,
we use several examples to show some principles for the construction of global function
fields with many rational places based on cyclotomic function fields. From now on, we will
often identify a monic irreducible polynomial $p(x)$ over $\mathbf{F}_q$ with its unique zero place of
the rational function field $\mathbf{F}_q(x)$.

Example 4.3.1 For $n \geq 2$ consider the cyclotomic function field $E_n := F(\Lambda_{x^n})$, where
$F = \mathbf{F}_2(x)$. Then the Galois group $G_n := \mathrm{Gal}(E_n/F)$ can be identified with $(\mathbf{F}_2[x]/(x^n))^*$.
Let C_n be the cyclic subgroup of G_n generated by $\overline{x+1}$, where $\overline{x+1}$ is the residue class
of $x + 1$ modulo x^n, and K_n the subfield of E_n/F fixed by C_n.

In order to calculate the genus and the number of rational places of K_n, we first have
to determine the order of C_n. Since $(\mathbf{F}_2[x]/(x^n))^*$ has order 2^{n-1}, we have $|C_n| = 2^k$ for
some integer $k \geq 1$. Note that for integers $h \geq 1$ we have $(x+1)^{2^h} - 1 = x^{2^h} \equiv 0 \pmod{x^n}$
if and only if $2^h \geq n$, that is, if and only if $h \geq \lfloor \log_2(n - 1) \rfloor + 1$. Hence the order of C_n
is equal to $2^{\lfloor \log_2(n-1) \rfloor + 1}$.

Next we determine the number of rational places of K_n. The infinite place of F splits
completely and the place x is totally ramified in the extension E_n/F, in view of Theorem
3.2.8 and Proposition 3.2.4(iii). Hence these two places of F contribute $[K_n : F]+1$ rational
places to the field K_n. By Theorem 3.2.6(iv), the Artin symbol

$$\left[\frac{E_n/F}{x+1} \right]$$

corresponds to the element $\overline{x+1} \in (\mathbf{F}_2[x]/(x^n))^*$, and so the place $x+1$ splits completely
in the extension K_n/F. Thus,

$$N(K_n) = 2[K_n : F] + 1 = 2\frac{[E_n : F]}{|C_n|} + 1 = 2^{n-\lfloor \log_2(n-1) \rfloor - 1} + 1.$$

Now we calculate the genus of K_n. Let P be the unique place of K_n lying over the
place x of F and let P_n be the unique place of E_n lying over P. In order to determine
$g(K_n)$, it suffices to compute the different exponent $d(P_n|P)$. Let $\lambda \in \Lambda_{x^n}$ be a root of
$f(z) = z^{x^n}/z^{x^{n-1}}$. Then λ is a local parameter at P_n by the proof of Proposition 3.2.4(i).
Furthermore, the minimal polynomial of λ over K_n is

$$m(z) = \prod_{\sigma \in C_n} (z + \sigma(\lambda)) \in K_n[z].$$

Since P_n is totally ramified in E_n/K_n, it follows from Proposition 1.3.13 that

$$d(P_n|P) = \nu_{P_n}(m'(\lambda)) = \sum_{\sigma \in C_n \setminus \{\mathrm{id}\}} \nu_{P_n}(\lambda + \sigma(\lambda)) = \sum_{i=1}^{r-1} \nu_{P_n}(\lambda + \lambda^{(x+1)^i}),$$

where $r = 2^{\lfloor \log_2(n-1) \rfloor + 1}$. For $1 \leq i \leq r - 1$ we have

$$\lambda + \lambda^{(x+1)^i} = \lambda + \sum_{j=0}^{i} \binom{i}{j} \lambda^{x^j} = \sum_{j=0}^{\min(i, n-1)} \binom{i}{j} \lambda^{x^j}.$$

By induction on j one shows that $\nu_{P_n}(\lambda^{x^j}) = 2^j$ for $0 \leq j \leq n-1$. Hence a straightforward computation yields

$$d(P_n | P) = \sum_{j=0}^{k-1} 2^{2^j + k - j - 1},$$

where $k = \lfloor \log_2(n-1) \rfloor + 1$. By the Hurwitz genus formula we get

$$2g(E_n) - 2 = [E_n : K_n](2g(K_n) - 2) + d(P_n | P).$$

Together with Theorem 3.2.9 this yields

$$g(K_n) = 1 + (n-3)2^{n-k-2} - \sum_{j=0}^{k-1} 2^{2^j - j - 2}.$$

This example stems from Niederreiter and Xing [103].

Example 4.3.2 In the above example, taking $n = 6, 7, 8$, respectively, we obtain three global function fields K_6, K_7, K_8 over $\mathbf{F}_2$ with

$$g(K_6) = 5, N(K_6) = 9; \quad g(K_7) = 15, N(K_6) = 17; \quad g(K_8) = 39, N(K_8) = 33.$$

All these three function fields are optimal.

Example 4.3.3 $g(K/\mathbf{F}_2) = 3$, $N(K/\mathbf{F}_2) = 7$, $K = F(\Lambda_{x^3+x+1}) = \mathbf{F}_2(x, y)$ with $F = \mathbf{F}_2(x)$ and

$$y^7 + (x^4 + x^2 + x)y^3 + (x^4 + x^3 + x^2 + 1)y + x^3 + x + 1 = 0.$$

The function field K is optimal and was considered in Niederreiter and Xing [103].

Example 4.3.4 $g(K/\mathbf{F}_2) = 9$, $N(K/\mathbf{F}_2) = 12$, $K = F(\Lambda_{(x^2+x+1)^2}) = \mathbf{F}_2(x, y)$ with $F = \mathbf{F}_2(x)$ and

$$z^3 + (x^2 + x + 1)z + x^2 + x + 1 = 0,$$

where $z = y^4 + (x^2 + x + 1)y^2 + (x^2 + x + 1)y$. This function field K from [103] is optimal.

Example 4.3.5 $g(K/\mathbf{F}_2) = 21$, $N(K/\mathbf{F}_2) = 21$, $K = F(\Lambda_{(x^2+x+1)(x^3+x+1)}) = \mathbf{F}_2(x, y)$ with $F = \mathbf{F}_2(x)$ and

$$y\{z^2(x + z + 1)^3 + z(x + z + 1)^2(y^2 + xy) + (x + z + 1)(y^4 + x^2 y^2)$$
$$+ (x^2 + x + 1)(x + z + 1)\} + y^3 + xy^2 + (x + 1)y + 1 = 0,$$

where $z = y^4 + (x^2 + x + 1)y^2 + (x^2 + x + 1)y$. This function field K from [103] is optimal.

Example 4.3.6 $g(K/\mathbf{F}_3) = 10$, $N(K/\mathbf{F}_3) = 19$. Consider the cyclotomic function field $E = F(\Lambda_M)$ with $F = \mathbf{F}_3(x)$ and $M = x^5$. The Galois group $\mathrm{Gal}(E/F) = (\mathbf{F}_3[x]/(x^5))^*$ has order 162. Let K be the subfield of E/F fixed by the subgroup $J = \mathbf{F}_3^* \cdot \langle \overline{x+1} \rangle$ of $\mathrm{Gal}(E/F)$, where $\langle \overline{x+1} \rangle$ is the cyclic subgroup generated by the residue class $\overline{x+1}$ of $x+1$ modulo x^5. Then $|J| = [E:K] = 18$ and $[K:F] = 9$. By Theorem 3.2.8, the place ∞ splits completely in K/F. Since $\overline{x+1}$ is contained in J, a consideration of the Artin symbol of the place $x+1$ shows that $x+1$ also splits completely in K/F. The place x is totally ramified in K/F by Proposition 3.2.4(iii). Therefore $N(K) = 19$. To calculate the genus of K, we consider the extension E/K. Let Q be the unique place of K lying over the place x and R the unique place of E lying over Q. Let $\lambda \in \Lambda_{x^5}$ be a root of $f(z) = z^{x^5}/z^{x^4}$. Then λ is a local parameter at R by the proof of Proposition 3.2.4(i). Furthermore, the minimal polynomial of λ over K is

$$m(z) = \prod_{\tau \in J} (z - \tau(\lambda)) \in K[z].$$

It follows then from Proposition 1.3.13 that the different exponent $d(R|Q)$ of R over Q is given by

$$d(R|Q) = \nu_R(m'(\lambda)) = \sum_{\tau \in J\backslash\{\mathrm{id}\}} \nu_R(\lambda - \tau(\lambda)).$$

The elements of J are the residue classes of $(-1)^i(x+1)^j \pmod{x^5}$, where $i \in \{0,1\}$ and $j \in \{0,1,\ldots,8\}$. A direct calculation shows that $d(R|Q) = 81$. The 81 places of E lying over ∞ are tamely ramified in E/K with ramification index 2 by Theorem 3.2.8. Altogether, the Hurwitz genus formula yields

$$18(2g(K) - 2) + 81 \cdot 1 + 81 \cdot (2-1) = 2g(E) - 2 = 486,$$

where the last identity is obtained from Theorem 3.2.9. Hence we get $g(K) = 10$. This example is due to Niederreiter and Xing [105].

Example 4.3.7 $g(K/\mathbf{F}_3) = 15$, $N(K/\mathbf{F}_3) = 28$. Consider the cyclotomic function field $E = F(\Lambda_M)$ with $F = \mathbf{F}_3(x)$ and $M = x^6$. The Galois group $\mathrm{Gal}(E/F) = (\mathbf{F}_3[x]/(x^6))^*$ has order 486. Let K be the subfield of E/F fixed by the subgroup J of $(\mathbf{F}_3[x]/(x^6))^*$ generated by $\overline{x+1}$ and $\overline{x-1}$, where the bar denotes residue classes modulo x^6. By noting that $(x-1)^6 \equiv (x+1)^3 \pmod{x^6}$, it is easily seen that $|J| = 54$. We also have $\mathbf{F}_3^* \subseteq J$ and $[K:F] = 9$. By Theorem 3.2.8, the place ∞ splits completely in K/F. By considering the Artin symbols, we see that the places $x+1$ and $x-1$ also split completely in K/F. Furthermore, the place x is totally ramified in K/F by Proposition 3.2.4(iii). Hence $N(K) = 28$. Let Q be the unique place of K lying over the place x and R the unique place of E lying over Q. Then, as in Example 4.3.6, we obtain that the different exponent $d(R|Q)$ of R over Q is given by

$$d(R|Q) = \sum_{\tau \in J\backslash\{\mathrm{id}\}} \nu_R(\lambda - \tau(\lambda)),$$

where $\lambda \in \Lambda_{x^6}$ is a root of $f(z) = z^{x^6}/z^{x^5}$. The elements of J are the residue classes of $(-1)^i(x+1)^j(x-1)^k \pmod{x^6}$, where $i \in \{0,1\}$, $j \in \{0,1,2\}$, and $k \in \{0,1,\ldots,8\}$. A direct calculation shows that $d(R|Q) = 189$. The 243 places of E lying over ∞ are tamely ramified in E/K with ramification index 2. Altogether, the Hurwitz genus formula yields

$$54(2g(K) - 2) + 189 \cdot 1 + 243 \cdot (2 - 1) = 2g(E) - 2 = 1944,$$

where the last identity is obtained from Theorem 3.2.9. Hence we get $g(K) = 15$. The function field K is optimal. This example was found by Niederreiter and Xing [105].

Example 4.3.8 $g(K/\mathbf{F}_3) = 40$, $N(K/\mathbf{F}_3) \geq 54$. Put $F = \mathbf{F}_3(x)$ and let E be the cyclo-tomic function field over F with modulus $M = (x+1)^3(x-1)^3$ and with the distinguished rational place ∞ of F. Then

$$[E : F] = \Phi_3((x+1)^3)\Phi_3((x-1)^3) = 324.$$

Since $x^6 \equiv 1 \pmod{M}$, we can find a subgroup G of $\mathrm{Gal}(E/F) = (\mathbf{F}_3[x]/(M))^*$ with $|G| = 12$ such that G contains the decomposition groups of the places x and ∞ in E/F. Let K be the subfield of E/F fixed by G. Then $[K : F] = 27$. By construction, x and ∞ split completely in K/F, and so $N(K) \geq 54$. Let E_1, respectively E_2, be the cyclotomic function field over F with modulus $(x+1)^3$, respectively $(x-1)^3$, and with the distinguished rational place ∞ of F. Then

$$[E_1 : F] = [E_2 : F] = 18$$

and E is the composite field of E_1 and E_2 by Lemma 3.2.5. To calculate $g(K)$, we first determine the ramification index $e_{x+1}(K/F)$ of the place $x+1$ in K/F. From $e_{x+1}(E_1/F) = 18$, $e_{x+1}(E_2/F) = 1$, and Abhyankar's lemma (see [152, Proposition III.8.9]) we deduce that $e_{x+1}(E/F) = 18$, and so $e_{x+1}(K/F)$ divides 9. We claim that $e_{x+1}(K/F) = 9$. If we had $e_{x+1}(K/F) = 1$ or 3, then the inertia field L of $x+1$ in K/F has degree $[L : F] = 27$ or 9. Since $x+1$ is unramified in L/F and E_2 is the inertia field of $x+1$ in E/F, we have $L \subseteq E_2$, and so $[E_2 : L] = 2$. From the fact that x splits completely in L/F it follows that the relative degree of x in E_2/F is 1 or 2. Thus, the Artin symbol of x in E_2/F has order 1 or 2, but this is impossible since $x^2 \not\equiv 1 \pmod{(x-1)^3}$; hence indeed $e_{x+1}(K/F) = 9$. By the proof of Theorem 3.2.9, the different exponent $d_{x+1}(E_1/F)$ of $x+1$ in E_1/F is given by

$$d_{x+1}(E_1/F) = 3 \cdot 2 \cdot 3^2 - 3^2 = 45.$$

Now the tower formula for different exponents (see Proposition 1.3.11) yields

$$2d_{x+1}(K/F) + 1 = d_{x+1}(E/F) = d_{x+1}(E_1/F) = 45,$$

that is, $d_{x+1}(K/F) = 22$. In the same way one shows that $e_{x-1}(K/F) = 9$ and $d_{x-1}(K/F) = 22$. Since $x+1$ and $x-1$ are the only ramified places in K/F, the Hurwitz genus formula yields $2g(K) - 2 = -2 \cdot 27 + 3 \cdot 22 + 3 \cdot 22$, that is, $g(K) = 40$. This example is due to Niederreiter and Xing [110].

Example 4.3.9 $g(K/\mathbf{F}_3) = 102$, $N(K/\mathbf{F}_3) = 104$. Put $F = \mathbf{F}_3(x)$ and let E be the cyclotomic function field over F with modulus $M = \sum_{i=0}^{6} x^i$ and with the distinguished rational place ∞ of F. Then $[E : F] = \Phi_3(M) = 3^6 - 1$. Since $x^7 \equiv 1 \pmod{M}$, the subgroup G of $\mathrm{Gal}(E/F) = (\mathbf{F}_3[x]/(M))^*$ generated by the decomposition groups of the places x and ∞ in E/F satisfies $|G| = 14$. Let K be the subfield of E/F fixed by G. Then $[K : F] = 52$. By construction, x and ∞ split completely in K/F, and so $N(K) = 104$. Since M is the only ramified place in K/F, the Hurwitz genus formula yields $2g(K) - 2 = -2 \cdot 52 + (52 - 1) \cdot 6$, that is, $g(K) = 102$. This example stems from Niederreiter and Xing [110].

Example 4.3.10 $g(K/\mathbf{F}_4) = 3$, $N(K/\mathbf{F}_4) = 14$. Consider the irreducible polynomial $P = x^3 + \alpha$ over $\mathbf{F}_4$ and the corresponding cyclotomic function field $F(\Lambda_P)$ with $F = \mathbf{F}_4(x)$, where α is a root of $x^2 + x + 1$. The Galois group $\mathrm{Gal}(F(\Lambda_P)/F) = (\mathbf{F}_4[x]/(P))^*$ has order 63. Let K be the subfield of $F(\Lambda_P)/F$ fixed by the subgroup of $(\mathbf{F}_4[x]/(P))^*$ generated by the residue class of x modulo P. Then $[K : F] = 7$. The place ∞ splits completely in K/F, and a consideration of the Artin symbol shows that the place x also splits completely in K/F. Hence $N(K) = 14$. By Proposition 3.2.4(iii), P is the unique ramified place in K/F and it is totally and tamely ramified. Thus, the Hurwitz genus formula yields $2g(K) - 2 = -2 \cdot 7 + 3 \cdot (7 - 1)$, i.e., $g(K) = 3$. The function field K is optimal. This example is due to Niederreiter and Xing [105].

Example 4.3.11 $g(K/\mathbf{F}_4) = 10$, $N(K/\mathbf{F}_4) = 27$. Consider the cyclotomic function field $E = F(\Lambda_{P_1 P_2})$ with $F = \mathbf{F}_4(x)$, with the irreducible polynomials $P_1 = x^3 + x + 1$ and $P_2 = x^3 + x^2 + 1$ in $\mathbf{F}_4[x]$, and the subfields $L_1 = F(\Lambda_{P_1})$ and $L_2 = F(\Lambda_{P_2})$ of E/F. For $i = 1, 2$ let K_i be a subfield of L_i/F such that K_i is contained in the decomposition field of ∞ in L_i/F and $[K_i : F] = 3$. Let K be the composite field of K_1 and K_2. Since L_1 and L_2 are linearly disjoint by the proof of Theorem 3.2.6(i), we have $[K : F] = 9$. The place ∞ splits completely in K/F by the construction and Corollary 1.4.9(i). Note that $G = (\mathbf{F}_2[x]/(P_1 P_2))^*$ is a subgroup of $\mathrm{Gal}(E/F) = (\mathbf{F}_4[x]/(P_1 P_2))^*$ of order 49 and is thus the 7-Sylow subgroup of $\mathrm{Gal}(E/F)$. Consequently, G is contained in the Galois group $\mathrm{Gal}(E/K)$ of order $9 \cdot 7^2$. By considering the Artin symbols, we see that the places x and $x + 1$ split completely in K/F, thus $N(K) \geq 3 \cdot 9 = 27$. In order to determine the genus of K, we observe that the places $x^3 + x + 1$ and $x^3 + x^2 + 1$ are tamely ramified in K/F, each with ramification index 3. The Hurwitz genus formula gives $2g(K) - 2 = -2 \cdot 9 + 9 \cdot (3 - 1) + 9 \cdot (3 - 1)$, that is, $g(K) = 10$. Since $N_4(10) \leq 28$, we must have $N(K) = 27$. This example stems from Niederreiter and Xing [105].

The idea of using subfields of cyclotomic function fields for the construction of global function fields with many rational places was first proposed by Quebbemann [129]. In a recent paper, Keller [58] carried out a systematic search for subfields of cyclotomic function fields with many rational places. In particular, the following data for global function fields $K/\mathbf{F}_2$ that improved on previous constructions were obtained:

$$g(K/\mathbf{F}_2) = 38, \quad N(K/\mathbf{F}_2) = 30;$$
$$g(K/\mathbf{F}_2) = 66, \quad N(K/\mathbf{F}_2) = 48;$$
$$g(K/\mathbf{F}_2) = 81, \quad N(K/\mathbf{F}_2) = 56.$$

4.4 Explicit Function Fields

The most common explicit equations defining function fields are Kummer and Artin-Schreier equations. In this section, we present some examples of explicit equations for Kummer and Artin-Schreier extensions which yield global function fields with many rational places. In fact, many of these global function fields are optimal.

Example 4.4.1 $g(K/\mathbf{F}_2) = 2$, $N(K/\mathbf{F}_2) = 6$, $K = \mathbf{F}_2(x,y)$ with

$$y^2 + y = \frac{x(x+1)}{x^3 + x + 1}.$$

The function field K is optimal and can be found in Serre [144].

Example 4.4.2 $g(K/\mathbf{F}_2) = 4$, $N(K/\mathbf{F}_2) = 8$, $K = \mathbf{F}_2(x, y_1, y_2)$ with

$$y_1^2 + y_1 = x^3 + x, \quad y_2^2 + y_2 = \frac{xy_1}{x+1}.$$

The function field K is optimal and was constructed in Niederreiter and Xing [103]. There are four rational places of K lying over x and three rational places of K lying over $x + 1$, and there is a totally ramified place of K lying over the infinite place of $\mathbf{F}_2(x)$. Note that of the two rational places of $L = \mathbf{F}_2(x, y_1)$ lying over $x + 1$, one splits completely in the extension K/L and one is totally ramified in K/L.

Example 4.4.3 $g(K/\mathbf{F}_2) = 5$, $N(K/\mathbf{F}_2) = 9$, $K = \mathbf{F}_2(x, y_1, y_2)$ with

$$y_1^2 + y_1 = x^3 + x, \quad y_2^2 + y_2 = (x^2 + x)y_1.$$

The function field K is optimal and can be found in Niederreiter and Xing [103]. There are four rational places of K lying over each of x and $x + 1$, and there is a totally ramified place of K lying over the infinite place of $\mathbf{F}_2(x)$.

Example 4.4.4 $g(K/\mathbf{F}_2) = 6$, $N(K/\mathbf{F}_2) = 10$, $K = \mathbf{F}_2(x, y_1, y_2)$ with

$$y_1^2 + y_1 = x^3 + x, \quad y_2^2 + y_2 = u := \frac{x^2(x+1)((x+1)y_1 + x^3)}{x^5 + x^4 + x^3 + x^2 + 1}.$$

The function field K is optimal and was constructed in Niederreiter and Xing [103]. There are four rational places of K lying over each of x and $x+1$. The unique rational place P of $L = \mathbf{F}_2(x, y_1)$ lying over the infinite place of $\mathbf{F}_2(x)$ splits completely in the extension K/L, thus yielding two more rational places of K. This is seen by a short calculation showing that

$$\nu_P\left(u + \left(\frac{y_1}{x}\right)^2 + \frac{y_1}{x}\right) = 3.$$

To calculate $g(K)$, one also has to study the splitting behavior of the place $p(x) = x^5 + x^4 + x^3 + x^2 + 1$ of $\mathbf{F}_2(x)$. There are two different places P_1, P_2 of L lying over $p(x)$ which can be arranged in such a way that

$$y_1 \equiv x^4 + x \pmod{P_1}, \quad y_1 \equiv x^4 + x + 1 \pmod{P_2}.$$

Then $\nu_{P_1}(u) = -1$, hence $m_{P_1} = 1$ in the notation of Section 3.1, and $\nu_{P_2}(u) \geq 0$, so that $m_{P_2} = -1$. Thus, P_1 is the only ramified place in the extension K/L and $\deg(P_1) = 5$. Hence it follows from the formula in Theorem 3.1.9 that $g(K) = 6$.

Example 4.4.5 $g(K/\mathbf{F}_2) = 7$, $N(K/\mathbf{F}_2) = 10$, $K = \mathbf{F}_2(x, y_1, y_2)$ with

$$y_1^2 + y_1 = x^3 + x, \quad y_2^2 + y_2 = \frac{x(x+1)}{x^3 + x^2 + 1}.$$

The function field K is optimal and can be found in an equivalent form in Niederreiter and Xing [103]. There are four rational places of K lying over each of x and $x + 1$. The unique rational place of $L = \mathbf{F}_2(x, y_1)$ lying over the infinite place of $\mathbf{F}_2(x)$ splits completely in the extension K/L, thus yielding two more rational places of K. A straightforward calculation using Theorem 3.1.9 shows that the genus of K is 7.

Example 4.4.6 $g(K/\mathbf{F}_2) = 8$, $N(K/\mathbf{F}_2) = 11$, $K = \mathbf{F}_2(x, y_1, y_2)$ with

$$y_1^2 + y_1 = \frac{x(x+1)}{x^3 + x + 1}, \quad y_2^2 + y_2 = u := \frac{x(x+1)(x^3 + x + 1)}{(x^2 + x + 1)^2} y_1 + \frac{x(x+1)}{x^2 + x + 1}.$$

The function field K is optimal and was constructed in Niederreiter and Xing [103]. There are four rational places of K lying over each of x and $x + 1$. Of the two rational places of $L = \mathbf{F}_2(x, y_1)$ lying over the infinite place of $\mathbf{F}_2(x)$, one splits completely in the extension K/L and the other is totally ramified in K/L, which yields three more rational places of K. The only other ramified place in K/L is the unique place P of L with $\deg(P) = 4$ lying over $x^2 + x + 1$. If we put

$$z = \frac{(x+1)y_1 + 1}{x^2 + x + 1} \in L,$$

then a straightforward calculation shows that $\nu_P(u + z^2 + z) = -1$, so that $m_P = 1$ in the notation of Section 3.1. Hence it follows from the formula in Theorem 3.1.9 that $g(K) = 8$.

Example 4.4.7 $g(K/\mathbf{F}_2) = 9$, $N(K/\mathbf{F}_2) = 12$, $K = \mathbf{F}_2(x, y_1, y_2)$ with

$$y_1^2 + y_1 = \frac{x(x+1)}{x^3 + x + 1}, \quad y_2^2 + y_2 = \frac{x(x+1)}{x^3 + x^2 + 1}.$$

The function field K is optimal and can be found in Niederreiter and Xing [103]. There are four rational places of K lying over each of the three rational places of $\mathbf{F}_2(x)$. The genus of K is obtained from Theorem 3.1.9.

Example 4.4.8 $g(K/\mathbf{F}_2) = 11$, $N(K/\mathbf{F}_2) = 14$, $K = \mathbf{F}_2(x, y_1, y_2, y_3)$ with

$$y_1^2 + y_1 = x^3 + x, \quad y_2^2 + y_2 = \frac{xy_1}{x+1}, \quad y_3^2 + y_3 = u := \frac{x^2 y_1}{(x+1)^3} + \frac{x}{x+1}.$$

The function field K is optimal and was constructed in Niederreiter and Xing [104]. The field $E = \mathbf{F}_2(x, y_1, y_2)$ was already considered in Example 4.4.2. The following were found to be the rational places of E: a totally ramified place P_∞ lying over the infinite place of $\mathbf{F}_2(x)$, four unramified places P_1, P_2, P_3, P_4 lying over x, and three places P_5, P_6, P_7 lying over $x + 1$, where P_5 has ramification index 2 over $\mathbf{F}_2(x)$ and P_6 and P_7 are unramified. We observe that $\nu_{P_\infty}(y_1) = -6$ and $\nu_{P_\infty}(y_2) = -3$. With $w = (x + 1)^{-2} y_1 y_2 \in E$ we get

$$u + w^2 + w = \frac{y_1(xy_1 + (x + 1)^3 y_2)}{(x + 1)^5} + \frac{y_1^2 y_2}{(x + 1)^4} + \frac{x}{x + 1},$$

hence $\nu_{P_\infty}(u + w^2 + w) = -1$. Thus, by the theory of Artin-Schreier extensions in Section 3.1, P_∞ is totally ramified in K/E. The places P_1, P_2, P_3, P_4 split completely in K/E. Next we have $\nu_{P_5}(y_1) = 0$ and $\nu_{P_5}(y_2) = -1$. With $v = (x + 1)^{-1} y_2 \in E$ we get

$$u + v^2 + v = \frac{x(y_1 + y_2)}{(x + 1)^2} + \frac{x}{x + 1},$$

hence $\nu_{P_5}(u + v^2 + v) = -5$. Thus, P_5 is totally ramified in K/E. If P stands for either P_6 or P_7, then $\nu_P(y_1) = 2$. Furthermore, $z = x + 1$ is a local parameter at P and we have a local expansion of the form $y_1 = z^2 + z^3 + \cdots$ at P. Therefore

$$u = \frac{(1 + z^2)(z^2 + z^3 + \cdots)}{z^3} + \frac{1 + z}{z},$$

and so $\nu_P(u) \geq 1$. It follows that P_6 and P_7 split completely in K/E. Altogether, we get 14 rational places of K. The genus of K is obtained from Theorem 3.1.9.

Example 4.4.9 $g(K/\mathbf{F}_2) = 13$, $N(K/\mathbf{F}_2) = 15$, $K = \mathbf{F}_2(x, y_1, y_2, y_3)$ with

$$y_1^2 + y_1 = x^3 + x, \quad y_2^2 + y_2 = \frac{xy_1}{x + 1}, \quad y_3^2 + y_3 = x(x + 1)y_2.$$

The function field K is optimal and was constructed in Niederreiter and Xing [103]. The field $E = \mathbf{F}_2(x, y_1, y_2)$ was already considered in Example 4.4.2. Using the notation of Example 4.4.8 for the rational places of E, the place P_∞ is totally ramified in the extension K/E, whereas the places $P_1, \ldots, P_7$ split completely in K/E. Furthermore, P_∞ is the only ramified place in K/E and $m_{P_\infty} = 11$ in the notation of Theorem 3.1.9. The genus of K is obtained from this theorem.

Example 4.4.10 $g(K/\mathbf{F}_2) = 19$, $N(K/\mathbf{F}_2) = 20$, $K = \mathbf{F}_2(x, y_1, y_2, y_3)$ with

$$y_1^2 + y_1 = x^3 + x, \quad y_2^2 + y_2 = \frac{x(x + 1)}{x^3 + x^2 + 1}, \quad y_3^2 + y_3 = u := \frac{x(x + 1)(y_1 + x^2 + 1)}{x^3 + x^2 + 1}.$$

The function field K is optimal and was constructed in Niederreiter and Xing [104]. The field $E = \mathbf{F}_2(x, y_1, y_2)$ was already considered in Example 4.4.5. If P is one of the places of E lying over the infinite place of $\mathbf{F}_2(x)$, then $\nu_P(y_1) = -3$. With $w = x^{-1} y_1 \in E$ we get

$$u + w^2 + w = \frac{(x + 1)^3 y_1 + x(x + 1)^2}{x^2(x^3 + x^2 + 1)},$$

and so $\nu_P(u + w^2 + w) = 1$. It follows that all rational places of E split completely in the extension K/E, hence $N(K) = 20$. The only ramified place in K/E is the unique place Q of E lying over the place $x^3 + x^2 + 1$ of $\mathbf{F}_2(x)$, which has $\deg(Q) = 6$ and ramification index 2 over $\mathbf{F}_2(x)$. With $v = (y_1 + x^3 + x)y_2 \in E$ we get

$$u + v^2 + v = x^2(x + 1)^4 y_2 + x(x + 1)^4,$$

hence $\nu_Q(u + v^2 + v) = -1$ since $\nu_Q(y_2) = -1$. This implies $g(K) = 19$ by Theorem 3.1.9.

Example 4.4.11 $g(K/\mathbf{F}_3) = 2$, $N(K/\mathbf{F}_3) = 8$, $K = \mathbf{F}_3(x, y)$ with

$$y^2 = x^6 - x^2 + 1.$$

The function field K is optimal and can be found in Niederreiter and Xing [105].

Example 4.4.12 $g(K/\mathbf{F}_3) = 9$, $N(K/\mathbf{F}_3) = 19$, $K = \mathbf{F}_3(x, y_1, y_2)$ with

$$y_1^3 - y_1 = x(x - 1), \quad y_2^3 - y_2 = \frac{x(x - 1)}{x + 1}.$$

The function field K is optimal and was constructed in Niederreiter and Xing [102]. The places x and $x - 1$ of $F = \mathbf{F}_3(x)$ split completely in K/F and the infinite place of F is totally ramified in K/F.

Example 4.4.13 $g(K/\mathbf{F}_3) = 17$, $N(K/\mathbf{F}_3) = 24$, $K = \mathbf{F}_3(x, y_1, y_2, y_3)$ with

$$y_1^2 = x^3 - x + 1, \quad y_2^2 = -x^3 + x + 1, \quad y_3^2 = x^5 - x + 1.$$

This function field and the following ones with constant field $\mathbf{F}_3$ are from Niederreiter and Xing [110]. The necessary results on Kummer extensions can be found in Section 3.1.

Example 4.4.14 $g(K/\mathbf{F}_3) = 23$, $N(K/\mathbf{F}_3) = 26$, $K = \mathbf{F}_3(x, y_1, y_2, y_3)$ with

$$y_1^2 = x^3 - x + 1, \quad y_2^3 - y_2 = \frac{(x + 1)(y_1 + x + 1)}{x}, \quad y_3^2 = -x^4 + x^2 + 1.$$

Note that $L = \mathbf{F}_3(x, y_1, y_2)$ satisfies $g(L) = 6$ and $N(L) = 14$. The 13 rational places of L lying over the finite places of $F = \mathbf{F}_3(x)$ split completely in the Kummer extension K/L, whereas the place of K lying over the infinite place of F has degree 2. Therefore $N(K) = 26$.

Example 4.4.15 $g(K/\mathbf{F}_3) = 26$, $N(K/\mathbf{F}_3) = 36$, $K = \mathbf{F}_3(x, y_1, y_2, y_3)$ with

$$y_1^3 - y_1 = x(x - 1), \quad y_2^3 - y_2 = \frac{x(x - 1)}{x + 1}, \quad y_3^2 = -x^2 + x + 1.$$

Example 4.4.16 $g(K/\mathbf{F}_3) = 28$, $N(K/\mathbf{F}_3) = 37$, $K = \mathbf{F}_3(x, y_1, y_2, y_3)$ with

$$y_1^3 - y_1 = x(x - 1), \quad y_2^3 - y_2 = \frac{x(x - 1)}{x + 1}, \quad y_3^2 = (x + 1)(x^2 + 1).$$

Example 4.4.17 $g(K/\mathbf{F}_4) = 4$, $N(K/\mathbf{F}_4) = 15$, $K = \mathbf{F}_4(x, y_1, y_2)$ with

$$y_1^2 + y_1 = x^3, \quad y_2^3 = (x^2 + x)y_1 + x^4 + 1.$$

The function field K is optimal and equivalent to one given by Voss and Høholdt [168].

Example 4.4.18 $g(K/\mathbf{F}_5) = 9$, $N(K/\mathbf{F}_5) = 26$, $K = \mathbf{F}_5(x, y_1, y_2)$ with

$$y_1^2 = x(x - 1)(x - 2), \quad y_2^5 - y_2 = (x + 2)y_1.$$

This function field and the following ones with constant field $\mathbf{F}_5$ are from Niederreiter and Xing [107]. Note that $L = \mathbf{F}_5(x, y_1)$ satisfies $g(L) = 1$ and $N(L) = 8$. If P_∞ is the place of L lying over the infinite place of $\mathbf{F}_5(x)$, then

$$\nu_{P_\infty}\left((x+2)y_1 - \left(\frac{y_1}{x}\right)^5 + \frac{y_1}{x}\right) = \nu_{P_\infty}(y_1) = -3,$$

and so P_∞ is totally ramified in the Artin-Schreier extension K/L. There are no other ramified places in K/L. Over each of $x, x - 1$, and $x - 2$ there is exactly one place of L, and each of these splits completely in K/L. The two places of L lying over $x + 2$ also split completely in K/L.

Example 4.4.19 $g(K/\mathbf{F}_5) = 11, N(K/\mathbf{F}_5) = 32, K = \mathbf{F}_5(x, y_1, y_2)$ with

$$y_1^2 = x(x^2 - 2), \qquad y_2^5 - y_2 = \frac{x^4 - 1}{y_1 - 1}.$$

Example 4.4.20 $g(K/\mathbf{F}_5) = 17, N(K/\mathbf{F}_5) = 42, K = \mathbf{F}_5(x, y_1, y_2)$ with

$$y_1^2 = x(x^2 - 2), \qquad y_2^5 - y_2 = \frac{x^4 - 1}{y_1}.$$

Example 4.4.21 $g(K/\mathbf{F}_5) = 22, N(K/\mathbf{F}_5) = 51, K = \mathbf{F}_5(x, y_1, y_2)$ with

$$y_1^2 = x^5 - x + 1, \qquad y_2^5 - y_2 = (x^5 - x)y_1.$$

We conclude this section by giving some pointers to further literature on the construction of global function fields with many rational places. An excellent survey of early developments in this area is presented in Garcia and Stichtenoth [29]. More recent survey articles are those of Blake [10], Niederreiter [96], Niederreiter and Xing [108], [119], and van der Geer and van der Vlugt [161]. We refer to Section 3.1 for a discussion of maximal function fields.

A large number of further examples of global function fields with many rational places has been constructed in the work of Niederreiter and Xing. The methods are those discussed in the present chapter. For $q = 2$ these examples can be found in [103], [104], [108], [117], [181]; for $q = 3$ in [105], [110]; for $q = 4$ in [105], [106]; for $q = 5$ in [105], [107], [111]; for $q = 8, 16$ in [116]; and for $q = 9, 27$ in [113]. The first systematic search for global function fields with many rational places was carried out by Serre [141], [142], [143], [144] who used class field theory and ray class fields. Manin and Vlăduţ [79] is another early paper. Serre's approach was pursued further by Lauter [65], [66], [67]. Considerable extensions of this work, and also of some of the work of Niederreiter and Xing, were achieved by Auer [5], [6], [7]. Following the papers of Quebbemann [129] and Niederreiter and Xing [103], [105], methods based on cyclotomic function fields were used further by Keller [58] and Lauter [67]. In a series of papers, van der Geer and van der Vlugt [158], [159], [160], [161], [162] developed various methods for the explicit construction of global function fields with many rational places. These techniques use Artin-Schreier extensions and interesting connections with coding theory as principal tools. A systematic way of finding explicit global function fields with many rational places by means of Artin-Schreier extensions was described by van der Geer and van der Vlugt [163] and it reduces the problem to one of linear algebra. Recent work has concentrated on the case of full constant fields $\mathbf{F}_q$ with composite q; see Beelen and Pellikaan [8], Garcia and Stichtenoth [33], Özbudak and Stichtenoth [122], and van der Geer and van der Vlugt [165].

4.5 Tables

At the end of this section we provide two tables of bounds for the quantity $N_q(g)$ introduced in Definition 1.6.14. Table 4.5.1 is for $q = 2, 3, 4, 5, 8, 9, 16, 27$ and $1 \leq g \leq 50$, and Table 4.5.2 extends Table 4.5.1 for the important case $q = 2$ to the range $51 \leq g \leq 95$. In each entry of the tables, the first number is a lower bound for $N_q(g)$ and the second an upper bound for $N_q(g)$. If only one number is given, then this is the exact value of $N_q(g)$.

For fixed g, a general formula for $N_q(g)$ that is valid for all prime powers q is known only in the cases $g = 0, 1, 2$. It is trivial that $N_q(0) = q + 1$. A formula for $N_q(1)$, i.e., for the case of elliptic function fields, follows from the work of Deuring [17] and Waterhouse [171]. We have $N_q(1) = q + 1 + \lfloor 2q^{1/2} \rfloor$, except in the case where $q = p^e$ with a prime p dividing $\lfloor 2q^{1/2} \rfloor$ and an odd integer $e \geq 3$, in which case $N_q(1) = q + \lfloor 2q^{1/2} \rfloor$. Serre [141], [142], [143], [144] determined the values of $N_q(2)$. Write again $q = p^e$ with a prime p and an integer $e \geq 1$. If e is even and $q \neq 4, 9$, then $N_q(2) = q + 1 + 4q^{1/2}$, whereas $N_4(2) = 10$ and $N_9(2) = 20$. If e is odd and q is nonspecial (see the definition in Section 4.2.3), then $N_q(2) = q + 1 + 2\lfloor 2q^{1/2} \rfloor$. If e is odd and q is not nonspecial, then $N_q(2) = q + 2\lfloor 2q^{1/2} \rfloor$ or $q - 1 + 2\lfloor 2q^{1/2} \rfloor$, depending on whether the fractional part $\{2q^{1/2}\}$ is greater than $(\sqrt{5} - 1)/2$ or not.

The first complete table of lower and upper bounds on $N_q(g)$ for the range considered in Table 4.5.1 was published by Niederreiter [96]. We took this table as our starting point and incorporated recent improvements to obtain Table 4.5.1. We emphasize that we used only data that we could verify because they were either published or made available to us in manuscript form. This explains the few discrepancies that occur between our table and the recent tables of van der Geer and van der Vlugt [164]. In the following we list the sources for the improved lower bounds in Table 4.5.1 relative to the table in Niederreiter [96]. For $q = 2$, $g = 27$, see Auer [5], and for $q = 2$, $g = 38$, see Keller [58]. For $q = 3$, $g = 19, 22, 24, 30, 33, 34, 39, 45, 47$, see Auer [5]. For $q = 4$, $g = 20, 22, 23, 24, 28, 48$, see Auer [5], and for $q = 4$, $g = 27$, see Schweizer [139]. For $q = 8$, $g = 25$, see van der Geer and van der Vlugt [165], and for $q = 8$, $g = 30, 48$, see Auer [5]. For $q = 9$, $g = 7, 9, 17, 21$, see Özbudak and Stichtenoth [122], and for $q = 9$, $g = 13, 33, 41$, see van der Geer and van der Vlugt [165]. For $q = 16$, $g = 12, 20, 40, 49$, see van der Geer and van der Vlugt [165], for $q = 16$, $g = 31$, see Garcia and Stichtenoth [33], and for $q = 16$, $g = 34$, see Schweizer [139]. For $q = 27$, $g = 24$, see Beelen and Pellikaan [8], and for $q = 27$, $g = 49$, see van der Geer and van der Vlugt [165]. For Table 4.5.2 we took Table 2 in Niederreiter and Xing [117] as our starting point and we incorporated the following improved lower bounds: for $g = 66, 81$ due to Keller [58], for $g = 68$ due to Auer [5], and for $g = 55, 58, 60, 61, 71, 74, 75, 83, 89, 91$

due to Auer [7].

The upper bounds on $N_q(g)$ in Tables 4.5.1 and 4.5.2 are obtained from the method in Theorem 1.6.18 (see also Example 1.6.19) due to Serre and Oesterlé. In some cases one can obtain improvements on these bounds by special arguments of algebraic geometry. This was already illustrated by Serre [144]. In Table 4.5.1 we have also incorporated recent improved upper bounds of this type which are due to Lauter. The bound $N_3(5) \leq 13$ is from Lauter [69] and the bound $N_3(7) \leq 16$ from Lauter [70]. Furthermore, the bounds $N_8(6) \leq 35$, $N_9(5) \leq 35$, $N_{16}(4) \leq 46$, $N_{16}(5) \leq 54$, and

$$N_q(g) \leq q - 1 + g\lfloor 2q^{1/2} \rfloor$$

if $q = 8$ and $g \geq 4$, or if $q = 27$ and $g \geq 3$, are from Lauter [68] (see also Lauter and Serre [71]).

Table 4.5.1. Bounds for $N_q(g)$

$g\backslash q$	2	3	4	5	8	9	16	27
1	5	7	9	10	14	16	25	38
2	6	8	10	12	18	20	33	48
3	7	10	14	16	24	28	38	56
4	8	12	15	18	25-27	30	45-46	64-66
5	9	12-13	17-18	20-22	29-32	32-35	49-54	55-76
6	10	14-15	20	21-25	33-35	35-40	65	76-86
7	10	16	21-22	22-27	33-39	40-43	63-70	64-96
8	11	15-18	21-24	22-29	34-43	38-47	61-76	92-106
9	12	19	26	26-32	45-47	48-51	72-81	82-116
10	13	19-21	27-28	27-34	38-50	54-55	81-87	91-126
11	14	20-22	26-30	32-36	48-54	55-59	80-92	96-136
12	14-15	22-24	29-31	30-38	49-57	55-63	83-97	109-146
13	15	24-25	33	36-40	50-61	64-66	97-103	136-156
14	15-16	24-26	32-35	39-43	65	56-70	97-108	84-164
15	17	28	33-37	35-45	54-68	64-74	98-113	136-171
16	17-18	27-29	36-38	40-47	56-71	74-78	93-118	136-178
17	17-18	24-30	40	42-49	61-74	64-82	96-124	128-185
18	18-19	26-31	41-42	32-51	65-77	46-85	113-129	94-192
19	20	28-32	37-43	45-54	58-80	84-88	121-134	126-199
20	19-21	30-34	40-45	30-56	68-83	48-91	127-140	133-207
21	21	32-35	41-47	50-58	72-86	88-95	129-145	163-214
22	21-22	30-36	41-48	51-60	66-89	78-98	129-150	112-221
23	22-23	26-37	45-50	55-62	68-92	92-101	126-155	114-228
24	20-23	31-38	49-52	46-64	66-95	91-104	129-161	212-235
25	24	36-40	51-53	52-66	86-97	64-108	144-166	196-242
26	24-25	36-41	55	45-68	72-100	110-111	150-171	108-249
27	24-25	39-42	50-56	52-70	96-103	60-114	145-176	114-256
28	25-26	37-43	53-58	54-71	97-106	105-117	136-181	108-263
29	25-27	42-44	49-60	56-73	97-109	104-120	161-187	114-270
30	25-27	37-46	53-61	58-75	96-112	60-123	161-192	117-277
31	27-28	40-47	60-63	72-77	72-115	84-127	165-197	114-284
32	26-29	38-48	57-65	62-79	72-118	81-130	132-202	126-291
33	28-29	46-49	65-66	64-81	92-121	128-133	193-207	220-298
34	27-30	45-50	57-68	76-83	80-124	111-136	161-213	135-305
35	29-31	47-51	58-69	68-85	106-127	84-139	144-218	126-312
36	30-31	46-52	64-71	64-87	105-130	110-142	185-223	244-319
37	29-32	48-54	66-72	72-89	121-132	120-145	208-228	162-326
38	30-33	36-55	56-74	78-91	129-135	105-149	193-233	144-333
39	33	48-56	65-75	76-93	117-138	84-152	160-239	271-340
40	32-34	54-57	75-77	65-94	100-141	90-155	225-244	244-346

Table 4.5.1. Bounds for $N_q(g)$ (cont.)

$g\backslash q$	2	3	4	5	8	9	16	27
41	33-35	50-58	65-78	80-96	112-144	128-158	216-249	153-353
42	33-35	39-59	66-80	60-98	129-147	90-161	209-254	280-360
43	33-36	55-60	72-81	84-100	100-150	120-164	226-259	196-367
44	33-37	42-61	68-83	60-102	129-153	90-167	162-264	153-374
45	32-37	54-62	80-84	88-104	144-156	112-170	242-268	171-381
46	34-38	55-63	81-86	75-106	129-158	138-173	243-273	162-388
47	36-38	54-65	73-87	92-108	120-161	154-177	176-277	174-395
48	34-39	55-66	80-89	82-110	128-164	163-180	184-282	325-402
49	36-40	63-67	81-90	96-111	130-167	168-183	213-286	314-409
50	40	56-68	91-92	70-113	130-170	182-186	225-291	180-416

Table 4.5.2. Bounds for $N_2(g)$

g	51	52	53	54	55	56	57	58	59
$N_2(g)$	36-41	34-42	40-42	42-43	41-43	38-44	40-45	41-45	40-46

g	60	61	62	63	64	65	66	67	68
$N_2(g)$	41-47	41-47	44-48	42-48	42-49	48-50	48-50	44-51	49-51

g	69	70	71	72	73	74	75	76	77
$N_2(g)$	49-52	46-53	49-53	48-54	48-54	49-55	49-56	50-56	52-57

g	78	79	80	81	82	83	84	85	86
$N_2(g)$	48-57	52-58	56-59	56-59	53-60	57-60	57-61	52-62	56-62

g	87	88	89	90	91	92	93	94	95
$N_2(g)$	56-63	56-63	57-64	56-65	57-65	60-66	56-66	56-67	65-68

Chapter 5

Asymptotic Results

In Definition 1.6.20 we introduced the quantity

$$A(q) = \limsup_{g \to \infty} \frac{N_q(g)}{g}$$

for any prime power q. It follows from the Serre bound that $A(q) \le \lfloor 2q^{1/2} \rfloor$, and an improvement on this upper bound was obtained in Theorem 1.6.21, namely

$$A(q) \le q^{1/2} - 1.$$

In this chapter we will establish further results on $A(q)$, in particular, lower bounds.

5.1 Asymptotic Behavior of Towers

Definition 5.1.1 A **tower of function fields** over $\mathbf{F}_q$ is a sequence $\mathcal{F} = (F_1, F_2, F_3, \ldots)$ of global function fields $F_i / \mathbf{F}_q$ having the following properties:

(i) $F_1 \subseteq F_2 \subseteq F_3 \subseteq \ldots$.

(ii) For each $i \ge 1$, the extension F_{i+1}/F_i is separable of degree $[F_{i+1} : F_i] > 1$.

(iii) $g(F_j) > 1$ for some $j \ge 1$.

Note that this definition and the Hurwitz genus formula imply that

$$\lim_{i \to \infty} g(F_i) = \infty.$$

Definition 5.1.2 Let $\mathcal{F} = (F_1, F_2, F_3, \ldots)$ be a tower of function fields over $\mathbf{F}_q$. Another such tower $\mathcal{E} = (E_1, E_2, E_3, \ldots)$ over $\mathbf{F}_q$ is said to be a **subtower** of $\mathcal{F}$ (written $\mathcal{E} \prec \mathcal{F}$) if there exists an embedding (over $\mathbf{F}_q$)

$$\iota : \bigcup_{i \ge 1} E_i \longrightarrow \bigcup_{i \ge 1} F_i.$$

In other words, for any $i \ge 1$ there is an index $m = m(i) \ge 1$ such that $\iota(E_i) \subseteq F_m$.

In this section, we put together some simple observations on the behavior of the sequence $(N(F_i)/g(F_i))_{i\geq 1}$, where $\mathcal{F} = (F_1, F_2, F_3, \ldots)$ is a tower of function fields over $\mathbf{F}_q$. The results in this section are due to Garcia and Stichtenoth [31]. Related work can be found in Garcia and Stichtenoth [32], [35], Garcia, Stichtenoth, and Thomas [36], Özbudak and Thomas [124], and Stepanov [151, Chapter 6].

Lemma 5.1.3 *Let E/F be a finite extension of global function fields over $\mathbf{F}_q$. Assume that $g(F) > 1$. Then*

$$\frac{N(E)}{g(E) - 1} \leq \frac{N(F)}{g(F) - 1}.$$

Proof. There is a subfield H of E/F such that H/F is separable and E/H is purely inseparable of degree $l = p^e$, where $p = \operatorname{char}(\mathbf{F}_q)$ and $e \geq 0$. Then $H = E^l$ is isomorphic to E, so $N(H) = N(E)$ and $g(H) = g(E)$. The Hurwitz genus formula for H/F gives

$$g(H) - 1 \ = [H : F](g(F) - 1) + \tfrac{1}{2}\deg(\operatorname{Diff}(H/F))$$

$$\geq [H : F](g(F) - 1).$$

Any rational place of F has at most $[H : F]$ rational places of H lying over it, hence

$$N(H) \leq [H : F]N(F).$$

It follows that

$$\frac{N(E)}{g(E) - 1} = \frac{N(H)}{g(H) - 1} \leq \frac{[H : F]N(F)}{[H : F](g(F) - 1)} = \frac{N(F)}{g(F) - 1}.$$

$\square$

Corollary 5.1.4 *For any tower $\mathcal{F} = (F_1, F_2, F_3, \ldots)$ of function fields over $\mathbf{F}_q$, the sequence*

$$(N(F_i)/g(F_i))_{i\geq 1}$$

is convergent.

Proof. We can assume that $g(F_i) > 1$ for all i (see Definition 5.1.1(iii)). By Lemma 5.1.3, the sequence

$$\left(\frac{N(F_i)}{g(F_i) - 1}\right)_{i\geq 1}$$

is nonincreasing, hence convergent. Since $\lim_{i\to\infty} g(F_i) = \infty$, the sequence $(N(F_i)/g(F_i))_{i\geq 1}$ is also convergent, and

$$\lim_{i\to\infty} \frac{N(F_i)}{g(F_i)} = \lim_{i\to\infty} \frac{N(F_i)}{g(F_i) - 1}.$$

$\square$

Definition 5.1.5 For a tower $\mathcal{F} = (F_1, F_2, F_3, \ldots)$ of function fields over $\mathbf{F}_q$, let

$$\lambda(\mathcal{F}) := \lim_{i \to \infty} \frac{N(F_i)}{g(F_i)}.$$

The tower $\mathcal{F}$ is said to be **asymptotically good** (respectively, **asymptotically bad**) if $\lambda(\mathcal{F}) > 0$ (respectively, $\lambda(\mathcal{F}) = 0$).

It is obvious that $\lambda(\mathcal{F}) \leq A(q)$. We call the tower $\mathcal{F}$ **optimal** if $\lambda(\mathcal{F}) = A(q)$.

Corollary 5.1.6 *Let $\mathcal{F}$ be a tower of function fields over $\mathbf{F}_q$ and let $\mathcal{E} \prec \mathcal{F}$ be a subtower. Then:*

(i) $\lambda(\mathcal{E}) \geq \lambda(\mathcal{F})$.

(ii) *If $\mathcal{E}$ is asymptotically bad, then $\mathcal{F}$ is also asymptotically bad.*

(iii) *If $\mathcal{F}$ is optimal, then $\mathcal{E}$ is also optimal.*

Proof. This follows easily from Lemma 5.1.3. $\qquad\square$

The following two propositions provide sufficient conditions for a tower of function fields to be asymptotically bad (respectively, asymptotically good) in terms of degrees of global different divisors.

Proposition 5.1.7 *Let $\mathcal{F} = (F_1, F_2, F_3, \ldots)$ be a tower of function fields over $\mathbf{F}_q$. Suppose that $\rho_2, \rho_3, \rho_4, \ldots$ is a sequence of real numbers with the following properties:*

(a) $\rho_2 > 0$;

(b) $\rho_{i+1} \leq \deg(\mathrm{Diff}(F_{i+1}/F_i))$ *for all $i \geq 1$;*

(c) $\rho_{i+1} \geq [F_{i+1} : F_i]\rho_i$ *for all $i \geq 2$.*

Then:

(i) *there is a constant $\rho > 0$ such that, for all $n \geq 1$,*

$$g(F_{n+1}) - 1 \geq [F_{n+1} : F_1]\left(g(F_1) - 1 + \rho n\right);$$

(ii) $\lambda(\mathcal{F}) = 0$, *i.e., the tower $\mathcal{F}$ is asymptotically bad.*

Proof. (i) We abbreviate $d_{i+1} := \deg(\mathrm{Diff}(F_{i+1}/F_i))$ for all $i \geq 1$. The assumption (c) implies, by induction, that $\rho_{i+1} \geq [F_{i+1} : F_2]\rho_2$ for all $i \geq 1$. By the tower formula for different exponents (see Proposition 1.3.11), the degree of the global different divisor of F_{n+1}/F_1 is given by

$$\deg(\mathrm{Diff}(F_{n+1}/F_1)) = \sum_{i=1}^{n}[F_{n+1} : F_{i+1}]d_{i+1} \qquad \text{for all } n \geq 1.$$

Now the Hurwitz genus formula for the extension F_{n+1}/F_1 yields

$$
\begin{aligned}
2g(F_{n+1}) - 2 &= [F_{n+1} : F_1](2g(F_1) - 2) + \deg(\mathrm{Diff}(F_{n+1}/F_1)) \\
&\geq [F_{n+1} : F_1](2g(F_1) - 2) + \sum_{i=1}^{n}[F_{n+1} : F_{i+1}]\rho_{i+1} \\
&\geq [F_{n+1} : F_1](2g(F_1) - 2) + \sum_{i=1}^{n}[F_{n+1} : F_2]\rho_2 \\
&= [F_{n+1} : F_1](2g(F_1) - 2) + \frac{\rho_2}{[F_2 : F_1]}[F_{n+1} : F_1]n.
\end{aligned}
$$

Setting $\rho := \rho_2/(2[F_2 : F_1])$, we obtain the desired inequality

$$g(F_{n+1}) - 1 \geq [F_{n+1} : F_1] \left(g(F_1) - 1 + \rho n\right).$$

(ii) Since $N(F_{n+1}) \leq [F_{n+1} : F_1] N(F_1)$, the assertion

$$\lambda(\mathcal{F}) = \lim_{n \to \infty} \frac{N(F_{n+1})}{g(F_{n+1}) - 1} = 0$$

follows immediately from (i). $\qquad\qquad\qquad\qquad\qquad\qquad\qquad\qquad\qquad\qquad\quad$ □

Remark 5.1.8 The conclusions of Proposition 5.1.7 also hold if condition (c) is replaced by the slightly weaker condition

(c') $\rho_{i+1} \geq [F_{i+1} : F_2] \rho_2$ for all $i \geq 2$.

Proposition 5.1.9 *Let $\mathcal{F} = (F_1, F_2, F_3, \ldots)$ be a tower of function fields over $\mathbf{F}_q$. Suppose that*

$$\deg(\mathrm{Diff}(F_{i+1}/F_i)) \leq \varepsilon\, [F_{i+1} : F_i] \deg(\mathrm{Diff}(F_i/F_{i-1}))$$

holds for all $i \geq 2$, where ε is a constant satisfying $0 \leq \varepsilon < 1$. Moreover, suppose that there exists a nonempty set S of rational places of $F_1/\mathbf{F}_q$ such that any $P \in S$ splits completely in all extensions F_n/F_1. Then the tower $\mathcal{F}$ is asymptotically good. Moreover, we have

$$\lambda(\mathcal{F}) \geq \frac{2(1-\varepsilon)\, [F_2 : F_1]\, |S|}{\deg(\mathrm{Diff}(F_2/F_1)) + (1-\varepsilon)\, [F_2 : F_1]\, (2g(F_1) - 2)}$$

provided that the denominator is positive.

Proof. We set $d_{i+1} := \deg(\mathrm{Diff}(F_{i+1}/F_i))$ for all $i \geq 1$. Assume first that

$$d_2 + (1 - \varepsilon)\, [F_2 : F_1]\, (2g(F_1) - 2) > 0.$$

The assumption $d_{i+1} \leq \varepsilon\, [F_{i+1} : F_i]\, d_i$ implies that the inequality

$$d_{i+1} \leq \varepsilon^{i-1}\, [F_{i+1} : F_2]\, d_2$$

holds for each $i \geq 1$. As in the proof of Proposition 5.1.7, we therefore obtain

$$
\begin{aligned}
2g(F_{n+1}) - 2 &= [F_{n+1} : F_1]\, (2g(F_1) - 2) + \deg(\mathrm{Diff}(F_{n+1}/F_1)) \\
&= [F_{n+1} : F_1]\, (2g(F_1) - 2) + \sum_{i=1}^{n} [F_{n+1} : F_{i+1}]\, d_{i+1} \\
&\leq [F_{n+1} : F_1]\, (2g(F_1) - 2) + \sum_{i=1}^{n} [F_{n+1} : F_2]\, \varepsilon^{i-1}\, d_2 \\
&= [F_{n+1} : F_1] \left(2g(F_1) - 2 + \frac{d_2}{[F_2 : F_1]} \cdot \frac{1 - \varepsilon^n}{1 - \varepsilon}\right) \\
&\leq [F_{n+1} : F_1] \left(2g(F_1) - 2 + \frac{d_2}{(1-\varepsilon)\, [F_2 : F_1]}\right).
\end{aligned}
$$

Since $N(F_{n+1}) \geq [F_{n+1} : F_1] |S|$, it follows that

$$\lambda(\mathcal{F}) \geq \frac{2|S|}{2g(F_1) - 2 + \frac{d_2}{(1-\varepsilon)[F_2:F_1]}} = \frac{2(1-\varepsilon)[F_2 : F_1]|S|}{d_2 + (1 - \varepsilon)[F_2 : F_1](2g(F_1) - 2)}.$$

Now if $d_2 + (1 - \varepsilon)[F_2 : F_1](2g(F_1) - 2) \leq 0$, we replace $\mathcal{F}$ by the subtower $\mathcal{F}' :=$ $(F_j, F_{j+1}, F_{j+2}, \ldots)$, where j is chosen such that $g(F_j) > 1$. Applying the same arguments as at the beginning of the proof to the tower $\mathcal{F}'$, we conclude that $\mathcal{F}'$ and, *a fortiori*, the tower $\mathcal{F}$ is asymptotically good.   □

Remark 5.1.10 In Proposition 5.1.9, the assumption $d_{i+1} \leq \varepsilon [F_{i+1} : F_i] d_i$ can be replaced by the weaker condition $d_{i+1} \leq \varepsilon^{i-1} [F_{i+1} : F_2] d_2$ for all $i \geq 2$.

It is in general hard to find asymptotically good towers of function fields. For instance, if $\mathcal{F} = (F_1, F_2, F_3, \ldots)$ is a tower of abelian extensions of F_1 (i.e., all extensions F_n/F_1 are abelian), then $\mathcal{F}$ is asymptotically bad, as was shown by Frey, Perret, and Stichtenoth [26].

5.2 The Lower Bound of Serre

Serre [141] first employed Hilbert class field towers to obtain lower bounds on $A(q)$. The bounds on $A(q)$ depend crucially on lower bounds for the ℓ-rank of fractional ideal class groups. We start by combining Theorems 2.7.6 and 2.7.7 to obtain the following result.

Proposition 5.2.1 *Let K be a global function field of genus $g(K) > 1$ with full constant field $\mathbf{F}_q$ and let S be a subset of $\mathbf{P}_K$ such that $S' := \mathbf{P}_K \setminus S$ is a nonempty set of rational places of K. Suppose that there exists a prime number ℓ such that*

$$d_\ell(\mathrm{Cl}(O_S)) \geq 2 + 2(|S'| + \varepsilon_\ell(q))^{1/2},$$

where $\varepsilon_\ell(q) = 1$ if $\ell|(q - 1)$ and $\varepsilon_\ell(q) = 0$ otherwise. Then we have

$$A(q) \geq \frac{|S'|}{g(K) - 1}.$$

From Proposition 5.2.1 it is clear that in order to get good lower bounds for $A(q)$, we need to have good lower bounds for $d_\ell(\mathrm{Cl}(O_S))$. The following proposition, which is an improved version of a result of Schoof [138] given in Niederreiter and Xing [112], provides such a bound on the ℓ-rank of fractional ideal class groups.

Proposition 5.2.2 *Let $F/\mathbf{F}_q$ be a global function field and K/F a finite abelian extension. Let $\mathcal{T}$ be a proper subset of $\mathbf{P}_F$ such that $\mathcal{T}' := \mathbf{P}_F \setminus \mathcal{T}$ is finite and let S be the over-set of $\mathcal{T}$ with respect to K/F. Then for any prime number ℓ we have*

$$d_\ell(\mathrm{Cl}(O_S)) \geq \sum_P d_\ell(G_P) - (|\mathcal{T}'| - 1 + \varepsilon_\ell(q)) - d_\ell(G),$$

where $\varepsilon_\ell(q)$ is defined as in Proposition 5.2.1, $G = \mathrm{Gal}(K/F)$, and G_P is the inertia group of the place P in K/F. The sum is extended over all places P of F.

Proof. We will make use of the principle noted by Perret [126, Proposition 6] that an exact sequence

$$B_1 \to B_2 \to B_3$$

of finitely generated abelian groups leads to the inequality

$$d_\ell(B_2) \le d_\ell(B_1) + d_\ell(B_3)$$

for ℓ-ranks. In particular, passing to a subgroup or to a factor group cannot increase the ℓ-rank.

Now we employ Tate cohomology and notation from Sections 2.4 and 2.7. First of all, since $\hat{H}^{-1}(G, \mathrm{Cl}(O_S))$ is a subgroup of a factor group of $\mathrm{Cl}(O_S)$, we get

$$d_\ell(\mathrm{Cl}(O_S)) \ge d_\ell\left(\hat{H}^{-1}(G, \mathrm{Cl}(O_S))\right).$$

The exact sequence

$$1 \to J_S/K_S^* \to C_K \to \mathrm{Cl}(O_S) \to 1$$

is obtained from Propositions 2.4.2 and 2.4.3 and yields the exact sequence

$$\hat{H}^{-1}(G, \mathrm{Cl}(O_S)) \to \hat{H}^0(G, J_S/K_S^*) \to \hat{H}^0(G, C_K).$$

Hence

$$d_\ell\left(\hat{H}^{-1}(G, \mathrm{Cl}(O_S))\right) \ge d_\ell\left(\hat{H}^0(G, J_S/K_S^*)\right) - d_\ell\left(\hat{H}^0(G, C_K)\right).$$

The trivial exact sequence

$$1 \to K_S^* \to J_S \to J_S/K_S^* \to 1$$

yields the exact sequence

$$\hat{H}^0(G, K_S^*) \to \hat{H}^0(G, J_S) \to \hat{H}^0(G, J_S/K_S^*)$$

and hence

$$d_\ell\left(\hat{H}^0(G, J_S/K_S^*)\right) \ge d_\ell\left(\hat{H}^0(G, J_S)\right) - d_\ell\left(\hat{H}^0(G, K_S^*)\right).$$

By combining these inequalities, we get

$$d_\ell(\mathrm{Cl}(O_S)) \ge d_\ell\left(\hat{H}^0(G, J_S)\right) - d_\ell\left(\hat{H}^0(G, K_S^*)\right) - d_\ell\left(\hat{H}^0(G, C_K)\right).$$

By Proposition 2.7.4,

$$\hat{H}^0(G, C_K) \simeq \hat{H}^{-2}(G, \mathbf{Z}) \simeq G.$$

Since $\hat{H}^0(G, K_S^*)$ is a factor group of F_T^*, we have

$$d_\ell\left(\hat{H}^0(G, K_S^*)\right) \le d_\ell\left(F_T^*\right).$$

Therefore

$$d_\ell(\mathrm{Cl}(O_S)) \ge d_\ell\left(\hat{H}^0(G, J_S)\right) - d_\ell\left(F_T^*\right) - d_\ell(G),$$

and in view of $d_\ell(F_T^*) = |T'| - 1 + \varepsilon_\ell(q)$ (see the proof of Theorem 2.7.7) we obtain

$$d_\ell(\mathrm{Cl}(O_S)) \geq d_\ell\left(\hat{H}^0(G, J_S)\right) - (|T'| - 1 + \varepsilon_\ell(q)) - d_\ell(G). \qquad (5.1)$$

Now let $\mathcal{R}$ be a finite set of places of F such that $\mathcal{R}$ is a subset of T and $\mathcal{R}$ contains all ramified places in K/F belonging to T. By Propositions 2.7.1 and 2.7.2 we have

$$\hat{H}^0(G, J_S) \simeq \prod_{P \in \mathcal{R}} \hat{H}^0(G^P, U^P) \times \prod_{P \in T'} \hat{H}^0(G^P, (K^P)^*),$$

where K^P is the completion of K at some place lying over P, U^P is the unit group of the valuation ring of K^P, and G^P is the decomposition group of P in K/F. Note that by Theorem 2.3.1(i), G^P is isomorphic to $\mathrm{Gal}(K^P/F_P)$ with F_P being the P-adic completion of F. Then by local class field theory we get

$$\hat{H}^0(G^P, U^P) \simeq G_P, \quad \hat{H}^0(G^P, (K^P)^*) \simeq G^P.$$

Since clearly $d_\ell(G^P) \geq d_\ell(G_P)$, we obtain

$$d_\ell\left(\hat{H}^0(G, J_S)\right) \geq \sum_{P \in \mathcal{R} \cup T'} d_\ell(G_P) = \sum_P d_\ell(G_P),$$

and together with (5.1) this yields the desired result. $\qquad\qquad\qquad\qquad\qquad\qquad \square$

The following result from Niederreiter and Xing [112] yields unramified abelian extensions for which the Galois group has a large ℓ-rank. This result is useful in connection with Corollary 2.7.8.

Proposition 5.2.3 *Let F be a global function field and let $P_1, \ldots, P_n$ be $n \geq 1$ distinct places of F. Let T be a subset of $\mathbf{P}_F$. Suppose that there exist a prime number ℓ and n Galois extensions $K_1/F, \ldots, K_n/F$ such that the following four conditions are satisfied:*

(i) *$[K_i : F] = \ell$ for $1 \leq i \leq n$;*

(ii) *for $1 \leq i, j \leq n$, the place P_i is ramified in K_j/F if and only if $i = j$;*

(iii) *any place of F can be ramified in at most one of the extensions $K_1/F, \ldots, K_n/F$;*

(iv) *all places in T split completely in K_i/F for $1 \leq i \leq n$.*

Then there exists a subfield K of the extension $K_1 \cdots K_n/F$ such that $[K : F] = \ell$, the extension $K_1 \cdots K_n/K$ is unramified abelian, all places in the over-set of T with respect to K/F split completely in $K_1 \cdots K_n/K$, and

$$d_\ell(\mathrm{Gal}(K_1 \cdots K_n/K)) = n - 1.$$

Proof. For any fixed i with $1 \leq i \leq n$, the place P_i is ramified in the extension K_i/F and unramified in the extension $K_1 \cdots K_{i-1}K_{i+1} \cdots K_n/F$, hence

$$K_i \cap (K_1 \cdots K_{i-1}K_{i+1} \cdots K_n) = F.$$

Therefore the extensions $K_1/F, \ldots, K_n/F$ are linearly disjoint. Thus,

$$\mathrm{Gal}(K_1 \cdots K_n/F) \simeq \prod_{i=1}^{n} \mathrm{Gal}(K_i/F) \simeq (\mathbf{Z}/\ell\mathbf{Z})^n.$$

We prove the proposition by induction on n. The case $n = 1$ is trivial. Suppose the proposition is correct for $n - 1$, where $n \geq 2$. Let the subfield L of $K_1 \cdots K_{n-1}/F$ be such that $[L : F] = \ell$ and $K_1 \cdots K_{n-1}/L$ is unramified. Since P_n is ramified in K_n/F and unramified in L/F, we have $L \cap K_n = F$. Thus,

$$\text{Gal}(LK_n/F) \simeq \text{Gal}(LK_n/L) \times \text{Gal}(LK_n/K_n) \simeq (\mathbf{Z}/\ell\mathbf{Z})^2.$$

Let σ and θ be generators of $\text{Gal}(LK_n/L)$ and $\text{Gal}(LK_n/K_n)$, respectively. Then $\langle \sigma\theta \rangle \simeq \mathbf{Z}/\ell\mathbf{Z}$, $\langle \sigma\theta \rangle \cap \text{Gal}(LK_n/L) = \{\text{id}\}$, and $\langle \sigma\theta \rangle \cap \text{Gal}(LK_n/K_n) = \{\text{id}\}$. Let K be the subfield of LK_n/F fixed by $\langle \sigma\theta \rangle$, then $[K : F] = \ell$. For any place R of K, let Q be the place of F lying under R. We now distinguish three cases.

Case 1: Q is unramified in both L/F and K_n/F. Then Q is unramified in $K_1 \cdots K_{n-1}/F$ since $K_1 \cdots K_{n-1}/L$ is an unramified extension. Thus, Q is unramified in $K_1 \cdots K_n/F$, and so R is unramified in $K_1 \cdots K_n/K$.

Case 2: Q is ramified in L/F. Then Q is ramified in K_i/F for some $1 \leq i \leq n - 1$. By condition (iii), Q is unramified in K_n/F. Hence the inertia group of Q in LK_n/F is $\text{Gal}(LK_n/K_n) = \langle \theta \rangle$. Since $\langle \theta \rangle \cap \text{Gal}(LK_n/K) = \{\text{id}\}$, R is unramified in LK_n/K, and so Q is ramified in K/F. By condition (iii), the ramification index of Q in $K_1 \cdots K_n/F$ is at most ℓ, hence R is unramified in $K_1 \cdots K_n/K$.

Case 3: Q is ramified in K_n/F. Then Q is unramified in L/F by condition (iii). Now the arguments in Case 2 can be applied *mutatis mutandis* to show that R is unramified in $K_1 \cdots K_n/K$.

Thus, we have proved in all cases that R is unramified in $K_1 \cdots K_n/K$, and so the extension $K_1 \cdots K_n/K$ is unramified. Since $\text{Gal}(K_1 \cdots K_n/F) \simeq (\mathbf{Z}/\ell\mathbf{Z})^n$ and $[K : F] = \ell$, we have

$$\text{Gal}(K_1 \cdots K_n/K) \simeq (\mathbf{Z}/\ell\mathbf{Z})^{n-1}.$$

By condition (iv) it is obvious that all places in the over-set of $\mathcal{T}$ with respect to K/F split completely in $K_1 \cdots K_n/K$, and the induction is complete. □

A general lower bound on $A(q)$ of the same order of magnitude as that in Theorem 5.2.9 below was first announced by Serre [141] and proved in detail in Serre [144]. In our proof we follow the method of Niederreiter and Xing [115].

Lemma 5.2.4 *We have $A(q) > \frac{1}{4}$ for all even prime powers q and $A(q) \geq \frac{2}{5}$ for all odd prime powers q.*

Proof. According to results in Section 5.5 which will be shown independently, we have $A(2) > \frac{1}{4}$ and $A(3) > \frac{2}{5}$. By considering constant field extensions, we see that $A(q) \geq A(2) > \frac{1}{4}$ for all even prime powers q. Thus, it remains to prove the lemma for odd prime powers $q \geq 5$.

For such a q, let $F = \mathbf{F}_q(x)$ be the rational function field over $\mathbf{F}_q$ and choose seven distinct monic irreducible polynomials $f_1, \ldots, f_7 \in \mathbf{F}_q[x]$ of degree 2. For $1 \leq i \leq 7$ define the Kummer extension $K_i = F(y_i)$ of F with $y_i^2 = f_i$. Then we can apply Proposition

5.2.3 with $\ell = 2$ and $\mathcal{T}$ consisting of the infinite place ∞ of F. This yields a subfield K of $K_1 \cdots K_7/F$ with $[K : F] = 2$ such that the extension $K_1 \cdots K_7/K$ is unramified abelian. The two places of K lying over ∞ split completely in $K_1 \cdots K_7/K$. In the present case, a suitable field K can in fact be described explicitly, namely $K = F(y)$ with $y^2 = f_1 \cdots f_7$. We can apply Corollary 2.7.8 since

$$d_2(\mathrm{Gal}(K_1 \cdots K_7/K)) = 7 - 1 = 6 \geq 2 + 2\sqrt{3}.$$

In this way we get

$$A(q) \geq \frac{2}{g(K) - 1},$$

and since $g(K) = 6$ by Corollary 3.1.5, we arrive at the desired result.   □

If q is odd, let η be the quadratic character of $\mathbf{F}_q$ defined for $a \in \mathbf{F}_q$ by

$$\eta(a) = \begin{cases} 1 & \text{if } a = b^2 \text{ for some } b \in \mathbf{F}_q^*, \\ 0 & \text{if } a = 0, \\ -1 & \text{otherwise.} \end{cases}$$

The following result can be viewed as a generalization of [76, Exercise 5.64].

Lemma 5.2.5 *Let $F/\mathbf{F}_q$, q odd, be a global function field with exactly $N \geq q + 1$ rational places $P_\infty, P_1, \ldots, P_{N-1}$. Suppose that the divisor class number $h = h(F)$ of F is odd. Let $f_1, \ldots, f_{N-1}$ be $N - 1$ elements in F satisfying*

$$\mathrm{div}(f_i) = hP_i - hP_\infty \qquad \text{for } 1 \leq i \leq N - 1.$$

For $1 \leq k \leq N - 1$ let M_k denote the number of rational places P ($\neq P_\infty$) of F with $\eta(f_i(P)) = 1$ for all $1 \leq i \leq k$. Then

$$\left| M_k - \frac{N-1}{2^k} \right| \leq N - q + (4g(F) + k - 1)q^{1/2} + \frac{k}{2},$$

where $g(F)$ is the genus of F.

Proof. For $1 \leq i_1 < i_2 < \cdots < i_s \leq k$ consider the function field $F_{i_1,i_2,\ldots,i_s} = F(y_{i_1,i_2,\ldots,i_s})$ defined by

$$y_{i_1,i_2,\ldots,i_s}^2 = f_{i_1} f_{i_2} \cdots f_{i_s}.$$

Then the genus of $F_{i_1,i_2,\ldots,i_s}$ is at most $2g(F) + (s-1)/2$. By the definition of the f_i we know that for $1 \leq i, j \leq N - 1$ we have $\eta(f_i(P_j)) = 0$ if and only if $i = j$, so $F_{i_1,i_2,\ldots,i_s}$ has

$$\sum_{j=1}^{N-1} (1 + \eta(f_{i_1} \cdots f_{i_s}(P_j))) + A_{i_1,i_2,\ldots,i_s} = N - 1 + \sum_{j=1}^{N-1} \eta(f_{i_1} \cdots f_{i_s}(P_j)) + A_{i_1,i_2,\ldots,i_s}$$

rational places, where $0 \leq A_{i_1,i_2,\ldots,i_s} \leq 2$. By the Hasse-Weil bound we get

$$\left| N - 1 + \sum_{j=1}^{N-1} \eta(f_{i_1} \cdots f_{i_s}(P_j)) + A_{i_1,i_2,\ldots,i_s} - q - 1 \right| \leq 2g(F_{i_1,i_2,\ldots,i_s})q^{1/2} \leq (4g(F)+s-1)q^{1/2}.$$

Therefore

$$\left| \sum_{j=1}^{N-1} \eta(f_{i_1} \cdots f_{i_s}(P_j)) \right| \le N - q + (4g(F) + s - 1)q^{1/2}.$$

Obviously (see also [76, Exercise 5.63]) M_k is equal to

$$M_k = \frac{1}{2^k} \sum_{j=1}^{N-1} (1 + \eta(f_1(P_j))) \cdots (1 + \eta(f_k(P_j))) - A,$$

where $0 \le A \le k/2$, i.e.,

$$M_k = \frac{1}{2^k} \sum_{j=1}^{N-1} \left(1 + \sum_{i=1}^{k} \eta(f_i(P_j)) + \sum_{1 \le i < l \le k} \eta(f_i f_l(P_j)) + \cdots + \eta(f_1 f_2 \cdots f_k(P_j)) \right) - A.$$

Thus,

$$\begin{aligned}
\left| M_k - \frac{N-1}{2^k} \right| &= \left| \frac{1}{2^k} \sum_{s=1}^{k} \sum_{1 \le i_1 < \cdots < i_s \le k} \sum_{j=1}^{N-1} \eta(f_{i_1} \cdots f_{i_s}(P_j)) - A \right| \\
&\le \frac{1}{2^k} \sum_{s=1}^{k} \sum_{1 \le i_1 < \cdots < i_s \le k} \left(N - q + (4g(F) + s - 1)q^{1/2} \right) + \frac{k}{2} \\
&\le \frac{1}{2^k} \sum_{s=1}^{k} \sum_{1 \le i_1 < \cdots < i_s \le k} (N - q + (4g(F) + k - 1)q^{1/2}) + \frac{k}{2} \\
&= \frac{2^k - 1}{2^k}(N - q + (4g(F) + k - 1)q^{1/2}) + \frac{k}{2} \\
&\le N - q + (4g(F) + k - 1)q^{1/2} + \frac{k}{2}.
\end{aligned}$$

$\square$

If q is even, let ψ be the additive character of $\mathbf{F}_q$ defined for $a \in \mathbf{F}_q$ by

$$\psi(a) = \begin{cases} 1 & \text{if } a \in \{b^2 + b : b \in \mathbf{F}_q\}, \\ -1 & \text{otherwise.} \end{cases}$$

Then we get the following analog of Lemma 5.2.5 for rational function fields.

Lemma 5.2.6 *Let $F/\mathbf{F}_q$, q even, be the rational function field with $q+1$ rational places $P_\infty, P_1, \ldots, P_q$. Let $g_1, \ldots, g_q$ be q elements in F satisfying*

$$\operatorname{div}(g_i) = P_\infty - P_i$$

for $1 \le i \le q$. Let ψ be the additive character of $\mathbf{F}_q$ defined above, and for $1 \le k \le q$ let M_k denote the number of rational places P ($\ne P_1, \ldots, P_k$) of F with $\psi(g_i(P)) = 1$ for all $1 \le i \le k$. Then

$$\left| M_k - \frac{q+1-k}{2^k} \right| \le (2k - 2)q^{1/2}.$$

Remark 5.2.7 The proof of Lemma 5.2.6 just repeats the steps of the proof of Lemma 5.2.5, but with Artin-Schreier extensions defined by

$$y_{i_1,i_2,...,i_s}^2 + y_{i_1,i_2,...,i_s} = g_{i_1} + g_{i_2} + \cdots + g_{i_s}$$

instead of Kummer extensions. The bound in Lemma 5.2.6 is a little different from that in Lemma 5.2.5 because the different exponent of $F_{i_1,i_2,...,i_s}/F$ at P_i in Lemma 5.2.6 is larger than that in Lemma 5.2.5.

Theorem 5.2.8 (i) *Let q be an odd prime power and $F/\mathbb{F}_q$ a global function field with exactly $N \geq q+1$ rational places. Let $n \geq 7$ be an odd integer. Suppose that the divisor class number of F is odd and that*

$$\frac{N-1}{2^n} - (N-q) - (4g(F)+n-1)q^{1/2} - \frac{n}{2} \geq \left\lfloor \frac{1}{2}\left(\frac{(n-3)^2}{4}-1\right)\right\rfloor .$$

Then

$$A(q) \geq \frac{4\lfloor \frac{1}{2}(\frac{(n-3)^2}{4}-1)\rfloor}{4g(F)+n-3}.$$

(ii) *If q is a power of 2 and $n \geq 6$ satisfies*

$$\frac{q+1-n}{2^n} - (2n-2)q^{1/2} \geq \left\lfloor \frac{(n-3)^2}{8} \right\rfloor ,$$

then

$$A(q) \geq \frac{2\lfloor \frac{(n-3)^2}{8}\rfloor}{n-2}.$$

Proof. For the sake of brevity we prove this theorem only for odd q. The same arguments apply to the case where q is even.

Let $P_\infty, P_1, \ldots, P_{N-1}$ be all rational places of F and let $f_i \in F$ satisfy

$$\mathrm{div}(f_i) = h(F)P_i - h(F)P_\infty \quad \text{for } 1 \leq i \leq N-1.$$

Put $t = \lfloor \frac{1}{2}(\frac{(n-3)^2}{4}-1)\rfloor \geq 1$. From Lemma 5.2.5 we know that

$$M_n \geq \frac{N-1}{2^n} - (N-q) - (4g(F)+n-1)q^{1/2} - \frac{n}{2} \geq t.$$

Thus, there exists a set $\mathcal{T}$ of rational places of F satisfying:
  (i) $|\mathcal{T}| = t$;
  (ii) $\{P_\infty, P_1, \ldots, P_n\} \cap \mathcal{T} = \emptyset$;
  (iii) $\eta(f_i(P)) = 1$ for all $1 \leq i \leq n$ and $P \in \mathcal{T}$.
For $1 \leq i \leq n$ let K_i be the function field $F(y_i)$ defined by $y_i^2 = f_i$. Let $y = y_1 \cdots y_n$ and $K = F(y)$. Then

$$y^2 = f_1 \cdots f_n.$$

It is obvious that $P_1, \ldots, P_n$ are ramified in K/F. Since n is odd, P_∞ is also ramified in K/F. Therefore $K_1 \cdots K_n/K$ is an unramified abelian extension with the 2-rank of its

Galois group being $n-1$. If S is the over-set of $\mathcal{T}$ with respect to K/F, then all places in S split completely in $K_1 \cdots K_n/K$. From our construction we know that $|S| = 2t$ and that all places in S are rational places of K. By the definition of t we have

$$d_2(\mathrm{Gal}(K_1 \cdots K_n/K)) = n - 1 \geq 2 + 2(2t+1)^{1/2} = 2 + 2(|S|+1)^{1/2}.$$

Thus, by Corollary 2.7.8 we get

$$A(q) \geq \frac{2t}{g(K) - 1}.$$

The genus of K is equal to $2g(F) + (n-1)/2$. The result follows. $\qquad\square$

Theorem 5.2.9 *For any prime power q we have*

$$A(q) > \frac{1}{96} \log_2 q,$$

where $\log_2$ denotes the logarithm to the base 2.

Proof. For $q \geq 2^{24}$ we take F to be the rational function field over $\mathbf{F}_q$ and n the odd integer $2\lfloor (\log_2 q)/6 \rfloor - 1$. Then all conditions in Theorem 5.2.8 are satisfied. Thus,

$$A(q) \geq \frac{2\lfloor \frac{1}{2}(\frac{(n-3)^2}{4} - 1) \rfloor}{n-2} \geq \frac{1}{75} \log_2 q.$$

For $q < 2^{24}$ we use Lemma 5.2.4, and this yields the desired result. $\qquad\square$

We have not attempted to optimize the constant in the lower bound in Theorem 5.2.9. Li and Maharaj [75] and Temkine [155] improved on the bound in Theorem 5.2.9 by showing that there exists an effective absolute constant $c > 0$ such that for any prime power q and any integer $n \geq 1$ we have

$$A(q^n) \geq \frac{cn^2(\log q)^2}{\log n + \log q}.$$

5.3 Further Lower Bounds for $A(q^m)$

The lower bounds on $A(q)$ in Section 5.2 can be improved in various cases. We focus here on the case where q is not a square. For squares q, and only for squares, the exact value of $A(q)$ is known (see Section 5.4). Zink [188] showed that for any prime p we have

$$A(p^3) \geq \frac{2(p^2 - 1)}{p + 2}. \tag{5.2}$$

Perret [126] proved that if ℓ is a prime and q is a prime power with $q > 4\ell + 1$ and $q \equiv 1 \pmod{\ell}$, then

$$A(q^\ell) \geq \frac{\ell^{1/2}(q-1)^{1/2} - 2\ell}{\ell - 1}.$$

A conjectured sufficient condition for infinite class field towers that was put forth by Perret in the same paper was disproved by Niederreiter and Xing [118] (see also Niederreiter and Xing [119]).

In this section we show the results of Niederreiter and Xing [112] on $A(q)$ for composite nonsquares q. We follow the proofs in Niederreiter and Xing [117] which construct the base fields of the required infinite class field towers explicitly. Refinements of these results can be found in Li and Maharaj [75] and in Niederreiter and Xing [115].

Theorem 5.3.1 *If q is an odd prime power and $m \geq 3$ is an integer, then*

$$A(q^m) \geq \frac{2q + 2}{\lceil 2(2q+3)^{1/2} \rceil + 1}.$$

Proof. Put $n = \lceil 2(2q+3)^{1/2} \rceil + 3$. Then n does not exceed the number of monic irreducible polynomials of degree m in $\mathbf{F}_q[x]$, except for $q = m = 3$, but in this case the result of the theorem is implied by (5.2).

Let $f_1, \ldots, f_n$ be n distinct monic irreducible polynomials of degree m in $\mathbf{F}_q[x]$ and for $1 \leq i \leq n$ let $\beta_i \in \mathbf{F}_{q^m}$ be a root of f_i. Put $F = \mathbf{F}_{q^m}(x)$ and let $K = F(y)$ with

$$y^2 = \prod_{i=1}^{n} (x - \beta_i)(x - \beta_i^q).$$

Then K/F is a Kummer extension with $g(K) = n - 1$ by Corollary 3.1.5. For $1 \leq i \leq n$ let $K_i = F(y_i)$ with

$$y_i^2 = (x - \beta_i)(x - \beta_i^q)$$

and let L be the composite field of $K_1, \ldots, K_n$. Since the extensions $K_1/F, \ldots, K_n/F$ are linearly disjoint, we have

$$\mathrm{Gal}(L/F) \simeq \prod_{i=1}^{n} \mathrm{Gal}(K_i/F) = (\mathbf{Z}/2\mathbf{Z})^n.$$

Note also that $K \subseteq L$ since we can take $y = y_1 \cdots y_n$. Thus,

$$\mathrm{Gal}(L/K) \simeq (\mathbf{Z}/2\mathbf{Z})^{n-1},$$

and so $d_2(\mathrm{Gal}(L/K)) = n - 1$. For $1 \leq i \leq n$ the places $x - \beta_i$ and $x - \beta_i^q$ of F have ramification index 2 in L/F and also ramification index 2 in K/F. Since there are no other ramified places in L/F, the extension L/K is unramified.

Now let $\mathcal{T}$ be the set of rational places of F given by

$$\mathcal{T} = \{P_b = x - b : b \in \mathbf{F}_q\} \cup \{\infty\},$$

where ∞ is the infinite place of F. For all $1 \leq i \leq n$ and $b \in \mathbf{F}_q$ we have

$$(x - \beta_i)(x - \beta_i^q) \equiv (b - \beta_i)(b - \beta_i^q) \equiv (b - \beta_i)(b^q - \beta_i^q) \equiv (b - \beta_i)^{q+1} \pmod{P_b}.$$

Since q is odd, the element $(b - \beta_i)^{q+1}$ is a nonzero square in the residue class field of P_b, and so P_b splits completely in K_i/F by Kummer's theorem. Moreover, ∞ splits completely in K_i/F. Thus, if S is the over-set of $\mathcal{T}$ with respect to K/F, then $|S| = 2q + 2$ and all places in S split completely in L/K. By the definition of n,

$$d_2(\mathrm{Gal}(L/K)) = n - 1 \geq 2 + 2(2q+3)^{1/2} = 2 + 2(|S|+1)^{1/2}.$$

Therefore, all conditions in Corollary 2.7.8 are satisfied with $\ell = 2$, and so we obtain the result of the theorem. □

Corollary 5.3.2 *If* $q = p^e$ *with an odd prime* p *and an odd integer* $e \geq 3$, *then*

$$A(q) \geq \frac{2q^{1/m} + 2}{\lceil 2(2q^{1/m} + 3)^{1/2} \rceil + 1},$$

where m *is the least prime dividing* e.

Theorem 5.3.3 *If* $q \geq 8$ *is a power of 2 and* $m \geq 3$ *is an odd integer, then*

$$A(q^m) \geq \frac{q+1}{\lceil 2(2q+2)^{1/2} \rceil + 2}.$$

Proof. Put $n = \lceil 2(2q+2)^{1/2} \rceil + 3$. Then n does not exceed the number of monic irreducible polynomials of degree 2 in $\mathbf{F}_q[x]$.

Let $f_1, \ldots, f_n$ be n distinct monic irreducible polynomials of degree 2 in $\mathbf{F}_q[x]$. Since m is odd, each f_i is irreducible over $\mathbf{F}_{q^m}$ and can thus be identified with a place Q_i of $F = \mathbf{F}_{q^m}(x)$ of degree 2. Choose $\beta \in \mathbf{F}_{q^m}^*$ with trace

$$\mathrm{Tr}_{\mathbf{F}_{q^m}/\mathbf{F}_q}(\beta) = 0$$

and let $K = F(y)$ with

$$y^2 + y = \sum_{i=1}^{n} \frac{\beta}{f_i(x)}.$$

Then K/F is an Artin-Schreier extension with $g(K) = 2n - 1$ by Theorem 3.1.9. For $1 \leq i \leq n$ let $K_i = F(y_i)$ with

$$y_i^2 + y_i = \frac{\beta}{f_i(x)},$$

and let L be the composite field of $K_1, \ldots, K_n$. We have $K \subseteq L$ since we can take $y = y_1 + \cdots + y_n$. The only ramified places in L/F are the Q_i, $1 \leq i \leq n$, and so we see, as in the proof of Theorem 5.3.1, that L/K is an unramified abelian extension with $d_2(\mathrm{Gal}(L/K)) = n - 1$.

Now let $\mathcal{T}$ be the set of rational places of F given by

$$\mathcal{T} = \{P_b = x - b : b \in \mathbf{F}_q\} \cup \{\infty\},$$

where ∞ is the infinite place of F. For all $1 \leq i \leq n$ and $b \in \mathbf{F}_q$ we have

$$\mathrm{Tr}_{\mathbf{F}_{q^m}/\mathbf{F}_q}\left(\frac{\beta}{f_i(b)}\right) = \frac{1}{f_i(b)} \mathrm{Tr}_{\mathbf{F}_{q^m}/\mathbf{F}_q}(\beta) = 0,$$

and so for the absolute trace we get

$$\mathrm{Tr}_{\mathbf{F}_{q^m}}\left(\frac{\beta}{f_i(b)}\right) = 0$$

by the transitivity of the trace (see [76, Theorem 2.26]). It follows then from Kummer's theorem and [76, Theorem 2.25] that P_b splits completely in K_i/F. Moreover, ∞ splits completely in K_i/F. Thus, if S is the over-set of $\mathcal{T}$ with respect to K/F, then $|S| = 2q+2$ and all places in S split completely in L/K. By the definition of n,

$$d_2(\mathrm{Gal}(L/K)) = n - 1 \geq 2 + 2(2q + 2)^{1/2} = 2 + 2|S|^{1/2}.$$

Therefore, all conditions in Corollary 2.7.8 are satisfied with $\ell = 2$, and so we obtain the result of the theorem. $\square$

Corollary 5.3.4 *If $q = 2^e$ with an odd composite integer $e \geq 3$, then*

$$A(q) \geq \frac{q^{1/m} + 1}{\lceil 2(2q^{1/m} + 2)^{1/2}\rceil + 2},$$

where m is the least prime dividing e.

5.4 Explicit Towers

Explicit examples of asymptotically good towers of function fields over $\mathbf{F}_q$ are of great interest for coding theory since they can be used for the explicit construction of asymptotically good families of linear codes over $\mathbf{F}_q$ (see Chapter 6). In this section we present several explicit towers $\mathcal{F}$ with $\lambda(\mathcal{F}) > 0$.

An important case is that of a prime power q that is a square. Here Ihara [56] showed $A(q) \geq q^{1/2} - 1$ by using sequences of modular and Shimura curves over $\mathbf{F}_q$. The basic ideas of the method go back to Ihara [55]. In the special cases $q = p^2$ and $q = p^4$ with a prime p, this lower bound on $A(q)$ was proved also by Tsfasman, Vlăduţ, and Zink [157]. In view of the Vlăduţ-Drinfeld bound (see Theorem 1.6.21), we obtain

$$A(q) = q^{1/2} - 1 \qquad \text{for all squares } q.$$

Surveys of the method of modular and Shimura curves can be found in Ihara [57], Moreno [82, Chapter 5], Stepanov [151, Chapter 9], and Tsfasman and Vlăduţ [156, Part 4]. The first two examples in this section demonstrate that Ihara's result $A(q) \geq q^{1/2} - 1$ for squares q can be proved also in a more elementary way by means of explicit towers of function fields over $\mathbf{F}_q$.

Example 5.4.1 We consider a tower of function fields over $\mathbf{F}_{q^2}$ which was introduced by Garcia and Stichtenoth [31] and meets the Vlăduţ-Drinfeld bound. The tower $\mathcal{K} = (K_1, K_2, K_3, \ldots)$ is given by $K_n := \mathbf{F}_{q^2}(x_1, \ldots, x_n)$, with

$$x_{i+1}^q + x_{i+1} = \frac{x_i^q}{x_i^{q-1} + 1} \qquad \text{for } i = 1, \ldots, n-1.$$

A careful analysis of the tower $\mathcal{K}$ shows that for the genus of K_n we have

$$g(K_n) = \begin{cases} (q^{n/2} - 1)^2 & \text{if } n \text{ is even,} \\ (q^{(n+1)/2} - 1)(q^{(n-1)/2} - 1) & \text{if } n \text{ is odd.} \end{cases}$$

Moreover, for $\alpha \in \mathbf{F}_{q^2} \setminus \{\beta \in \mathbf{F}_{q^2} : \beta^q + \beta = 0\}$, the zero place of $x_1 - \alpha$ in the rational function field $K_1 = \mathbf{F}_{q^2}(x_1)$ splits completely in all extensions K_n/K_1. The infinite place of K_1 is totally ramified in all extensions K_n/K_1. Hence the number of rational places of K_n satisfies

$$N(K_n) \geq q^{n-1}(q^2 - q) + 1.$$

Therefore

$$\lambda(\mathcal{K}) \geq q - 1.$$

Applying the Vlăduţ-Drinfeld bound, we obtain $\lambda(\mathcal{K}) = q - 1$.

Example 5.4.2 In fact, the tower $\mathcal{K}$ considered in Example 5.4.1 is a subtower of the following tower $\mathcal{E} = (E_1, E_2, E_3, \ldots)$ of function fields over $\mathbf{F}_{q^2}$ constructed earlier by Garcia and Stichtenoth [30]. Let $E_1 := \mathbf{F}_{q^2}(x_1)$ be the rational function field and, for $n \geq 1$, let $E_{n+1} := E_n(z_{n+1})$ with

$$z_{n+1}^q + z_{n+1} = x_n^{q+1} \qquad \text{and} \qquad x_{n+1} = \frac{z_{n+1}}{x_n}.$$

Then we note that for $n \geq 2$,

$$z_{n+1}^q + z_{n+1} = x_n^{q+1} = \frac{z_n^{q+1}}{x_{n-1}^{q+1}} = \frac{z_n^{q+1}}{z_n^q + z_n} = \frac{z_n^q}{z_n^{q-1} + 1}.$$

It follows that the subfield $\mathbf{F}_{q^2}(z_2, \ldots, z_{n+1}) \subseteq E_{n+1}$ is isomorphic to the field K_n in the tower $\mathcal{K}$, and hence $\mathcal{K}$ is a subtower of $\mathcal{E}$. In fact, the tower $\mathcal{E}$ is also optimal, i.e., $\lambda(\mathcal{E}) = q - 1$. By results in Garcia and Stichtenoth [30] we have

$$g(E_n) \leq q^n + q^{n-1} - q^{(n+1)/2} - 2q^{(n-1)/2} + 1 \qquad \text{for all } n \geq 1,$$

$$N(E_n) \geq (q^2 - 1)q^{n-1} + 1 \qquad \text{for all } n \geq 1.$$

In Examples 5.4.1 and 5.4.2 above, the extensions K_n/K_1 and E_n/E_1 have wildly ramified places for $n > 1$. Now we give two examples where F_n/F_1 has no wildly ramified places. These examples provide asymptotically good towers of Kummer extensions and are due to Garcia and Stichtenoth [32] (see also Garcia, Stichtenoth, and Thomas [36]).

Example 5.4.3 Assume that $q = p^e$ with a prime p and $e > 1$, and set $m := (q-1)/(p-1)$. Let $F_n := \mathbf{F}_q(x_1, \ldots, x_n)$, with

$$x_{i+1}^m + (x_i + 1)^m = 1 \qquad \text{for } i = 1, \ldots, n - 1.$$

This tower $\mathcal{F} = (F_1, F_2, F_3, \ldots)$ has the following properties:

- for all $n \geq 1$, the extension F_{n+1}/F_n is cyclic of degree $[F_{n+1} : F_n] = m$;

- if P is a place of F_1 that ramifies in F_n/F_1 for some $n > 1$, then P is the zero of $x_1 - \alpha$ for some $\alpha \in \mathbf{F}_q$;
- the infinite place of F_1 splits completely in F_n/F_1 for all $n > 1$.

These properties readily imply that

$$\lambda(\mathcal{F}) \geq \frac{2}{q-2}.$$

Hence the tower is optimal for $q = 4$. Note that this example yields a very simple and elementary proof that $A(q) > 0$ if q is not a prime number.

Example 5.4.4 Assume that $q = r^2$ with a prime power $r > 2$. Define $F_n := \mathbf{F}_q(x_1, \ldots, x_n)$ by the equations

$$x_{i+1}^{r-1} + (x_i + 1)^{r-1} = 1 \qquad \text{for } i = 1, \ldots, n-1.$$

For this tower $\mathcal{F} = (F_1, F_2, F_3, \ldots)$ we have

$$\lambda(\mathcal{F}) \geq \frac{2}{r-2} = \frac{2}{\sqrt{q}-2}.$$

The tower is optimal for $q = 9$.

5.5   Lower Bounds on $A(2)$, $A(3)$, and $A(5)$

We derive lower bounds on $A(q)$ for small primes q. The following bound for $q = 2$ is due to Niederreiter and Xing [112] and improves on earlier results by Serre [141], [144] and Schoof [138].

Theorem 5.5.1 *We have*

$$A(2) \geq \frac{81}{317} = 0.2555\ldots.$$

Proof. Let F be the rational function field $\mathbf{F}_2(x)$. Put $M = (x^2+x+1)(x^6+x^3+1) \in \mathbf{F}_2[x]$ and let K be the subfield of the cyclotomic function field $F_M := F(\Lambda_M)$ fixed by the cyclic subgroup $\langle \overline{x} \rangle$ of $\mathrm{Gal}(F_M/F) = (\mathbf{F}_2[x]/(M))^*$. Then $[K : F] = 21$ and the places ∞ and x split completely in K/F. Furthermore, $x^2 + x + 1$ is tamely ramified in K/F with ramification index 3 and $x^6 + x^3 + 1$ is totally ramified in K/F, hence $g(K) = 54$.

Put $N = x^4 \in \mathbf{F}_2[x]$ and let L be the subfield of the cyclotomic function field $F_N := F(\Lambda_N)$ fixed by the cyclic subgroup $\langle \overline{x^2+1} \rangle$ of $\mathrm{Gal}(F_N/F) = (\mathbf{F}_2[x]/(N))^*$. Then $[L : F] = 4$ and $\mathrm{Gal}(L/F) \simeq (\mathbf{Z}/2\mathbf{Z})^2$. Let P be the place of F_N lying over x and λ a generator of the cyclic $\mathbf{F}_2[x]$-module Λ_N. If ν_P is the normalized discrete valuation corresponding to P, then $\nu_P(\lambda) = 1$ by the proof of Proposition 3.2.4(i). Thus, by Proposition 1.3.13 the different exponent of P in F_N/L is given by

$$d_P(F_N/L) = \nu_P\left(\lambda - \lambda^{x^2+1}\right) = 4.$$

Furthermore, $d_P(F_N/F) = 24$ by the proof of Theorem 3.2.9. If Q is the place of L lying over x, then the tower formula for different exponents shows that

$$d_Q(L/F) = \frac{1}{2}(d_P(F_N/F) - d_P(F_N/L)) = 10.$$

Now we consider the composite field KL of K and L. It is clear that

$$\mathrm{Gal}(KL/K) \simeq \mathrm{Gal}(L/F) \simeq (\mathbf{Z}/2\mathbf{Z})^2$$

since $K \cap L = F$. The only ramified places in KL/K are those places of K lying over x. Any such place is totally ramified in KL/K, and so its inertia group in KL/K is the whole group $G = \mathrm{Gal}(KL/K)$. Let $\mathcal{T}'$ be a set of rational places of K consisting of 20 places lying over ∞ and one place lying over x. Let S be the over-set of $\mathcal{T} = \mathbf{P}_K \setminus \mathcal{T}'$ and S' the over-set of $\mathcal{T}'$ with respect to KL/K. Then $|S'| = 4 \cdot 20 + 1 = 81$. By Proposition 5.2.2, applied to KL/K, we get

$$d_2(\mathrm{Cl}(O_S)) \geq \sum_R d_2(G_R) - (|\mathcal{T}'| - 1) - d_2(G) = 21 \cdot 2 - 20 - 2 = 20,$$

where the sum is over all places R of K. Thus, the condition on $d_2(\mathrm{Cl}(O_S))$ in Proposition 5.2.1 is satisfied. It remains to calculate the genus of KL. The ramified places in KL/K are exactly the 21 places of K lying over x. For any such place, the different exponent in KL/K is equal to $d_Q(L/F) = 10$. Hence by the Hurwitz genus formula,

$$2g(KL) - 2 = 4(2g(K) - 2) + 21 \cdot 10 = 634,$$

and so

$$A(2) \geq \frac{|S'|}{g(KL) - 1} = \frac{81}{317}.$$

$\square$

The following two results were shown independently by Anglès and Maire [3] and Temkine [155] and improve on earlier bounds due to Xing [174] and Niederreiter and Xing [112].

Theorem 5.5.2 *We have*

$$A(3) \geq \frac{8}{17} = 0.4705\ldots.$$

Proof. Let

$$
\begin{aligned}
f(x) \ = \ & (x^2 + 1)(x^2 + x + 2)(x^2 + 2x + 2)(x^3 + 2x + 1)(x^3 + 2x + 2) \\
& (x^3 + x^2 + 2)(x^3 + x^2 + x + 2)(x^3 + x^2 + 2x + 1) \\
& (x^3 + 2x^2 + 1)(x^3 + 2x^2 + x + 1)(x^3 + 2x^2 + 2x + 2) \\
& (x^4 + x^2 + x + 1)(x^4 + x^3 + x^2 + 1) \in \mathbf{F}_3[x],
\end{aligned}
$$

written as a product of 13 irreducible polynomials over $\mathbf{F}_3$. Let F be the rational function field $\mathbf{F}_3(x)$ and $K = F(y)$ with $y^2 = f(x)$. Furthermore, let $\mathcal{T}'$ be the set of rational places of F given by

$$\mathcal{T}' = \{P_b = x - b : b \in \mathbf{F}_3\} \cup \{\infty\},$$

where ∞ is the infinite place of F. Then, with S being the over-set of $\mathcal{T} = \mathbf{P}_F \setminus \mathcal{T}'$ with respect to K/F, we get by Proposition 5.2.2,

$$d_2(\mathrm{Cl}(O_S)) \geq 13 - 4 - 1 = 8.$$

For any $b \in \mathbf{F}_3$ we have $f(b) = 1$, and so P_b splits completely in K/F. Moreover, since f is monic and $\deg(f) = 38$ is even, the place ∞ splits completely in K/F. Hence for $S' = \mathbf{P}_K \setminus S$ we have $|S'| = 8$, and so the condition on $d_2(\mathrm{Cl}(O_S))$ in Proposition 5.2.1 is satisfied. From $\deg(f) = 38$ we get $g(K) = 18$ by Corollary 3.1.5, and so

$$A(3) \geq \frac{8}{17}.$$

□

Theorem 5.5.3 *We have*
$$A(5) \geq \frac{8}{11} = 0.7272\ldots.$$

Proof. We proceed as in the proof of Theorem 5.5.2 and use similar notation. Let

$$\begin{aligned} f(x) = \ & (x+1)(x+2)(x^2+2)(x^2+3)(x^2+x+1) \\ & (x^2+x+2)(x^2+2x+3)(x^2+2x+4)(x^2+3x+3) \\ & (x^2+3x+4)(x^2+4x+1)(x^2+4x+2)(x^4+x^2+2) \in \mathbf{F}_5[x], \end{aligned}$$

written as a product of 13 irreducible polynomials over $\mathbf{F}_5$. Let F be the rational function field $\mathbf{F}_5(x)$, $K = F(y)$ with $y^2 = f(x)$, and

$$\mathcal{T}' = \{x,\ x-1,\ x-2,\ \infty\} \subseteq \mathbf{P}_F.$$

Then by Proposition 5.2.2,

$$d_2(\mathrm{Cl}(O_S)) \geq 13 - 4 - 1 = 8.$$

The place ∞ splits completely in K/F, and so do the places $x, x-1, x-2$ since $f(0) = f(1) = f(2) = 4 = 2^2$. Hence $|S'| = 8$, and so the condition on $d_2(\mathrm{Cl}(O_S))$ in Proposition 5.2.1 is satisfied. From $\deg(f) = 26$ we get $g(K) = 12$ by Corollary 3.1.5, and so

$$A(5) \geq \frac{8}{11}.$$

□

Chapter 6

Applications to Algebraic Coding Theory

Goppa's celebrated construction of algebraic-geometry codes uses algebraic curves over finite fields with many rational points or, equivalently, global function fields with many rational places. This construction was a breakthrough in algebraic coding theory because it yields sequences of linear codes beating the asymptotic Gilbert-Varshamov bound. We describe this construction and its consequences, but also recent work which shows that improvements on Goppa's construction can be obtained by other constructions that also employ places of higher degree. As basic references for algebraic coding theory we recommend the books of MacWilliams and Sloane [77] and van Lint [166].

6.1 Goppa's Algebraic-Geometry Codes

We start with a brief recapitulation of the theory of linear codes. Recall that a code is a scheme for detecting and correcting transmission errors in noisy communication channels. A code operates by adding redundant information to messages. As the signal alphabet we always use $\mathbf{F}_q$, where $q = 2$ is naturally an important special case.

A **linear code** over $\mathbf{F}_q$ is a nonzero linear subspace of the vector space $\mathbf{F}_q^n$ for some $n \geq 1$. If $C \subseteq \mathbf{F}_q^n$ is a linear code over $\mathbf{F}_q$, then n is the **length** of C and $k := \dim(C)$ is the **dimension** of C. We express these facts by saying that C is a linear $[n, k]$ code over $\mathbf{F}_q$. Note that by definition we have $1 \leq k \leq n$. The **(Hamming) weight** $w(\mathbf{x})$ is the number of nonzero coordinates of $\mathbf{x} \in \mathbf{F}_q^n$. For a linear code C over $\mathbf{F}_q$, its **minimum distance** (or **minimum weight**) is defined to be the smallest weight of any nonzero vector in C. We say that C is a linear $[n, k, d]$ code over $\mathbf{F}_q$ if C has length n, dimension k, and minimum distance d.

If C is a linear $[n, k]$ code over $\mathbf{F}_q$, then a **generator matrix** of C is any $k \times n$ matrix over $\mathbf{F}_q$ whose row space is equal to C. If M is a generator matrix of C, then the coding scheme afforded by C proceeds by viewing messages as row vectors in $\mathbf{F}_q^k$ and mapping a row vector $\mathbf{a} \in \mathbf{F}_q^k$ into the coded message $\mathbf{c} = \mathbf{a}M \in \mathbf{F}_q^n$. The image of this map, i.e., the set of all coded messages, is the code C. If C has minimum distance d, then this coding scheme can correct up to $\lfloor (d-1)/2 \rfloor$ transmission errors in each coded message block of length n. Thus, d is the crucial parameter for the error-correction capability of C. It is a common aim of constructions of linear $[n, k]$ codes over $\mathbf{F}_q$ to maximize d for given n, k, and q.

We can always put a generator matrix M of C in row-reduced echelon form by using elementary row operations on M. If we further apply suitable column permutations, then we get a $k \times n$ matrix of the special form $(I_k|A)$, where I_k is the $k \times k$ identity matrix over $\mathbf{F}_q$ and A is a suitable matrix over $\mathbf{F}_q$. A matrix of this special form is called a **canonical generator matrix**. Note that column permutations may change C, but they do not change the basic parameters n, k, d. If C allows a canonical generator matrix M, then the encoding rule $\mathbf{c} = \mathbf{a}M$ means that the first k coordinates of $\mathbf{c}$ are the original message coordinates of $\mathbf{a}$ and an additional $n - k$ coordinates are appended as control coordinates. These observations also yield the following simple bound on the minimum distance.

Proposition 6.1.1 (Singleton Bound) *For any linear $[n, k, d]$ code over $\mathbf{F}_q$ we have*

$$d \le n - k + 1.$$

Proof. We can assume that the given code allows a canonical generator matrix M. Then it suffices to note that any row of M has weight at most $n - k + 1$. $\square$

A linear code over $\mathbf{F}_q$ can also be viewed as the null space of a suitable matrix over $\mathbf{F}_q$. If C is a linear $[n, k]$ code over $\mathbf{F}_q$, then any matrix H over $\mathbf{F}_q$ with n columns which has C as its null space, i.e.,

$$C = \{\mathbf{x} \in \mathbf{F}_q^n : H\mathbf{x}^\top = \mathbf{0}\},$$

is called a **parity-check matrix** of C. Note that H must have rank $n - k$. Often, but not always, it is assumed that H is an $(n - k) \times n$ matrix. The minimum distance of C can be read off from any parity-check matrix of C.

Lemma 6.1.2 *Let H be any parity-check matrix of the linear code C over $\mathbf{F}_q$. Then C has minimum distance d if and only if any $d - 1$ columns of H are linearly independent and some d columns of H are linearly dependent.*

Proof. There exists a $\mathbf{c} \in C$ of weight $w \ge 1$ if and only if $H\mathbf{c}^\top = \mathbf{0}$ for some $\mathbf{c} \in \mathbf{F}_q^n$ of weight w, and this is in turn equivalent to some w columns of H being linearly dependent. $\square$

Remark 6.1.3 Lemma 6.1.2 yields another simple proof of the Singleton bound in Proposition 6.1.1 since $n - k$, being the rank of H, is the maximum number of linearly independent columns of H.

We are now ready to introduce Goppa's construction of algebraic-geometry codes. Let $F/\mathbf{F}_q$ be a global function field over $\mathbf{F}_q$ of genus g with $N(F) \geq 1$. Choose distinct rational places $P_1, \ldots, P_n$ of F, where $n > g$, and let G be a divisor of F with $\mathrm{supp}(G) \cap \{P_1, \ldots, P_n\} = \emptyset$. Consider the Riemann-Roch space $\mathcal{L}(G)$ and note that $\nu_{P_i}(f) \geq 0$ for $1 \leq i \leq n$ and all $f \in \mathcal{L}(G)$, i.e.,

$$\mathcal{L}(G) \subseteq \bigcap_{i=1}^{n} O_{P_i}.$$

Thus, it is meaningful to define the $\mathbf{F}_q$-linear map $\psi : \mathcal{L}(G) \longrightarrow \mathbf{F}_q^n$ by

$$\psi(f) = (f(P_1), \ldots, f(P_n)) \qquad \text{for all } f \in \mathcal{L}(G),$$

where $f(P)$ denotes, as usual, the residue class of $f \in O_P$ modulo the place P. The image of ψ is a linear subspace of $\mathbf{F}_q^n$ which is denoted by $C(P_1, \ldots, P_n; G)$ and called an **algebraic-geometry code** (or **AG code**). The standard result on the parameters of AG codes is the following one.

Theorem 6.1.4 *Let $F/\mathbf{F}_q$ be a global function field of genus g and let $P_1, \ldots, P_n$ be distinct rational places of F. Choose a divisor G of F with $g \leq \deg(G) < n$ and $\mathrm{supp}(G) \cap \{P_1, \ldots, P_n\} = \emptyset$. Then $C(P_1, \ldots, P_n; G)$ is a linear $[n, k, d]$ code over $\mathbf{F}_q$ with*

$$k = \ell(G) \geq \deg(G) - g + 1, \qquad d \geq n - \deg(G).$$

Moreover, $k = \deg(G) - g + 1$ if $\deg(G) \geq 2g - 1$.

Proof. For any nonzero $f \in \mathcal{L}(G)$ we consider the weight of $\psi(f)$. We have $w(\psi(f)) = n - r$, where r is the number of zeros of f from $\{P_1, \ldots, P_n\}$. If $P_{i_1}, \ldots, P_{i_r}$ are these distinct zeros, then

$$f \in \mathcal{L}(G - P_{i_1} - \cdots - P_{i_r}).$$

Since $f \neq 0$, this implies

$$0 \leq \deg(G - P_{i_1} - \cdots - P_{i_r}) = \deg(G) - r,$$

and so $w(\psi(f)) \geq n - \deg(G) > 0$. This shows not only the desired lower bound on d, but also that ψ is injective. Hence $k = \ell(G)$, and the rest follows from the Riemann-Roch theorem. $\square$

Remark 6.1.5 Theorem 6.1.4 implies that $k + d \geq n + 1 - g$. This should be compared with the Singleton bound $k + d \leq n + 1$ in Proposition 6.1.1. Thus, the genus of F controls, in a sense, the deviation of $k + d$ from the Singleton bound. If $g = 0$, i.e., if F is the rational function field over $\mathbf{F}_q$, then we must have $k = \deg(G) + 1$ and $d = n - \deg(G)$. If $g = 0$ and $n \leq q$, then the resulting code is actually a generalized Reed-Solomon code. If $g = 0$ and $n = q + 1$, then we get an extended generalized Reed-Solomon code.

Remark 6.1.6 The condition on the support of G in Theorem 6.1.4 is a conventional one in the area. However, we can get rid of this condition as follows. By the approximation theorem, we can choose $u \in F$ such that $\nu_{P_i}(u) = \nu_{P_i}(G)$ for $1 \leq i \leq n$. Then we apply Theorem 6.1.4 with G replaced by $G - \operatorname{div}(u)$. If we drop the condition on $\deg(G)$ in Theorem 6.1.4, then ψ has the kernel $\mathcal{L}(G - D)$ with the divisor $D = \sum_{i=1}^{n} P_i$ of F. Thus, the dimension k of $C(P_1, \ldots, P_n; G)$ is given by $\ell(G) - \ell(G - D)$.

Remark 6.1.7 The construction of AG codes can be generalized in an obvious manner. Let $P_1, \ldots, P_n$ again be distinct rational places of $F/\mathbf{F}_q$ and for an integer $k \geq 1$ let V be a k-dimensional $\mathbf{F}_q$-linear subspace of F satisfying the following two conditions:

$$V \subseteq \bigcap_{i=1}^{n} O_{P_i} \quad \text{and} \quad \bigcap_{i=1}^{n} (V \cap \mathsf{M}_{P_i}) = \{0\}.$$

Then the $\mathbf{F}_q$-linear map $\psi : V \longrightarrow \mathbf{F}_q^n$ defined by

$$\psi(f) = (f(P_1), \ldots, f(P_n)) \qquad \text{for all } f \in V$$

is injective, and so its image is a linear $[n, k]$ code over $\mathbf{F}_q$. There is no standard terminology for these codes. In this book we call such a code a **function-field code** and we denote it by $C_V(P_1, \ldots, P_n)$. These codes will be used in Section 6.5.

The idea of an algebraic-geometry code goes back to Goppa [38], [39], [40] who actually introduced the dual code of $C(P_1, \ldots, P_n; G)$ which can be described in terms of local components of Weil differentials. The construction of AG codes inspired a lot of work on algebraic curves over finite fields and global function fields, particularly on the question of the number of rational points, respectively places, which is clearly relevant in this context. We will say more about this connection in Section 6.2. Algebraic-geometry codes have been treated in detail in several books, see e.g. Goppa [41], Moreno [82], Pretzel [127], [128], Stepanov [151], Stichtenoth [152], and Tsfasman and Vlăduţ [156]. There is also an extensive survey article on AG codes by Høholdt, van Lint, and Pellikaan [53] which discusses, among other things, alternative approaches to AG codes and the important practical issue of decoding algorithms for AG codes. For the latter topic we also refer to the earlier survey by Høholdt and Pellikaan [52] and the book of Stepanov [151, Chapter 12].

For the concrete implementation of AG codes it is desirable to know how to produce generator matrices of AG codes. It is obvious that if $\{f_1, \ldots, f_k\}$ is a basis of the vector space $\mathcal{L}(G)$ over $\mathbf{F}_q$, then under the conditions of Theorem 6.1.4 the matrix

$$\begin{pmatrix} f_1(P_1) & f_1(P_2) & \cdots & f_1(P_n) \\ \vdots & \vdots & & \vdots \\ f_k(P_1) & f_k(P_2) & \cdots & f_k(P_n) \end{pmatrix}$$

is a generator matrix of the code $C(P_1, \ldots, P_n; G)$. The problem of constructing generator matrices of AG codes is thus reduced to that of finding bases of Riemann-Roch spaces.

Algorithms for computing bases of Riemann-Roch spaces were developed e.g. by Haché [44], Heß [51], Le Brigand and Risler [74], and Matsumoto and Miura [80].

We now present a way of introducing algebraic-geometry codes via parity-check matrices and we follow here the paper of Xing, Niederreiter, and Lam [183]. Let $F/\mathbf{F}_q$ be a global function field of genus g and let $P_\infty, P_1, \ldots, P_n$ be $n+1$ distinct rational places of F, where $n > g$. Choose a positive nonspecial divisor D of F with $\ell(D) = 1$. Then $\ell(D + P_i) = 2$ for $1 \le i \le n$ by Corollary 1.1.4, and so for each $1 \le i \le n$ we can choose a function

$$f_i \in \mathcal{L}(D + P_i)\backslash\mathcal{L}(D).$$

Let t be a local parameter at P_∞ and consider the local expansions of the $f_i, 1 \le i \le n$, at P_∞ which are of the form

$$f_i = t^{-v} \sum_{r=0}^{\infty} b_{r,i} t^r, \tag{6.1}$$

where $v := \nu_{P_\infty}(D) \ge 0$ and all coefficients $b_{r,i} \in \mathbf{F}_q$. For each $i = 1, \ldots, n$ we define

$$c_{r,i} = \begin{cases} b_{r-1,i} & \text{for } 1 \le r \le v, \\ b_{r,i} & \text{for } r \ge v+1. \end{cases}$$

Now we choose a positive integer m with $g \le m < n$ and define the $m \times n$ matrix H over $\mathbf{F}_q$ by

$$H = \begin{pmatrix} c_{1,1} & c_{1,2} & \cdots & c_{1,n} \\ \vdots & \vdots & & \vdots \\ c_{m,1} & c_{m,2} & \cdots & c_{m,n} \end{pmatrix}.$$

We denote by $C(P_\infty, P_1, \ldots, P_n; D; m)$ the linear code over $\mathbf{F}_q$ of length n with parity-check matrix H. This code is equivalent to an AG code, in the sense of the following definition.

Definition 6.1.8 Two linear codes C_1 and C_2 over $\mathbf{F}_q$ of length n are **equivalent** if there exist nonzero elements $\lambda_1, \ldots, \lambda_n$ of $\mathbf{F}_q$ such that

$$C_2 = \{(\lambda_1 a_1, \ldots, \lambda_n a_n) \in \mathbf{F}_q^n : (a_1, \ldots, a_n) \in C_1\}.$$

Theorem 6.1.9 *Let $F/\mathbf{F}_q$ be a global function field of genus g and let $P_\infty, P_1, \ldots, P_n$ be $n+1$ distinct rational places of F, where $n > g$. Let D be a positive nonspecial divisor of F with $\ell(D) = 1$ and let m be a positive integer with $g \le m < n$. Then the code $C(P_\infty, P_1, \ldots, P_n; D; m)$ is equivalent to the AG code $C(P_1, \ldots, P_n; G)$, where G is a divisor of F that is equivalent to the divisor $D + \sum_{i=1}^n P_i - (m+1)P_\infty$ and satisfies $\text{supp}(G) \cap \{P_1, \ldots, P_n\} = \emptyset$.*

Proof. Put $C_1 = C(P_\infty, P_1, \ldots, P_n; D; m)$. Note that the conditions on D imply that $\deg(D) = g$ and $\mathcal{L}(D) = \mathbf{F}_q$. In particular, we get

$$0 \le v = \nu_{P_\infty}(D) \le g \le m.$$

By the construction of C_1 we know that a vector $(x_1, \ldots, x_n) \in \mathbf{F}_q^n$ belongs to C_1 if and only if

$$\sum_{i=1}^{n} x_i b_{r,i} = 0 \qquad \text{for all } r \in \{0, 1, \ldots, m\}\backslash\{v\},$$

that is, if and only if

$$\sum_{i=1}^{n} x_i f_i - \sum_{i=1}^{n} x_i b_{v,i} \in \mathcal{L}(D + \sum_{i=1}^{n} P_i - (m+1)P_\infty).$$

It is easily seen from the construction that $1, f_1, \ldots, f_n$ form a basis of $\mathcal{L}(D + \sum_{i=1}^{n} P_i)$ over $\mathbf{F}_q$, hence for any element

$$f \in \mathcal{L}(D + \sum_{i=1}^{n} P_i - (m+1)P_\infty) \subseteq \mathcal{L}(D + \sum_{i=1}^{n} P_i)$$

there exist elements $y_1, \ldots, y_n, \lambda \in \mathbf{F}_q$ such that

$$f = \sum_{i=1}^{n} y_i f_i + \lambda.$$

We denote the vector $(y_1, \ldots, y_n)$ by $\mathbf{y}_f$. It is uniquely determined by f. Consider the local expansion of f at P_∞, then from (6.1) we obtain

$$
\begin{aligned}
f &= t^{-v} \sum_{r=0}^{\infty} \left(\sum_{i=1}^{n} y_i b_{r,i} \right) t^r + \lambda \\
&= t^{-v} \sum_{\substack{r=0 \\ r \neq v}}^{\infty} \left(\sum_{i=1}^{n} y_i b_{r,i} \right) t^r + \lambda + \sum_{i=1}^{n} y_i b_{v,i}.
\end{aligned}
$$

Since $\nu_{P_\infty}(f) \geq m + 1 - \nu_{P_\infty}(D) = m + 1 - v$, we get

$$\sum_{i=1}^{n} y_i b_{r,i} = 0 \qquad \text{for all } r \in \{0, 1, \ldots, m\}\backslash\{v\},$$

and so $\mathbf{y}_f \in C_1$. Altogether, we have shown that

$$C_1 = \{\mathbf{y}_f \in \mathbf{F}_q^n : f \in \mathcal{L}(D + \sum_{i=1}^{n} P_i - (m+1)P_\infty)\}.$$

By the approximation theorem, there exists an $h \in F$ such that

$$\nu_{P_i}(h) = \nu_{P_i}(D) + 1 \geq 1 \qquad \text{for } 1 \leq i \leq n.$$

Then $\nu_{P_i}(f_i h) = 0$ for $1 \leq i \leq n$, and so the elements

$$\lambda_i = (f_i h)(P_i) \qquad \text{for } 1 \leq i \leq n$$

are in $\mathbf{F}_q^*$. Put

$$G = D + \sum_{i=1}^{n} P_i - (m+1)P_\infty - \text{div}(h),$$

then $\operatorname{supp}(G) \cap \{P_1, \ldots, P_n\} = \emptyset$ and G is equivalent to

$$D + \sum_{i=1}^{n} P_i - (m+1)P_\infty.$$

Furthermore, we have

$$\mathcal{L}(G) = \{fh : f \in \mathcal{L}(D + \sum_{i=1}^{n} P_i - (m+1)P_\infty)\}.$$

The codes C_1 and $C_2 := C(P_1, \ldots, P_n; G)$ have the same dimension

$$\ell(D + \sum_{i=1}^{n} P_i - (m+1)P_\infty).$$

Thus, in order to prove that C_1 and C_2 are equivalent, it suffices to verify that any vector in C_2 is equal to $(\lambda_1 a_1, \ldots, \lambda_n a_n)$ for some $(a_1, \ldots, a_n) \in C_1$. By construction, any vector in C_2 is of the form

$$((fh)(P_1), \ldots, (fh)(P_n))$$

with some $f \in \mathcal{L}(D + \sum_{i=1}^{n} P_i - (m+1)P_\infty)$. We can write

$$f = \sum_{i=1}^{n} y_i f_i + \lambda$$

for some $y_1, \ldots, y_n, \lambda \in \mathbf{F}_q$. Then $(y_1, \ldots, y_n) = \mathbf{y}_f \in C_1$ and

$$(fh)(P_j) = \left(\sum_{i=1}^{n} y_i f_i h + \lambda h\right)(P_j) = y_j(f_j h)(P_j) = \lambda_j y_j \qquad \text{for } 1 \leq j \leq n,$$

which is the desired statement. $\qquad\qquad\qquad\qquad\qquad\qquad\qquad\qquad\qquad\qquad\qquad\square$

Remark 6.1.10 Obviously, equivalent codes have the same parameters. Thus, it follows from Theorems 6.1.4 and 6.1.9 that $C(P_\infty, P_1, \ldots, P_n; D; m)$ is a linear $[n, k, d]$ code over $\mathbf{F}_q$ with

$$k = \ell\left(D + \sum_{i=1}^{n} P_i - (m+1)P_\infty\right) \geq n - m, \qquad d \geq m - g + 1.$$

The bound $k \geq n - m$ is, in fact, trivial by the construction of the code since its parity-check matrix is an $m \times n$ matrix over $\mathbf{F}_q$. The bound $d \geq m - g + 1$ can also be proved directly by using the form of the parity-check matrix and Lemma 6.1.2 (see the proof of [183, Theorem 5.3]).

Remark 6.1.11 A divisor D of F as in Theorem 6.1.9, i.e., a positive nonspecial divisor with $\ell(D) = 1$, exists in each of the following cases: (i) $g = 0$ (take D to be the zero divisor); (ii) $g = 1$ (take D to be a rational place); (iii) $q = 2$ and $N(F/\mathbf{F}_q) \geq 4$; (iv) $q \geq 3$ and $N(F/\mathbf{F}_q) \geq 2$. The validity of (iii) and (iv) was proved in Niederreiter and Xing [102, Lemma 6]. Thus, a suitable divisor D exists in all cases that are of interest in Theorem 6.1.9.

The paper of Xing, Niederreiter, and Lam [183] contains another method of defining algebraic-geometry codes via parity-check matrices, called the "first construction" in [183]. It was later shown by Özbudak and Stichtenoth [123] that the "first construction" yields linear codes that are also equivalent to AG codes. In fact, Özbudak and Stichtenoth [123] introduced a more general construction (though not by means of parity-check matrices) which contains the construction of the codes $C(P_\infty, P_1, \ldots, P_n; D; m)$ and the "first construction" as special cases. They proved not only that every linear code resulting from their more general construction can be viewed as an AG code, but also the converse that every AG code can be obtained by their construction. We remark that the construction of Özbudak and Stichtenoth [123] is, in turn, a special case of the construction to be presented in Section 6.3.

Another construction of linear codes due to Xing, Niederreiter, and Lam [183] uses again parity-check matrices, but is not equivalent to the construction of AG codes. Let s be a prime power, put $q = s^2$, and let $F/\mathbf{F}_q$ be a maximal function field of genus g (see Section 3.1), that is, we have

$$N(F/\mathbf{F}_q) = q + 1 + 2gs.$$

It is a standard fact (see e.g. Rück and Stichtenoth [132]) that for a maximal function field $F/\mathbf{F}_q$ the group $\mathrm{Cl}(F)$ of divisor classes of degree zero of F has a nice structure as an abelian group, namely

$$\mathrm{Cl}(F) \simeq (\mathbf{Z}/(s+1)\mathbf{Z})^{2g}.$$

Actually, all we need to know is that $\mathrm{Cl}(F)$ has exponent $s + 1$. Now let $P_\infty, P_1, \ldots, P_n$ be $n + 1$ distinct rational places of F. Then, by what we have just noted, $(s+1)P_\infty - (s+1)P_i$ is a principal divisor of F for $1 \le i \le n$, and so for suitable $h_i \in F$ we have

$$\mathrm{div}(h_i) = (s+1)P_\infty - (s+1)P_i \qquad \text{for } 1 \le i \le n.$$

Next, we choose a local parameter t at P_∞ and consider the local expansions of the h_i, $1 \le i \le n$, at P_∞ which are of the form

$$h_i = t^s \sum_{r=1}^{\infty} e_{r,i} t^r, \tag{6.2}$$

where all coefficients $e_{r,i} \in \mathbf{F}_q$. Finally, we select an integer m with $1 \le m < n$ and define the $m \times n$ matrix H over $\mathbf{F}_q$ by

$$H = \begin{pmatrix} e_{1,1} & e_{1,2} & \cdots & e_{1,n} \\ \vdots & \vdots & & \vdots \\ e_{m,1} & e_{m,2} & \cdots & e_{m,n} \end{pmatrix}.$$

We denote by $C(P_\infty, P_1, \ldots, P_n; m)$ the linear code over $\mathbf{F}_q$ of length n with parity-check matrix H. We have the following parameter bounds for this code.

Theorem 6.1.12 *Let $F/\mathbf{F}_q$ with $q = s^2$ be a maximal function field, let $P_\infty, P_1, \ldots, P_n$ be $n + 1$ distinct rational places of F, and let m be an integer with $1 \leq m < n$. Then $C(P_\infty, P_1, \ldots, P_n; m)$ is a linear $[n, k, d]$ code over $\mathbf{F}_q$ with*

$$k \geq n - m, \qquad d \geq \left\lceil \frac{m}{s+1} \right\rceil + 1.$$

Proof. The parity-check matrix H of $C(P_\infty, P_1, \ldots, P_n; m)$ has m rows, and so it is trivial that $k \geq n - m$.

To obtain $d \geq \lceil m/(s+1) \rceil + 1$, we note that by Lemma 6.1.2 it suffices to prove that any $l := \lceil m/(s+1) \rceil$ columns of H are linearly independent. Let $\mathbf{e}_i$ denote the ith column of H and consider l vectors $\mathbf{e}_{i_1}, \mathbf{e}_{i_2}, \ldots, \mathbf{e}_{i_l}$ with $1 \leq i_1 < i_2 < \ldots < i_l \leq n$. Suppose that

$$\sum_{j=1}^{l} b_j \mathbf{e}_{i_j} = 0 \in \mathbf{F}_q^m$$

for some $b_j \in \mathbf{F}_q$, that is,

$$\sum_{j=1}^{l} b_j e_{r,i_j} = 0 \qquad \text{for } 1 \leq r \leq m. \tag{6.3}$$

Now we consider

$$f = \sum_{j=1}^{l} b_j h_{i_j}$$

and its local expansion at P_∞ obtained from (6.2) and (6.3) as

$$\begin{aligned}
f &= \sum_{j=1}^{l} b_j t^s \sum_{r=1}^{\infty} e_{r,i_j} t^r = t^s \sum_{r=1}^{\infty} \left(\sum_{j=1}^{l} b_j e_{r,i_j} \right) t^r \\
&= t^s \sum_{r=m+1}^{\infty} \left(\sum_{j=1}^{l} b_j e_{r,i_j} \right) t^r.
\end{aligned}$$

As a consequence we get

$$\nu_{P_\infty}(f) \geq s + m + 1.$$

By the definition of the h_i we have

$$f \in \mathcal{L}\left((s+1) \sum_{j=1}^{l} P_{i_j} \right).$$

If we had $f \neq 0$, then

$$\deg((f)_\infty) \leq \deg\left((s+1) \sum_{j=1}^{l} P_{i_j} \right) = (s+1)l < s + m + 1,$$

and so

$$s + m + 1 \leq \nu_{P_\infty}(f) \leq \deg((f)_0) = \deg((f)_\infty) < s + m + 1,$$

a contradiction. Hence $f = 0$, and since $h_1, \ldots, h_n$ are linearly independent over $\mathbf{F}_q$, we obtain $b_j = 0$ for $1 \leq j \leq l$. $\square$

If g is the genus of the maximal function field F and $m < (s+1)g/s$, then it follows immediately from Theorem 6.1.12 that the linear code $C(P_\infty, P_1, \ldots, P_n; m)$ satisfies

$$k + d > n + 1 - g.$$

This is better than the bound $k + d \geq n + 1 - g$ for AG codes of the same length noted in Remark 6.1.5. An example in [183] shows that the linear codes $C(P_\infty, P_1, \ldots, P_n; m)$ can be better than AG codes not only in terms of bounds, but also in terms of exact values.

Theorem 6.1.12 suggests that there are other interesting ways of obtaining good linear codes from global function fields besides Goppa's method. We mention also the work of Feng and Rao [23], [24] in this direction. Even more powerful constructions that contain Goppa's construction as a simple special case will be presented in Sections 6.3 and 6.4.

6.2   Beating the Asymptotic Gilbert-Varshamov Bound

In a practical engineering context, the problem of constructing good linear codes often poses itself in the following way: for a given q and a given **relative minimum distance** d/n (which may be suggested by physical properties of the communication channel), find linear $[n, k, d]$ codes over $\mathbf{F}_q$ for which the **information rate** k/n is as large as possible. The asymptotic theory of linear codes treats this problem for sufficiently long linear codes. The basic object of study in this theory is the following set of ordered pairs of asymptotic relative minimum distances and information rates. For a given prime power q, let U_q be the set of points (δ, R) in the unit square $[0, 1]^2$ for which there exists a sequence of linear $[n_i, k_i, d_i]$ codes over $\mathbf{F}_q$ with $i = 1, 2, \ldots$ such that $n_i \to \infty$ as $i \to \infty$ and

$$\delta = \lim_{i \to \infty} \frac{d_i}{n_i}, \qquad R = \lim_{i \to \infty} \frac{k_i}{n_i}.$$

In view of the problem posed above, it is reasonable to introduce the following function.

Definition 6.2.1 For a given prime power q put

$$\alpha_q(\delta) = \sup\{R \in [0, 1] : (\delta, R) \in U_q\} \qquad \text{for } 0 \leq \delta \leq 1.$$

It was shown by Manin [78] using standard coding-theoretic arguments that α_q is a nonincreasing continuous function on $[0, 1]$ and that

$$U_q = \{(\delta, R) \in [0, 1]^2 : 0 \leq R \leq \alpha_q(\delta)\}.$$

Thus, U_q is completely determined by the function α_q. However, the function α_q is not known explicitly, and it may in fact be a very complicated function. Therefore, at present we have to be satisfied with partial results and bounds on α_q. We note the trivial result $\alpha_q(0) = 1$.

Bounds for linear codes often imply bounds on α_q. For instance, the Singleton bound in Proposition 6.1.1 immediately yields

$$\alpha_q(\delta) \leq 1 - \delta \qquad \text{for } 0 \leq \delta \leq 1.$$

A better result is obtained from the Plotkin bound.

Proposition 6.2.2 (Plotkin Bound) *For any linear $[n, k, d]$ code over $\mathbf{F}_q$ we have*

$$d \leq \frac{nq^{k-1}(q-1)}{q^k - 1}.$$

Proof. Let C be a given linear $[n, k, d]$ code over $\mathbf{F}_q$. For $j = 1, \ldots, n$ consider the $\mathbf{F}_q$-linear map

$$\pi_j : (c_1, \ldots, c_n) \in C \mapsto c_j \in \mathbf{F}_q.$$

We have either $\text{im}(\pi_j) = \{0\}$ or $|\ker(\pi_j)| = q^{k-1}$. By counting nonzero entries along the coordinates, we then get

$$\sum_{\mathbf{c} \in C} w(\mathbf{c}) \leq nq^{k-1}(q-1).$$

On the other hand, it is trivial that

$$\sum_{\mathbf{c} \in C} w(\mathbf{c}) \geq (q^k - 1)d,$$

and so the desired result follows. □

Corollary 6.2.3 *For any prime power q we have*

$$\alpha_q(\delta) = 0 \qquad for \ \frac{q-1}{q} \leq \delta \leq 1.$$

Proof. If we assume $(\delta, R) \in U_q$ with $R > 0$, then Proposition 6.2.2 implies $\delta \leq (q-1)/q$. Thus $\alpha_q(\delta) = 0$ for $(q-1)/q < \delta \leq 1$. The continuity of α_q shows $\alpha_q((q-1)/q) = 0$. □

By combining Proposition 6.2.2 with an argument of Manin [78] (see also [166, Chapter 5]), we obtain

$$\alpha_q(\delta) \leq 1 - \frac{q}{q-1}\delta \qquad \text{for } 0 \leq \delta < \frac{q-1}{q}.$$

We refer to [77, Section 17.7] and [166, Chapter 5] for further upper bounds on α_q.

Lower bounds on α_q are of great practical relevance since they can be viewed as existence theorems for good linear codes over $\mathbf{F}_q$ of large length. Note that because of $\alpha_q(0) = 1$ and Corollary 6.2.3, it suffices to consider the function α_q on the open interval $(0, (q-1)/q)$. The following standard result from coding theory yields an important lower bound on α_q.

Proposition 6.2.4 (Gilbert-Varshamov Bound) *There exists a linear $[n, k]$ code over $\mathbf{F}_q$ with minimum distance at least d, provided that*

$$q^{n-k} > \sum_{i=0}^{d-2} \binom{n-1}{i}(q-1)^i.$$

Proof. We construct a suitable $(n - k) \times n$ parity-check matrix H over $\mathbf{F}_q$. We choose the first column of H as any nonzero vector from $\mathbf{F}_q^{n-k}$. The second column is any vector from $\mathbf{F}_q^{n-k}$ that is not a scalar multiple of the first column. In general, suppose $j - 1$ columns (with $j \leq n$) have been chosen so that any $d - 1$ of them are linearly independent. There are at most

$$\sum_{i=0}^{d-2} \binom{j-1}{i}(q-1)^i \leq \sum_{i=0}^{d-2} \binom{n-1}{i}(q-1)^i$$

vectors obtained by linear combinations of $d - 2$ or fewer of these $j - 1$ columns. If the inequality of the proposition holds, then it will be possible to choose a jth column that is linearly independent of any $d - 2$ of the first $j - 1$ columns. The null space of H is a linear code over $\mathbf{F}_q$ of length n, of minimum distance at least d by Lemma 6.1.2, and of dimension at least k. By turning to a k-dimensional subspace we get a linear code of the desired type. $\square$

Let $\log_q$ denote the logarithm to the base q. Then for $0 \leq \delta < 1$ we define the q-ary **entropy function** H_q by $H_q(0) = 0$ and

$$H_q(\delta) = \delta \log_q(q-1) - \delta \log_q \delta - (1-\delta) \log_q(1-\delta) \qquad \text{for } 0 < \delta < 1.$$

Note that $H_q(\delta)$ increases from 0 to 1 as δ runs from 0 to $(q-1)/q$.

Corollary 6.2.5 (Asymptotic Gilbert-Varshamov Bound) *For any prime power q we have*

$$\alpha_q(\delta) \geq R_{GV}(q, \delta) := 1 - H_q(\delta) \qquad \text{for } 0 \leq \delta \leq \frac{q-1}{q}.$$

Proof. Apply Proposition 6.2.4 with $d = \lfloor \delta n \rfloor + 2$ and note that

$$\sum_{i=0}^{\lfloor \delta n \rfloor} \binom{n-1}{i}(q-1)^i \leq (\lfloor \delta n \rfloor + 1)\binom{n}{\lfloor \delta n \rfloor}(q-1)^{\lfloor \delta n \rfloor}$$

since the largest term of the sum on the left-hand side is the last one. The rest follows by straightforward arguments involving Stirling's formula. $\square$

The asymptotic Gilbert-Varshamov bound has been known since the 1950s and for a long time there was no improvement on this bound. It was believed by some that this bound may even be best possible. Thus, it came as a major breakthrough when, shortly after Goppa's invention of AG codes, it was demonstrated that there are sequences of AG codes that beat the asymptotic Gilbert-Varshamov bound for certain sufficiently large q and certain ranges of the parameter δ.

The crucial fact is the following lower bound on α_q which can be derived from the theory of AG codes and involves the quantity

$$A(q) = \limsup_{g \to \infty} \frac{N_q(g)}{g}$$

from the asymptotic theory of the number of rational places of global function fields over $\mathbf{F}_q$ (see Chapter 5).

Theorem 6.2.6 *For any prime power q we have*

$$\alpha_q(\delta) \geq R_{AG}(q, \delta) := 1 - \frac{1}{A(q)} - \delta \qquad for \ 0 \leq \delta \leq 1.$$

Proof. We can assume that $A(q) > 1$ and $0 \leq \delta \leq 1 - A(q)^{-1}$, for otherwise the result is trivial. Let $F_1, F_2, \ldots$ be a sequence of global function fields over $\mathbf{F}_q$ such that $g_i := g(F_i)$ and $n_i := N(F_i)$ satisfy

$$\lim_{i \to \infty} g_i = \infty \quad and \quad \lim_{i \to \infty} \frac{n_i}{g_i} = A(q).$$

For sufficiently large i we can choose integers r_i with $g_i \leq r_i < n_i$ and

$$\lim_{i \to \infty} \frac{r_i}{n_i} = 1 - \delta.$$

Then Theorem 6.1.4 yields a sequence of linear $[n_i, k_i, d_i]$ codes over $\mathbf{F}_q$ with

$$k_i \geq r_i - g_i + 1 \quad and \quad d_i \geq n_i - r_i.$$

By passing, if necessary, to a subsequence, we can assume that the limits

$$R := \lim_{i \to \infty} \frac{k_i}{n_i} \quad and \quad \delta' := \lim_{i \to \infty} \frac{d_i}{n_i}$$

exist. It follows that

$$R \geq 1 - \delta - \frac{1}{A(q)} \quad and \quad \delta' \geq \delta.$$

Furthermore $(\delta', R) \in U_q$, and so

$$\alpha_q(\delta) \geq \alpha_q(\delta') \geq R \geq 1 - \frac{1}{A(q)} - \delta$$

since α_q is nonincreasing. $\qquad\qquad\qquad\qquad\qquad\qquad\qquad\qquad\qquad\qquad\qquad\qquad$ □

If q is a square, then we know from Section 5.4 that $A(q) = q^{1/2} - 1$. It is now a matter of comparing the two explicitly given functions $R_{AG}(q, \delta)$ and $R_{GV}(q, \delta)$ on the interval $[0, (q-1)/q]$ to determine whether AG codes can yield an improvement on the asymptotic Gilbert-Varshamov bound. For the following theorem, the weaker result $A(q) \geq q^{1/2} - 1$ established by Ihara [56] and Tsfasman, Vlăduţ, and Zink [157] is sufficient.

Theorem 6.2.7 *Let $q \geq 49$ be the square of a prime power. Then there exists an open interval $(\delta_1, \delta_2) \subseteq [0, (q-1)/q]$ containing $(q-1)/(2q-1)$ such that*

$$R_{AG}(q, \delta) > R_{GV}(q, \delta) \qquad for \ all \ \delta \in (\delta_1, \delta_2).$$

Proof. It suffices to prove the inequality for $\delta_0 = (q-1)/(2q-1)$. Note that

$$H_q(\delta_0) - \delta_0 = \log_q\left(2 - \frac{1}{q}\right).$$

Thus, it is enough to show that

$$\log_q\left(2 - \frac{1}{q}\right) > \frac{1}{q^{1/2} - 1},$$

and it is easily checked that this holds if and only if $q \geq 49$. $\qquad\square$

For a given square $q \geq 49$ it is an exercise in elementary calculus to find the precise interval on which $R_{AG}(q, \delta)$ exceeds $R_{GV}(q, \delta)$. We recall from Example 5.4.1 that for any square q there exists the Garcia-Stichtenoth tower $K_1 \subseteq K_2 \subseteq \ldots$ of function fields over $\mathbf{F}_q$ for which

$$\lim_{i \to \infty} \frac{N(K_i)}{g(K_i)} = A(q) = q^{1/2} - 1.$$

Since the construction of this tower is explicit, there is the potential of getting excellent long AG codes from this tower in a concrete fashion. A problem that has to be overcome is that of the explicit determination of bases of Riemann-Roch spaces in the function fields K_i. First steps towards this goal were made by Aleshnikov *et al.* [2] and Voss and Høholdt [169].

For composite nonsquares q we can use the lower bounds on $A(q)$ in Corollaries 5.3.2 and 5.3.4 to obtain an analog of Theorem 6.2.7. Let $q = p^h$ with an integer $h > 1$ and a prime p, and assume that h is composite if $p = 2$. Let l be the least prime dividing h. Then for $0 \leq \delta \leq 1$ we define the functions

$$R_{NX}(q, \delta) = 1 - \frac{\lceil 2(2q^{1/l} + 3)^{1/2}\rceil + 1}{2q^{1/l} + 2} - \delta \qquad \text{if } q \text{ is odd,}$$

$$R_{NX}(q, \delta) = 1 - \frac{\lceil 2(2q^{1/l} + 2)^{1/2}\rceil + 2}{q^{1/l} + 1} - \delta \qquad \text{if } q \text{ is even.}$$

Theorem 6.2.6 yields

$$\alpha_q(\delta) \geq R_{NX}(q, \delta) \quad \text{for } 0 \leq \delta \leq 1.$$

The following improvement on the asymptotic Gilbert-Varshamov bound was shown by Niederreiter and Xing [112].

Theorem 6.2.8 *Let $m \geq 3$ be an odd integer and let r be a prime power with $r \geq 100m^3$ for odd r and $r \geq 576m^3$ for even r. Then for $q = r^m$ there exists an open interval $(\delta_1, \delta_2) \subseteq [0, (q-1)/q]$ containing $(q-1)/(2q-1)$ such that*

$$R_{NX}(q, \delta) > R_{GV}(q, \delta) \qquad \text{for all } \delta \in (\delta_1, \delta_2).$$

Proof. As in the proof of Theorem 6.2.7 we see that for odd r it is enough to show that

$$\log_q\left(2 - \frac{1}{q}\right) \geq \frac{(2r + 3)^{1/2} + 1}{r + 1},$$

or equivalently,

$$\frac{r+1}{m} \log\left(2 - \frac{1}{q}\right) \geq ((2r+3)^{1/2} + 1)\log r. \tag{6.4}$$

We have

$$\log t \leq \frac{\log c}{c^{1/2}} t^{1/2} \qquad \text{for } t \geq c \geq e^2.$$

Hence for $r \geq 100m^3$ and $m \geq 3$ we get

$$
\begin{aligned}
\log r &\leq \frac{\log(100m^3)}{10m^{3/2}} r^{1/2} \leq \frac{4.61 + 3M(m)m^{1/2}}{10m^{3/2}} r^{1/2} \\
&\leq \frac{4.61 + 3M(m)m^{1/2}}{10m^{3/2}}(0.72)((2r+3)^{1/2} - 1)
\end{aligned}
$$

with $M(m) = m^{-1/2}\log m$ for $m = 3, 5, 7$ and $M(m) = 2e^{-1}$ for $m \geq 9$. It follows for $m \geq 9$ that

$$
\begin{aligned}
((2r+3)^{1/2} + 1)\log r &\leq \frac{0.77m^{1/2} + 3e^{-1}m^{1/2}}{5m^{3/2}}(1.44)(r+1) \\
&\leq 0.6\frac{r+1}{m} \leq \frac{r+1}{m}\log\left(2 - \frac{1}{q}\right).
\end{aligned}
$$

The last bound is easily checked for $m = 3, 5, 7$, and so (6.4) is established.

For even r it suffices to show that

$$\log_q\left(2 - \frac{1}{q}\right) \geq \frac{2(2r+2)^{1/2} + 3}{r+1},$$

or equivalently,

$$\frac{r+1}{m}\log\left(2 - \frac{1}{q}\right) \geq (2(2r+2)^{1/2} + 3)\log r. \tag{6.5}$$

For $r \geq 576m^3$ and $m \geq 3$ we obtain

$$
\begin{aligned}
\log r &\leq \frac{\log(576m^3)}{24m^{3/2}} r^{1/2} \leq \frac{6.36 + 3M(m)m^{1/2}}{24m^{3/2}} r^{1/2} \\
&\leq \frac{2.12 + M(m)m^{1/2}}{8m^{3/2}}(0.36)(2(2r+2)^{1/2} - 3).
\end{aligned}
$$

It follows for $m \geq 9$ that

$$
\begin{aligned}
(2(2r+2)^{1/2} + 3)\log r &\leq \frac{0.71m^{1/2} + 2e^{-1}m^{1/2}}{8m^{3/2}}(2.88)r \\
&\leq 0.6\frac{r}{m} \leq \frac{r}{m}\log\left(2 - \frac{1}{q}\right).
\end{aligned}
$$

The last bound is easily checked for $m = 3, 5, 7$, and so (6.5) is established. □

Remark 6.2.9 For small m the condition on r in Theorem 6.2.8 can be weakened by using explicit computations. For $m = 3$ we can take $r > 2294$ if r is odd and $r > 13822$ if r is even. For $m = 5$ we can take $r > 8666$ if r is odd and $r > 48999$ if r is even.

A recent paper of Xing [175] contains an improved version of Theorem 6.2.6 and, as a consequence, a further improvement on the asymptotic Gilbert-Varshamov bound. A key step is to show that in certain cases and for special choices of the divisor G, it is possible to have better parameter bounds for AG codes than those in Theorem 6.1.4.

6.3 NXL Codes

We have seen in the previous section that Goppa's construction of AG codes yields excellent linear codes over $\mathbf{F}_q$ in an asymptotic sense, provided that q is large enough. What happens for small q? Return to Theorem 6.1.4 and note that the genus g of the chosen global function field $F/\mathbf{F}_q$ must be smaller than the number n of the chosen rational places of F (and n is the length of the resulting linear code). Consequently, we must have $g < N(F)$, and so

$$N_q(g) > g.$$

If, for fixed q, the construction should work for infinitely many values of n, then infinitely many values of g must be allowed, and so it follows that necessarily $A(q) \geq 1$. However, for $q = 2$ and $q = 3$ this lower bound on $A(q)$ is impossible in view of Theorem 1.6.21. If we consider, for instance, the case $q = 2$, then Table 4.5.1 for $q = 2$ suggests that $N_2(g) \leq g$ for all $g \geq 21$, and this can indeed be proved by the Serre-Oesterlé method: use the result of Example 1.6.19 for $g \geq 32$ and Table 4.5.1 for $21 \leq g \leq 31$. This has the consequence that Goppa's construction can be applied only for $g \leq 20$. But even in this range, the designed distance $n - \deg(G)$ of the AG code, being bounded above by $N_2(g) - g$, can be at most 4. The upshot of this argument is that Goppa's construction is practically useless for the most important case $q = 2$.

The trouble with Goppa's construction of linear codes over $\mathbf{F}_q$ is that it uses only the rational places of a global function field over $\mathbf{F}_q$, and there are just too few of those relative to the genus for small q. The crucial step of devising constructions that employ places of arbitrary degree was taken by Niederreiter, Xing, and Lam [121].

The general principle of the method in [121] is quite simple. Let $F/\mathbf{F}_q$ be a global function field, let G_1 and G_2 be two divisors of F with $G_1 \leq G_2$, and consider the corresponding Riemann-Roch spaces $\mathcal{L}(G_1)$ and $\mathcal{L}(G_2)$. Note that $\mathcal{L}(G_1)$ is a linear subspace of the vector space $\mathcal{L}(G_2)$ over $\mathbf{F}_q$. Thus, if we choose an ordered basis of $\mathcal{L}(G_2)$, then the coordinate vectors of the elements of $\mathcal{L}(G_1)$ form a linear code over $\mathbf{F}_q$ of length $n = \ell(G_2)$ and dimension $k = \ell(G_1)$, provided that $\ell(G_1) \geq 1$. We call such a linear code an **NXL code**. In the family of NXL codes we include also the linear codes that are obtained from the above linear codes by simple manipulations such as puncturing.

As is to be expected, some care has to be taken in the construction of NXL codes in order to obtain good codes. First of all, a lot depends on the choice of the ordered basis of $\mathcal{L}(G_2)$. For instance, if we construct this ordered basis by starting from an ordered basis of the linear subspace $\mathcal{L}(G_1)$ and extending it to an ordered basis of $\mathcal{L}(G_2)$, then we get an uninteresting linear code of minimum distance 1. Clearly, the choice of the divisors G_1 and G_2 will also be decisive.

We now describe a special family of NXL codes that is slightly more general than that in [121] (see also [123]). This construction uses the following ingredients:

$F/\mathbf{F}_q$: a global function field over $\mathbf{F}_q$ of genus g;

$P_1, \ldots, P_r$: distinct places of F;

D: a nonspecial divisor of F;

E: a positive divisor of F with $\mathrm{supp}(E) \cap \{P_1, \ldots, P_r\} = \emptyset$ and

$$1 \le \deg(E - D) \le \sum_{i=1}^{r} \deg(P_i) - g.$$

We introduce the notation

$$s_i = \deg(P_i) \quad \text{for } 1 \le i \le r, \quad n = \sum_{i=1}^{r} s_i.$$

We emphasize that the degrees $s_1, \ldots, s_r$ of the chosen places of F can be completely arbitrary, except for the condition $n > g$. We observe now that

$$\ell(D + P_i) = \ell(D) + s_i \quad \text{for } 1 \le i \le r$$

since D is nonspecial (see Corollary 1.1.4). For each $i = 1, \ldots, r$ we choose a basis

$$\{f_{i,j} + \mathcal{L}(D) : 1 \le j \le s_i\}$$

of the factor space $\mathcal{L}(D+P_i)/\mathcal{L}(D)$. The n-dimensional factor space $\mathcal{L}(D+\sum_{i=1}^r P_i)/\mathcal{L}(D)$ has then the basis

$$\{f_{i,j} + \mathcal{L}(D) : 1 \le j \le s_i, 1 \le i \le r\}$$

which we order in a lexicographic manner. The latter basis property is proved by observing that if $\sum_{i=1}^r h_i \in \mathcal{L}(D)$ with $h_i \in \mathcal{L}(D + P_i)$ for $1 \le i \le r$ and some $h_j \notin \mathcal{L}(D)$, then

$$\nu_{P_j}(h_j) = -\nu_{P_j}(D) - 1,$$

but also

$$h_j \in \mathcal{L}\left(D + \sum_{\substack{i=1 \\ i \ne j}}^{r} P_i\right),$$

which implies

$$\nu_{P_j}(h_j) \ge -\nu_{P_j}(D),$$

a contradiction.

Now a linear code is constructed as follows. Every

$$f \in \mathcal{L}\left(D + \sum_{i=1}^{r} P_i - E\right) \subseteq \mathcal{L}\left(D + \sum_{i=1}^{r} P_i\right)$$

has a unique representation

$$f = \sum_{i=1}^{r} \sum_{j=1}^{s_i} c_{i,j} f_{i,j} + u \tag{6.6}$$

with all $c_{i,j} \in \mathbf{F}_q$ and $u \in \mathcal{L}(D)$. Define the $\mathbf{F}_q$-linear map

$$\eta : f \in \mathcal{L}\left(D + \sum_{i=1}^{r} P_i - E\right) \mapsto (c_{1,1}, \ldots, c_{1,s_1}, \ldots, c_{r,1}, \ldots, c_{r,s_r}) \in \mathbf{F}_q^n.$$

The image of η is the linear code $C(P_1, \ldots, P_r; D, E)$ over $\mathbf{F}_q$ of length n.

The linear code $C(P_1, \ldots, P_r; D, E)$ belongs to the family of NXL codes since it is obtained from the general construction principle of these codes by putting

$$G_1 = D + \sum_{i=1}^{r} P_i - E, \quad G_2 = D + \sum_{i=1}^{r} P_i,$$

and puncturing so that only the n coordinates corresponding to the $f_{i,j}$ are kept. The following theorem provides bounds for the dimension and the minimum distance of the linear code $C(P_1, \ldots, P_r; D, E)$.

Theorem 6.3.1 *Let $F/\mathbf{F}_q$ be a global function field of genus g, let $P_1, \ldots, P_r$ be distinct places of F with degrees $s_1, \ldots, s_r$, respectively, and let D be a nonspecial divisor of F. Furthermore, let E be a positive divisor of F with $\operatorname{supp}(E) \cap \{P_1, \ldots, P_r\} = \emptyset$ such that $m := \deg(E - D)$ satisfies*

$$1 \le m \le \sum_{i=1}^{r} s_i - g.$$

Then $C(P_1, \ldots, P_r; D, E)$ is a linear $[n, k, d]$ code over $\mathbf{F}_q$ with

$$n = \sum_{i=1}^{r} s_i, \quad k = \ell\left(D + \sum_{i=1}^{r} P_i - E\right) \ge n - m - g + 1, \quad d \ge d_0,$$

where d_0 is the least cardinality of a subset R of $\{1, \ldots, r\}$ for which $\sum_{i \in R} s_i \ge m$. Moreover, we have $k = n - m - g + 1$ if $n - m \ge 2g - 1$.

Proof. For any nonzero $f \in \mathcal{L}(D + \sum_{i=1}^{r} P_i - E)$ we consider the weight $w(\eta(f))$. We have the unique representation (6.6) for f. Define

$$R = \{1 \le i \le r : \exists j, 1 \le j \le s_i, \text{ with } c_{i,j} \ne 0\}$$

and note that $|R| \le w(\eta(f))$. We can then write

$$f = \sum_{i \in R} h_i + u$$

with $h_i \in \mathcal{L}(D + P_i)$ for $i \in R$. Therefore

$$f \in \mathcal{L}(D + \sum_{i \in R} P_i).$$

But we also have $f \in \mathcal{L}(D + \sum_{i=1}^{r} P_i - E)$ and $\operatorname{supp}(E) \cap \{P_1, \ldots, P_r\} = \emptyset$, hence

$$f \in \mathcal{L}\left(D + \sum_{i \in R} P_i - E\right).$$

Since $f \neq 0$, this implies

$$\deg\left(D + \sum_{i \in R} P_i - E\right) \geq 0,$$

that is, $\sum_{i \in R} s_i \geq m$. It follows that

$$w(\eta(f)) \geq |R| \geq d_0 > 0.$$

This shows not only the desired lower bound on d, but also that η is injective. Hence

$$k = \ell\left(D + \sum_{i=1}^{r} P_i - E\right),$$

and the rest follows from the Riemann-Roch theorem. $\square$

Remark 6.3.2 It was observed by Özbudak and Stichtenoth [123] that any AG code $C(P_1, \ldots, P_n; G)$ can be represented as an NXL code $C(P_1, \ldots, P_n; D, E)$ with suitable divisors D and E, where $P_1, \ldots, P_n$ are distinct rational places and $\mathrm{supp}(G) \cap \{P_1, \ldots, P_n\} = \emptyset$, as usual in the theory of AG codes. Indeed, choose $D \geq G$ of sufficiently large degree such that D and $D - G$ are nonspecial. In combination with [152, Theorem I.5.4] and the approximation theorem, we can then find $z \in \mathcal{L}(D + \sum_{i=1}^{n} P_i - G)$ with

$$\nu_{P_i}(z) = -\nu_{P_i}(D) - 1 \qquad \text{for } 1 \leq i \leq n.$$

Put

$$E = D + \sum_{i=1}^{n} P_i - G + \mathrm{div}(z),$$

then $E \geq 0$ and $\mathrm{supp}(E) \cap \{P_1, \ldots, P_n\} = \emptyset$. We choose $f_i \in \mathcal{L}(D + P_i) \backslash \mathcal{L}(D)$ for $1 \leq i \leq n$, then

$$\nu_{P_i}(f_i) = -\nu_{P_i}(D) - 1 \qquad \text{for } 1 \leq i \leq n$$

and $\{f_1 + \mathcal{L}(D), \ldots, f_n + \mathcal{L}(D)\}$ is an ordered basis of $\mathcal{L}(D + \sum_{i=1}^{n} P_i)/\mathcal{L}(D)$. By multiplying $f_1, \ldots, f_n$ by suitable elements of $\mathbf{F}_q^*$, we can assume that

$$(f_i z^{-1})(P_i) = 1 \qquad \text{for } 1 \leq i \leq n.$$

This completes the construction of the NXL code $C(P_1, \ldots, P_n; D, E)$. Note that

$$\mathcal{L}\left(D + \sum_{i=1}^{n} P_i - E\right) = \mathcal{L}(G - \mathrm{div}(z)) = \{hz : h \in \mathcal{L}(G)\},$$

and so we can define the $\mathbf{F}_q$-linear isomorphism

$$\varphi : f \in \mathcal{L}\left(D + \sum_{i=1}^{n} P_i - E\right) \mapsto f z^{-1} \in \mathcal{L}(G).$$

It is then easy to check that if $\eta : \mathcal{L}(D + \sum_{i=1}^{n} P_i - E) \longrightarrow \mathbf{F}_q^n$ and $\psi : \mathcal{L}(G) \longrightarrow \mathbf{F}_q^n$ are the $\mathbf{F}_q$-linear maps in the constructions of $C(P_1, \ldots, P_n; D, E)$ and $C(P_1, \ldots, P_n; G)$, respectively, then $\eta = \psi \circ \varphi$. Thus,

$$C(P_1, \ldots, P_n; G) = \mathrm{im}(\psi) = \mathrm{im}(\eta) = C(P_1, \ldots, P_n; D, E).$$

Remark 6.3.3 The construction of the NXL codes $C(P_1, \ldots, P_r; D, E)$ can be generalized in an obvious manner to the case where $P_1, \ldots, P_r$ are replaced by positive divisors $D_1, \ldots, D_r$ of F whose supports are pairwise disjoint and for which $s_i := \deg(D_i) \geq 1$ for $1 \leq i \leq r$. The condition on E now says that E is a positive divisor of F with $\operatorname{supp}(E) \cap \operatorname{supp}(D_i) = \emptyset$ for $1 \leq i \leq r$. Then the conclusions of Theorem 6.3.1 hold again, only the interpretation of the s_i has to be changed.

Example 6.3.4 Let $q = 2$ and let F be the rational function field over $\mathbf{F}_2$. Put $r = 6$ and for $P_1, \ldots, P_6$ choose three rational places, a place of degree 2, and two places of degree 3 of F. Let D be the zero divisor and E be a place of degree 7 of F. Then Theorem 6.3.1 shows that $C(P_1, \ldots, P_6; D, E)$ is a linear $[n, k, d]$ code over $\mathbf{F}_2$ with $n = 11, k = 5$, and $d \geq 3$, hence $k + d \geq 8$. By comparison, the best lower bound on $k + d$ for a linear code over $\mathbf{F}_2$ of length 11 from Theorem 6.1.4, i.e., for an AG code, is obtained by taking $g = 8$ (compare with Table 4.5.1), and then $k + d \geq 4$.

Example 6.3.5 Let $q = 3$ and let F be the rational function field over $\mathbf{F}_3$. Put $r = 13$ and for $P_1, \ldots, P_{13}$ choose four rational places, three places of degree 2, and six places of degree 3 of F. Let D be the zero divisor and E be a place of degree 7 of F. Then Theorem 6.3.1 shows that $C(P_1, \ldots, P_{13}; D, E)$ is a linear $[n, k, d]$ code over $\mathbf{F}_3$ with $n = 28, k = 22$, and $d \geq 3$, hence $k + d \geq 25$. By comparison, the best lower bound on $k + d$ for a linear code over $\mathbf{F}_3$ of length 28 from Theorem 6.1.4, i.e., for an AG code, is obtained by taking $g = 15$ (compare with Table 4.5.1), and then $k + d \geq 14$.

6.4   XNL Codes

We have argued in the previous section that using only rational places of global function fields for the construction of linear codes is too restrictive. The family of NXL codes overcomes this restriction. Another method of constructing linear codes from places of arbitrary degree of global function fields was introduced by Xing, Niederreiter, and Lam [184]. This method provides a considerable generalization of Goppa's construction of AG codes.

The idea is to use not only data from a global function field, but also (short) linear codes as inputs, and the output will then be a longer linear code. The construction uses the following ingredients:

$F/\mathbf{F}_q$: a global function field over $\mathbf{F}_q$ of genus g;

$P_1, \ldots, P_r$: distinct places of F;

G: a divisor of F with
$$\operatorname{supp}(G) \cap \{P_1, \ldots, P_r\} = \emptyset;$$

C_i: linear $[n_i, k_i \geq \deg(P_i), d_i]$ codes over $\mathbf{F}_q$ for $1 \leq i \leq r$;

ϕ_i: fixed $\mathbf{F}_q$-linear monomorphisms from the residue class field of P_i to the linear code C_i for $1 \leq i \leq r$.

We emphasize that for each $i = 1, \ldots, r$ the dimension k_i of the linear code C_i must be greater than or equal to the degree of the place P_i. This condition guarantees that there exists an $\mathbf{F}_q$-linear monomorphism ϕ_i mapping the residue class field of P_i into C_i. We put

$$n = \sum_{i=1}^{r} n_i,$$

where we repeat that n_i is the length of the linear code C_i. Now we define the $\mathbf{F}_q$-linear map

$$\beta : f \in \mathcal{L}(G) \mapsto (\phi_1(f(P_1)), \ldots, \phi_r(f(P_r))) \in \mathbf{F}_q^n,$$

where on the right-hand side we use concatenation of vectors. The image of β is the linear code $C(P_1, \ldots, P_r; G; C_1, \ldots, C_r)$ over $\mathbf{F}_q$ of length n. Such a linear code is called an **XNL code**.

The following result of Xing, Niederreiter, and Lam [184] gives bounds for the dimension and the minimum distance of XNL codes, where the bound for the minimum distance is stated in the improved form pointed out later by Özbudak and Stichtenoth [123].

Theorem 6.4.1 Let $F/\mathbf{F}_q$ be a global function field of genus g, let $P_1, \ldots, P_r$ be distinct places of F, and let $C_i, 1 \le i \le r$, be linear $[n_i, k_i, d_i]$ codes over $\mathbf{F}_q$ with $k_i \ge \deg(P_i)$. Furthermore, let G be a divisor of F with $\mathrm{supp}(G) \cap \{P_1, \ldots, P_r\} = \emptyset$ and

$$g \le \deg(G) < \sum_{i=1}^{r} k_i.$$

Then $C(P_1, \ldots, P_r; G; C_1, \ldots, C_r)$ is a linear $[n, k, d]$ code over $\mathbf{F}_q$ with

$$n = \sum_{i=1}^{r} n_i, \qquad k = \ell(G) \ge \deg(G) - g + 1, \qquad d \ge d_0,$$

where d_0 is the minimum of $\sum_{i \in M'} d_i$ taken over all subsets M of $\{1, \ldots, r\}$ for which $\sum_{i \in M} k_i \le \deg(G)$, with M' denoting the complement of M in $\{1, \ldots, r\}$. Moreover, we have $k = \deg(G) - g + 1$ if $\deg(G) \ge 2g - 1$.

Proof. For any nonzero $f \in \mathcal{L}(G)$ we consider the weight $w(\beta(f))$. Define

$$M = \{1 \le i \le r : f(P_i) = 0\}.$$

Then we get

$$w(\beta(f)) = \sum_{i \in M'} w(\phi_i(f(P_i))) \ge \sum_{i \in M'} d_i.$$

Furthermore, we have

$$f \in \mathcal{L}\left(G - \sum_{i \in M} P_i\right).$$

Since $f \ne 0$, this implies

$$\deg\left(G - \sum_{i \in M} P_i\right) \ge 0,$$

that is, $\sum_{i \in M} k_i \leq \deg(G)$. Hence $w(\beta(f)) \geq d_0 > 0$. This shows not only the desired lower bound on d, but also that β is injective. Consequently $k = \ell(G)$, and the rest follows from the Riemann-Roch theorem. $\square$

Corollary 6.4.2 *In the special case where $k_i \geq d_i$ for $1 \leq i \leq r$, the minimum distance d of the linear code $C(P_1, \ldots, P_r; G; C_1, \ldots, C_r)$ satisfies*

$$d \geq \sum_{i=1}^{r} d_i - \deg(G).$$

Proof. Proceed as in the proof of Theorem 6.4.1 and note that if $\sum_{i \in M} k_i \leq \deg(G)$, then

$$\sum_{i \in M'} d_i = \sum_{i=1}^{r} d_i - \sum_{i \in M} d_i \geq \sum_{i=1}^{r} d_i - \sum_{i \in M} k_i \geq \sum_{i=1}^{r} d_i - \deg(G).$$

This implies $d_0 \geq \sum_{i=1}^{r} d_i - \deg(G)$. $\square$

Remark 6.4.3 If for each $i = 1, \ldots, r$ we choose C_i to be the trivial linear $[1, 1, 1]$ code over $\mathbf{F}_q$ and $P_1, \ldots, P_r$ are distinct rational places of F, then the construction of XNL codes reduces to Goppa's construction of AG codes. Theorem 6.1.4 is then a special case of Theorem 6.4.1. Note that by the argument in Remark 6.1.6, the condition on $\mathrm{supp}(G)$ in Theorem 6.4.1 may be dropped.

Remark 6.4.4 It was pointed out by Özbudak and Stichtenoth [123] that the linear codes $C(P_1, \ldots, P_r; D, E)$ constructed in Section 6.3 can also be viewed as special XNL codes. We use the same notation as in the construction in Section 6.3. By the approximation theorem there exists an element $z \in F$ with

$$\nu_{P_i}(z) = \nu_{P_i}(D) + 1 \qquad \text{for } 1 \leq i \leq r.$$

Note that $f_{i,j} \in \mathcal{L}(D + P_i) \backslash \mathcal{L}(D)$, and so

$$\nu_{P_i}(f_{i,j}z) = 0 \qquad \text{for } 1 \leq j \leq s_i, 1 \leq i \leq r.$$

For $1 \leq i \leq r$ the residue classes

$$(f_{i,j}z)(P_i), \quad 1 \leq j \leq s_i,$$

form a basis of the residue class field R_i of P_i over $\mathbf{F}_q$. We choose $C_i = \mathbf{F}_q^{s_i}$, i.e., the trivial linear $[s_i, s_i, 1]$ code over $\mathbf{F}_q$. Then an $\mathbf{F}_q$-linear isomorphism $\phi_i : R_i \longrightarrow C_i$ is defined by requiring that $(f_{i,j}z)(P_i)$ be mapped to the jth unit vector $(0, \ldots, 1, \ldots, 0) \in \mathbf{F}_q^{s_i}$ for $1 \leq j \leq s_i$. Put

$$G = D + \sum_{i=1}^{r} P_i - E - \mathrm{div}(z).$$

Then G satisfies the conditions in Theorem 6.4.1. Define the $\mathbf{F}_q$-linear isomorphism

$$\varphi : f \in \mathcal{L}\left(D + \sum_{i=1}^{r} P_i - E\right) \mapsto fz \in \mathcal{L}(G).$$

It is then easy to check that if $\eta : \mathcal{L}(D + \sum_{i=1}^{r} P_i - E) \longrightarrow \mathbf{F}_q^n$ and $\beta : \mathcal{L}(G) \longrightarrow \mathbf{F}_q^n$ are the $\mathbf{F}_q$-linear maps in the constructions of $C(P_1, \ldots, P_r; D, E)$ and $C(P_1, \ldots, P_r; G; C_1, \ldots, C_r)$, respectively, then $\eta = \beta \circ \varphi$. Thus,

$$C(P_1, \ldots, P_r; D, E) = \mathrm{im}(\eta) = \mathrm{im}(\beta) = C(P_1, \ldots, P_r; G; C_1, \ldots, C_r).$$

Furthermore, Theorem 6.3.1 is then a special case of Theorem 6.4.1.

Remark 6.4.5 Özbudak and Stichtenoth [123] pointed out also that *any* linear code over $\mathbf{F}_q$ can be obtained from the construction of XNL codes. Indeed, let C be an arbitrary linear $[n, k]$ code over $\mathbf{F}_q$. Choose F to be the rational function field $\mathbf{F}_q(x)$ and let P_∞ be the pole of x. Put $r = 1$, let $P_1 \neq P_\infty$ be a place of F of degree k, and set $C_1 = C$. With the choice $G = (k - 1)P_\infty$ it is then clear that the XNL code $C(P_1; G; C_1)$ is equal to the given code C.

Although in the sense of Remark 6.4.5 any linear code is an XNL code with $r = 1$, in practice the construction of XNL codes is applied with $r \geq 2$ and short (and often trivial) linear codes $C_1, \ldots, C_r$. This will be illustrated by the examples given below. Several examples of good XNL codes are listed in the paper of Xing, Niederreiter, and Lam [184], and a more systematic search for good XNL codes was carried out by Ding, Niederreiter, and Xing [19]. Many excellent examples were found that improved on previous constructions, and some of these XNL codes are optimal in the sense of the following definition.

Definition 6.4.6 A linear $[n, k, d]$ code over $\mathbf{F}_q$ is **optimal** if any linear $[n, k]$ code over $\mathbf{F}_q$ has minimum distance at most d.

In the following examples in this section we write B_k for the number of places of degree k of the chosen global function field F. We think of all places of F as being listed according to nondecreasing degrees and we always take $P_1, \ldots, P_r$ to be the first r places in this list. Note that for sufficiently large b there exist places P and Q of F of degree b and $b + 1$, respectively, for instance by Lemma 1.6.13. We always choose $G = m(Q - P)$ with a positive integer m, so that $\mathrm{supp}(G) \cap \{P_1, \ldots, P_r\} = \emptyset$ for sufficiently large b and $\deg(G) = m$. In all examples we have $k_i = d_i$ for $1 \leq i \leq r$, so that the lower bound on the minimum distance is obtained from Corollary 6.4.2.

Example 6.4.7 Let $q = 2$ and let $F = \mathbf{F}_2(x, y)$ be the elliptic function field defined by

$$y^2 + y = x + \frac{1}{x}.$$

Then $B_1 = 4, B_2 = 2, B_3 = 0, B_4 = 2$. Take $r = 6, [n_i, k_i, d_i] = [1, 1, 1]$ for $1 \leq i \leq 4$, and $[n_i, k_i, d_i] = [3, 2, 2]$ for $i = 5, 6$. Then we obtain linear $[n, k, d]$ codes over $\mathbf{F}_2$ with parameters

$$n = 10, \quad k = m, \quad d \geq 8 - m \quad \text{for } 1 \leq m \leq 7.$$

The linear codes with $m = 2, 3, 4$ and $d = 8 - m$ are optimal.

Example 6.4.8 Let F be as in Example 6.4.7 and take $r = 7, [n_i, k_i, d_i] = [1, 1, 1]$ for $1 \leq i \leq 4, [n_i, k_i, d_i] = [3, 2, 2]$ for $i = 5, 6$, and $[n_7, k_7, d_7] = [8, 4, 4]$. Then we obtain linear $[n, k, d]$ codes over $\mathbf{F}_2$ with parameters

$$n = 18, \quad k = m, \quad d \geq 12 - m \quad \text{for } 1 \leq m \leq 11.$$

The linear code with $m = 4$ and $d = 8$ is optimal.

Example 6.4.9 Let $q = 2$ and let $F = \mathbf{F}_2(x, y)$ be the elliptic function field defined by

$$y^2 + y = x^3.$$

Then $B_1 = B_2 = 3$. Take $r = 6, [n_i, k_i, d_i] = [1, 1, 1]$ for $1 \leq i \leq 3$, and $[n_i, k_i, d_i] = [3, 2, 2]$ for $4 \leq i \leq 6$. Then we obtain linear $[n, k, d]$ codes over $\mathbf{F}_2$ with parameters

$$n = 12, \quad k = m, \quad d \geq 9 - m \quad \text{for } 1 \leq m \leq 8.$$

The linear codes with $m = 3, 5$ and $d = 9 - m$ are optimal.

Example 6.4.10 Let $q = 3$ and let $F = \mathbf{F}_3(x, y)$ be the elliptic function field defined by

$$y^2 = x(x^2 + x - 1).$$

Then $B_1 = 6$ and $B_2 = 3$. Take $r = 9, [n_i, k_i, d_i] = [1, 1, 1]$ for $1 \leq i \leq 6$, and $[n_i, k_i, d_i] = [3, 2, 2]$ for $7 \leq i \leq 9$. Then we obtain linear $[n, k, d]$ codes over $\mathbf{F}_3$ with parameters

$$n = 15, \quad k = m, \quad d \geq 12 - m \quad \text{for } 1 \leq m \leq 11.$$

The linear code with $m = 3$ and $d = 9$ is optimal.

A construction of linear codes over $\mathbf{F}_q$ using places of degree 2 of a global function field $F/\mathbf{F}_q$ was noted by Xing and Niederreiter [182]. This construction is related to a method of Xing and Ling [177], [178] in which rational places of a global function field $F/\mathbf{F}_{q^2}$ are used to construct linear codes over $\mathbf{F}_q$.

6.5   A Propagation Rule for Linear Codes

By a **propagation rule** for linear codes we mean a theorem or a construction which, from a collection of arbitrary linear codes, produces a linear code with new code parameters. Some classical examples of propagation rules for linear codes can be found in the book of

MacWilliams and Sloane [77, Chapter 18]. The construction of XNL codes in Section 6.4 may also be viewed as a propagation rule which uses short linear codes as inputs and yields as the output a linear code whose length is the sum of the lengths of the input codes.

In this section we describe a propagation rule for linear codes due to Niederreiter and Xing [120]. This propagation rule applies to a single linear code and produces a linear code of larger length and higher dimension. The key idea is to represent the given linear code as a function-field code (see Remark 6.1.7) and to use suitable extensions of the underlying global function field in this representation in order to construct new linear codes. The following lemma, which establishes that any linear code can be represented as a function-field code, is a special case of a result of Pellikaan, Shen, and van Wee [125].

Lemma 6.5.1 *Let C be an arbitrary linear $[n, k]$ code over $\mathbf{F}_q$. Then there exist a global function field $F/\mathbf{F}_q$, distinct rational places $P_1, \ldots, P_n$ of F, and a k-dimensional $\mathbf{F}_q$-linear subspace V of F such that C is equal to the function-field code $C_V(P_1, \ldots, P_n)$.*

Proof. Since $A(q) > 0$ by Lemma 5.2.4, there exists a global function field $F/\mathbf{F}_q$ possessing n distinct rational places $P_1, \ldots, P_n$. Let $\mathbf{c}_1, \ldots, \mathbf{c}_k$ form a basis of C and put

$$\mathbf{c}_j = (c_{j,1}, \ldots, c_{j,n}) \in \mathbf{F}_q^n \qquad \text{for } 1 \le j \le k.$$

By the approximation theorem, for each $1 \le j \le k$ we can find an $f_j \in F$ such that

$$\nu_{P_i}(f_j - c_{j,i}) \ge 1 \qquad \text{for } 1 \le i \le n.$$

Let V be the $\mathbf{F}_q$-linear subspace of F spanned by $f_1, \ldots, f_k$. From the construction of the f_j we infer

$$V \subseteq \bigcap_{i=1}^n O_{P_i}$$

and also

$$f_j(P_i) = c_{j,i} \qquad \text{for } 1 \le j \le k, 1 \le i \le n.$$

Let $\psi : V \longrightarrow \mathbf{F}_q^n$ be the $\mathbf{F}_q$-linear map defined in Remark 6.1.7. Then the identity above yields

$$\psi(f_j) = \mathbf{c}_j \qquad \text{for } 1 \le j \le k.$$

Since $\mathbf{c}_1, \ldots, \mathbf{c}_k$ are linearly independent over $\mathbf{F}_q$, this implies that $f_1, \ldots, f_k$ are linearly independent over $\mathbf{F}_q$ and also

$$\bigcap_{i=1}^n (V \cap M_{P_i}) = \{0\}.$$

Thus, V satisfies the conditions in Remark 6.1.7 and it is immediate that $C = C_V(P_1, \ldots, P_n)$. $\square$

Theorem 6.5.2 *Suppose that there exists a linear $[n, k, d]$ code over $\mathbf{F}_q$. Then for any integers b, r, s with $2 \le b \le q$, $1 \le r < b$, and $0 \le s \le r$ there exists a linear $[N, K, D]$ code over $\mathbf{F}_q$ with*

$$N = bn, \qquad K = k(s+1) + r - s, \qquad D \ge \min((b-s)d, (b-r)n).$$

Proof. Let C be a linear $[n, k, d]$ code over $\mathbf{F}_q$. By Lemma 6.5.1 there exist a global function field $F/\mathbf{F}_q$, distinct rational places $P_1, \ldots, P_n$ of F, and a k-dimensional $\mathbf{F}_q$-linear subspace V of F such that $C = C_V(P_1, \ldots, P_n)$. By the approximation theorem we can find an element $z \in F$ such that $\nu_{P_i}(z) \geq 1$ for $1 \leq i \leq n$ and $\nu_Q(z) = -1$ for some place Q of F. For a given b, let B be a subset of $\mathbf{F}_q$ with $|B| = b$ and $0 \in B$. Consider the simple algebraic extension $E = F(y)$ defined by

$$\prod_{\beta \in B} (y - \beta) = z.$$

Then $[E : F] = b$ by a variant of the Eisenstein theorem (see [152, Proposition III.1.14.(2)]) and the places $P_1, \ldots, P_n$ split completely in the extension E/F by Kummer's theorem [152, Theorem III.3.7]. Consequently, $\mathbf{F}_q$ is the full constant field of E. For $1 \leq i \leq n$ and $\beta \in B$, let $R_i^{(\beta)}$ be the unique place of E lying over P_i with

$$y(R_i^{(\beta)}) = \beta.$$

For given r and s put

$$U = \left\{ \sum_{j=0}^{r} v_j y^j \in E : v_j \in V \text{ for } 0 \leq j \leq s, \ v_j \in \mathbf{F}_q \text{ for } s+1 \leq j \leq r \right\}.$$

Then U is an $\mathbf{F}_q$-linear subspace of E. Since $\dim(V) = k$ and the elements $y^j, 0 \leq j \leq r$, are linearly independent over F, it is obvious that

$$\dim(U) = k(s+1) + r - s.$$

By the definitions of y and z we have $\nu_R(y) \geq 0$ for all places $R \in \{R_i^{(\beta)} : 1 \leq i \leq n, \ \beta \in B\}$. Furthermore, by the definition of the function-field code $C_V(P_1, \ldots, P_n)$ we have $\nu_{P_i}(f) \geq 0$ for all $f \in V$ and $1 \leq i \leq n$. Consequently, the analogous condition

$$\nu_R(u) \geq 0 \quad \text{for all } u \in U \text{ and } R \in \{R_i^{(\beta)} : 1 \leq i \leq n, \ \beta \in B\}$$

is satisfied. With $N = bn$ we consider the map $\psi : U \longrightarrow \mathbf{F}_q^N$ defined by

$$\psi(u) = \left(u(R_i^{(\beta)}) \right)_{1 \leq i \leq n, \ \beta \in B} \quad \text{for all } u \in U.$$

Since ψ is $\mathbf{F}_q$-linear, its image $\mathrm{im}(\psi)$ is a linear code over $\mathbf{F}_q$ of length N. We will prove that the linear code $\mathrm{im}(\psi)$ has the properties claimed in the theorem.

For an arbitrary nonzero $u \in U$ we bound the weight $w(\psi(u))$ from below. We have

$$u = \sum_{j=0}^{r} v_j y^j \quad \text{with } v_j \in V \text{ for } 0 \leq j \leq s, \ v_j \in \mathbf{F}_q \text{ for } s+1 \leq j \leq r,$$

where not all $v_j = 0$. Let t be the largest index with $v_t \neq 0$. We first consider the case $0 \leq t \leq s$. Then

$$u = \sum_{j=0}^{t} v_j y^j \quad \text{with } v_t \neq 0.$$

Thus, for all $1 \leq i \leq n$ and $\beta \in B$ we obtain

$$u\left(R_i^{(\beta)}\right) = \sum_{j=0}^{t} v_j(P_i)\beta^j.$$

Since $v_t \in V$ and $C = C_V(P_1, \ldots, P_n)$, we have

$$w\left((v_t(P_i))_{1 \leq i \leq n}\right) \geq d,$$

and so there are d choices of i with $v_t(P_i) \neq 0$. For each such i there are at most t, and so at most s, values of $\beta \in B$ with $u(R_i^{(\beta)}) = 0$. Consequently, there exist $b - s$ elements $\beta \in B$ with $u(R_i^{(\beta)}) \neq 0$, and so

$$w(\psi(u)) \geq (b - s)d.$$

Now we consider the remaining case $s + 1 \leq t \leq r$. Then again

$$u = \sum_{j=0}^{t} v_j y^j \qquad \text{with } v_t \neq 0.$$

But now $v_t \in \mathbf{F}_q$, and so for all $1 \leq i \leq n$ and $\beta \in B$ we obtain

$$u\left(R_i^{(\beta)}\right) = v_t \beta^t + \sum_{j=0}^{t-1} v_j(P_i)\beta^j.$$

Therefore, for each $1 \leq i \leq n$ there are at most t, and so at most r, values of $\beta \in B$ with $u(R_i^{(\beta)}) = 0$. Consequently, there exist $b - r$ elements $\beta \in B$ with $u(R_i^{(\beta)}) \neq 0$, and so

$$w(\psi(u)) \geq (b - r)n.$$

This proves not only the claimed lower bound on the minimum distance of $\mathrm{im}(\psi)$, but also that ψ is injective. Hence

$$\dim(\mathrm{im}(\psi)) = \dim(U) = k(s + 1) + r - s,$$

and so all the desired properties are shown. $\qquad\qquad\qquad\qquad\qquad\qquad\square$

Corollary 6.5.3 *Suppose that there exists a linear $[n, k, d]$ code over $\mathbf{F}_q$. Then for any integers b and r with $2 \leq b \leq q$ and $1 \leq r < b$ there exists a linear $[N, K, D]$ code over $\mathbf{F}_q$ with*

$$N = bn, \quad K = k + r, \quad D = \min(bd, (b - r)n).$$

Proof. This is basically the special case $s = 0$ of Theorem 6.5.2, but it remains to show that in the lower bound for D in Theorem 6.5.2 we have equality. First note that there exists an element $v_0 \in V$ with

$$w\left((v_0(P_i))_{1 \leq i \leq n}\right) = d.$$

Then with $u_1 = v_0 \in U$ we get $w(\psi(u_1)) = bd$. Next we choose r distinct elements $\beta_1, \ldots, \beta_r \in B$ with $\beta_1 = 0$ and put

$$u_2 = \prod_{j=1}^{r} (y - \beta_j) \in U.$$

Then we have

$$u_2\left(R_i^{(\beta)}\right) = \prod_{j=1}^{r}(\beta - \beta_j) \qquad \text{for all } 1 \le i \le n \text{ and } \beta \in B,$$

and so $w(\psi(u_2)) = (b - r)n$. □

It should be noted that the propagation rule in Theorem 6.5.2 contains one free parameter in the case $q = 2$ and three free parameters in the case $q > 2$, and so several linear codes are obtained from a single input code. Furthermore, the propagation rule in Theorem 6.5.2 and its special case in Corollary 6.5.3 can obviously be iterated. This yields the following results which are proved by straightforward induction. We use the standard convention that an empty product has the value 1.

Corollary 6.5.4 *Suppose that there exists a linear $[n, k, d]$ code over $\mathbf{F}_q$ and let m be an arbitrary positive integer. For $1 \le l \le m$ let b_l, r_l, s_l be integers with $2 \le b_l \le q$, $1 \le r_l < b_l$, and $0 \le s_l \le r_l$. Then there exists a linear $[N, K, D]$ code over $\mathbf{F}_q$ with*

$$N = n \prod_{l=1}^{m} b_l, \qquad K = k \prod_{l=1}^{m}(s_l + 1) + \sum_{l=1}^{m}(r_l - s_l) \prod_{j=l+1}^{m}(s_j + 1),$$

$$D \ge \min\left(d \prod_{l=1}^{m}(b_l - s_l),\ n \min_{0 \le l \le m-1} b_1 \cdots b_l(b_{l+1} - r_{l+1}) \prod_{j=l+2}^{m}(b_j - s_j)\right).$$

Corollary 6.5.5 *Suppose that there exists a linear $[n, k, d]$ code over $\mathbf{F}_q$ and let m be an arbitrary positive integer. For $1 \le l \le m$ let b_l and r_l be integers with $2 \le b_l \le q$ and $1 \le r_l < b_l$. Then there exists a linear $[N, K, D]$ code over $\mathbf{F}_q$ with*

$$N = n \prod_{l=1}^{m} b_l, \qquad K = k + \sum_{l=1}^{m} r_l,$$

$$D = \min\left(d \prod_{l=1}^{m} b_l,\ nb_1 \cdots b_m \min_{1 \le l \le m}\left(1 - \frac{r_l}{b_l}\right)\right).$$

Example 6.5.6 In Corollary 6.5.5 we take an arbitrary q, an arbitrary $m \ge 1$, and $b_l = q$ and $r_l = 1$ for $1 \le l \le m$. Then for any given linear $[n, k, d]$ code over $\mathbf{F}_q$ we obtain linear codes over $\mathbf{F}_q$ with parameters

$$[q^m n,\ k + m,\ \min\left(q^m d,\ q^{m-1}(q - 1)n\right)].$$

In particular, if for any $n \geq 1$ we start from the obvious linear $[n, 1, n]$ code over $\mathbf{F}_q$, then we get linear codes over $\mathbf{F}_q$ with parameters

$$[q^m n, \, m + 1, \, q^{m-1}(q-1)n].$$

For $n \leq q$ all these linear codes are optimal by the Plotkin bound in Proposition 6.2.2. If we start from the known linear $[q + 1, 2, q]$ code over $\mathbf{F}_q$, then we get linear codes over $\mathbf{F}_q$ with parameters

$$[q^m(q+1), \, m + 2, \, q^{m-1}(q^2 - 1)].$$

All these linear codes are again optimal by the Plotkin bound.

Chapter 7

Applications to Cryptography

It has been known at least since the invention of elliptic-curve cryptosystems in 1985 that algebraic curves over finite fields can be very useful for cryptography. Elliptic-curve cryptosystems have already been treated extensively in several books (see for instance Blake, Seroussi, and Smart [11], Enge [21], Koblitz [59], and Menezes [81]), and so we focus here on other, and more recent, applications of algebraic curves over finite fields to cryptography in which rational points of the curve play a role. Specifically, we discuss applications to stream ciphers, to the construction of hash functions, and to the construction of authentication schemes. As usual in this book, we prefer to use the equivalent language of algebraic function fields rather than the language of algebraic curves.

7.1 Background on Stream Ciphers and Linear Complexity

A cryptosystem is a basic tool for data security that transforms data in original form (**plaintexts**) into protected data in encrypted form (**ciphertexts**) and, vice versa, recovers plaintexts from ciphertexts by decryption. The encryption and decryption schemes depend on the choice of parameters (**keys**) from an extremely large set of possibilities. A fundamental distinction is made between public-key cryptosystems (such as the elliptic-curve cryptosystem) and symmetric cryptosystems. In a public-key cryptosystem, the encryption key K is public knowledge and only the decryption key K' is kept secret from unauthorized users. In a symmetric cryptosystem, both keys K and K' have to be kept secret since they are computationally equivalent in the sense that one can easily be obtained from the other; in many cases we even have $K = K'$.

A **stream cipher** is a symmetric cryptosystem in which plaintexts and ciphertexts are strings (or, in other words, finite sequences) of elements of a finite field $\mathbf{F}_q$ and encryption and decryption proceed by termwise addition, respectively subtraction, of the same string of elements of $\mathbf{F}_q$. The latter string serves as the (encryption and decryption) key; it is called the **keystream** in the context of stream ciphers and is supposed to be known only to authorized users.

In an ideal world, the keystream could be chosen to be a truly random string of elements of $\mathbf{F}_q$, and then the stream cipher would be perfectly secure since the ciphertexts will carry absolutely no information. In the real world, sources of true randomness are hard to come by, and so keystreams are generated from certain secret seed data by a (perhaps even publicly available) deterministic algorithm. Useful keystreams must meet requirements such as possessing good statistical randomness properties and a high complexity (in a suitable sense), so that the keystream cannot be inferred from a small portion of its terms. Many practical keystream generators use linear recurrence relations over $\mathbf{F}_q$ as basic steps in their algorithms. We refer to Rueppel [134] for a survey of algorithms for keystream generation and also for general background on stream ciphers.

Definition 7.1.1 Let k be a positive integer. Then the sequence $s_1, s_2, \ldots$ of elements of $\mathbf{F}_q$ is said to satisfy a **linear recurrence relation** over $\mathbf{F}_q$ (of order k) if there exist $a_0, a_1, \ldots, a_{k-1} \in \mathbf{F}_q$ such that

$$s_{i+k} = \sum_{h=0}^{k-1} a_h s_{i+h} \qquad \text{for } i = 1, 2, \ldots .$$

The **linear recurring sequence** $s_1, s_2, \ldots$ in Definition 7.1.1 is uniquely determined by the linear recurrence relation and by the initial values $s_1, \ldots, s_k$. By convention, the zero sequence $0, 0, \ldots$ in $\mathbf{F}_q$ is also meant to satisy a linear recurrence relation over $\mathbf{F}_q$ of order 0. The following notion of complexity is fundamental for the system-theoretic approach to the assessment of keystreams.

Definition 7.1.2 Let n be a positive integer and let S be a finite or infinite sequence of elements of $\mathbf{F}_q$ containing at least n terms. Then the **linear complexity** $L_n(S)$ is the least k such that the first n terms of S can be generated by a linear recurrence relation over $\mathbf{F}_q$ of order k.

It is clear that we always have $0 \leq L_n(S) \leq n$. The extreme values of $L_n(S)$ correspond to highly nonrandom behavior, for if $s_1, \ldots, s_n$ are the first n terms of S, then $L_n(S) = 0$ if and only if $s_i = 0$ for $1 \leq i \leq n$, whereas $L_n(S) = n$ if and only if $s_i = 0$ for $1 \leq i \leq n - 1$ and $s_n \neq 0$. We will usually consider $L_n(S)$ for (infinite) sequences S.

For a sequence S with terms $s_1, s_2, \ldots$ in $\mathbf{F}_q$, we introduce its **generating function**

$$G := \sum_{i=1}^{\infty} s_i x^{-i} \in \mathbf{F}_q((x^{-1})),$$

where $\mathbf{F}_q((x^{-1}))$ is the field of formal Laurent series over $\mathbf{F}_q$ in the variable x^{-1} or, equivalently, the completion of the rational function field $\mathbf{F}_q(x)$ with respect to the infinite place ∞. Let ν_∞ be the normalized discrete valuation of $\mathbf{F}_q((x^{-1}))$ extending ∞. Note that G is **rational**, i.e., G belongs to $\mathbf{F}_q(x)$, if and only if S is a linear recurring sequence. Otherwise, we say that G is **irrational**. If G is rational with denominator of degree k, then S satisfies a linear recurrence relation over $\mathbf{F}_q$ of order k, and vice versa.

Every element of $\mathbf{F}_q((x^{-1}))$ has a unique continued fraction expansion. For $G \in \mathbf{F}_q((x^{-1}))$ with $\nu_\infty(G) > 0$, i.e., for a generating function G, this continued fraction expansion has the form

$$G = 1/(A_1 + 1/(A_2 + \cdots)) =: [A_1, A_2, \ldots], \qquad (7.1)$$

where the partial quotients A_j are polynomials over $\mathbf{F}_q$ of positive degree. This expansion is finite if G is rational and infinite if G is irrational. The theory of continued fractions in $\mathbf{F}_q((x^{-1}))$ proceeds in analogy with the classical theory of continued fractions for real numbers. A systematic account of the theory of continued fractions in $\mathbf{F}_q((x^{-1}))$, and more generally in local fields, is given in de Mathan [16].

If the continued fraction expansion (7.1) is broken off after the term $A_j, j \geq 1$, then it represents a rational function p_j/q_j in reduced form. The polynomials $p_j, q_j \in \mathbf{F}_q[x]$ can be computed recursively by

$$p_{-1} = 1, \; p_0 = 0, \; p_j = A_j p_{j-1} + p_{j-2} \qquad \text{for } j \geq 1,$$

$$q_{-1} = 0, \; q_0 = 1, \; q_j = A_j q_{j-1} + q_{j-2} \qquad \text{for } j \geq 1.$$

Then we have

$$\deg(q_j) = \sum_{h=1}^{j} \deg(A_h) \qquad \text{for } j \geq 1, \qquad (7.2)$$

$$\nu_\infty(q_j G - p_j) = -\nu_\infty(q_{j+1}) \qquad \text{for } j \geq 0. \qquad (7.3)$$

For rational G we interpret $\deg(A_j) = \deg(q_j) = -\nu_\infty(q_j) = \infty$ whenever A_j and q_j do not exist.

The following theorem of Niederreiter [89] provides an explicit formula for the linear complexity $L_n(S)$ in Definition 7.1.2 in terms of the continued fraction expansion of the generating function G of S. For the proof we first need an auxiliary result.

Lemma 7.1.3 *If $f, g \in \mathbf{F}_q[x]$ are such that $\nu_\infty(fG - g) > 0$, then*

$$f = \sum_{h=0}^{j} c_h q_h \qquad \text{and} \qquad g = \sum_{h=0}^{j} c_h p_h$$

for some $j \geq 0$ and $c_h \in \mathbf{F}_q[x]$ with $\deg(c_h) < \deg(A_{h+1})$ for $0 \leq h \leq j$. If in addition $f \neq 0$, then

$$\nu_\infty(fG - g) = \deg(q_{i+1}) - \deg(c_i),$$

where i is the least index with $c_i \neq 0$.

Proof. Using (7.2) we see that every $f \in \mathbf{F}_q[x]$ can be represented in the indicated form. By (7.3) we have

$$\nu_\infty(c_h(q_h G - p_h)) = \nu_\infty(c_h) - \nu_\infty(q_{h+1}) > 0 \qquad \text{for } 0 \leq h \leq j,$$

hence

$$\nu_\infty \left(\sum_{h=0}^{j} c_h (q_h G - p_h) \right) > 0.$$

Using $\nu_\infty(fG - g) > 0$ we get

$$\nu_\infty \left(\sum_{h=0}^{j} c_h p_h - g \right) = \nu_\infty \left(fG - g - \sum_{h=0}^{j} c_h (q_h G - p_h) \right) > 0.$$

But $\sum_{h=0}^{j} c_h p_h - g$ is a polynomial, hence it must be 0.

To prove the second part, we note that if $f \neq 0$, then there exists a least index i with $c_i \neq 0$. From (7.3) we obtain

$$\nu_\infty \left(c_i (q_i G - p_i) \right) = \deg(q_{i+1}) - \deg(c_i) \leq \deg(q_{i+1}),$$

$$\nu_\infty \left(c_h (q_h G - p_h) \right) > \deg(q_{h+1}) - \deg(A_{h+1}) = \deg(q_h) \geq \deg(q_{i+1})$$

for $i < h \leq j$, and the desired formula for $\nu_\infty(fG - g)$ follows. □

Theorem 7.1.4 *Let S be an arbitrary sequence of elements of $\mathbf{F}_q$ and let G be its generating function with continued fraction expansion (7.1). Put $w_j = \deg(q_j)$ for $j \geq 0$ and $w_{-1} = 1$. Then for every integer $n \geq 1$ we have $L_n(S) = w_j$, where $j = j(n) \geq 0$ is uniquely determined by the condition*

$$w_{j-1} + w_j \leq n < w_j + w_{j+1}.$$

Proof. From (7.3) we get

$$\nu_\infty \left(G - \frac{p_j}{q_j} \right) = w_j + w_{j+1} \qquad \text{for } j \geq 0.$$

Hence if $s_1, s_2, \ldots$ are the terms of S and

$$\frac{p_j}{q_j} = \sum_{i=1}^{\infty} t_i x^{-i},$$

then $s_i = t_i$ for $1 \leq i < w_j + w_{j+1}$. Since the sequence $t_1, t_2, \ldots$ satisfies a linear recurrence relation over $\mathbf{F}_q$ of order $\deg(q_j) = w_j$, it follows that

$$L_n(S) \leq w_j \qquad \text{for } 1 \leq n < w_j + w_{j+1}. \tag{7.4}$$

Now let $w_{j-1} + w_j \leq n < w_j + w_{j+1}$. By the definition of $L_n(S)$ there exists a sequence $u_1, u_2, \ldots$ generated by a linear recurrence relation over $\mathbf{F}_q$ of order $L_n(S)$ such that $s_i = u_i$ for $1 \leq i \leq n$. The generating function of the sequence $u_1, u_2, \ldots$ is thus rational of the form g_n/f_n with $g_n, f_n \in \mathbf{F}_q[x]$ and $\deg(f_n) = L_n(S)$. We get

$$\nu_\infty(f_n G - g_n) = \nu_\infty \left(G - \frac{g_n}{f_n} \right) - L_n(S) \geq n + 1 - L_n(S) > 0. \tag{7.5}$$

Since $f_n \neq 0$, Lemma 7.1.3 yields the formulas

$$f_n = \sum_{h=i}^{j} c_h q_h \qquad \text{with } c_i \neq 0,$$

$$\nu_\infty(f_n G - g_n) = w_{i+1} - \deg(c_i).$$

Together with (7.5) this implies

$$L_n(S) \geq n + 1 - w_{i+1} + \deg(c_i) \geq w_{j-1} + w_j + 1 - w_{i+1}.$$

In view of (7.4) we conclude that $i \geq j - 1$. If we had $c_j = 0$, then $i = j - 1$ and

$$\deg(f_n) = L_n(S) \geq n + 1 - w_j + \deg(c_{j-1}) \geq w_{j-1} + 1 + \deg(c_{j-1}),$$

which contradicts $f_n = c_{j-1} q_{j-1}$. Thus $c_j \neq 0$ and

$$L_n(S) = \deg(f_n) = \deg(c_j q_j) \geq w_j,$$

hence $L_n(S) = w_j$ by (7.4). $\square$

Corollary 7.1.5 *If $L_n(S) > n/2$, then $L_{n+1}(S) = L_n(S)$. If $L_n(S) \leq n/2$, then $L_{n+1}(S) = L_n(S)$ for exactly one choice of $s_{n+1} \in \mathbf{F}_q$ and $L_{n+1}(S) = n + 1 - L_n(S)$ for exactly $q - 1$ choices of $s_{n+1} \in \mathbf{F}_q$.*

Proof. We use the notation of Theorem 7.1.4 and its proof. If $L_n(S) = w_j > n/2$, then $n + 1 \leq 2w_j < w_j + w_{j+1}$, and so $L_{n+1}(S) = w_j$ by Theorem 7.1.4.

If $L_n(S) = w_j \leq n/2$, then it follows from the proof of Theorem 7.1.4 that $f_n = c_j q_j + c_{j-1} q_{j-1}$ with $c_j \in \mathbf{F}_q^*$. If we had $c_{j-1} \neq 0$, then

$$w_j - \deg(c_{j-1}) = \nu_\infty(f_n G - g_n) \geq n + 1 - w_j,$$

where we used (7.5) in the second step. Hence $\deg(c_{j-1}) \leq 2w_j - n - 1 < 0$, a contradiction. Thus $c_{j-1} = 0$, and so the linear recurrence relation over $\mathbf{F}_q$ of order w_j generating $s_1, \ldots, s_n$ is uniquely determined. Thus, there exists a unique $s_{n+1} \in \mathbf{F}_q$ with $L_{n+1}(S) = L_n(S)$. For the other $q - 1$ choices of $s_{n+1} \in \mathbf{F}_q$ we have $L_{n+1}(S) > L_n(S)$. Then necessarily $L_{n+1}(S) = w_{j+1}$ and $n + 1 = w_j + w_{j+1}$ by Theorem 7.1.4, hence $L_{n+1}(S) = n + 1 - L_n(S)$. $\square$

For integers $n \geq 1$ and $L \geq 0$ let $N_n^{(q)}(L)$ denote the number of $S \in \mathbf{F}_q^n$ with $L_n(S) = L$. Note that $N_n^{(q)}(L) = 0$ for $L > n$. The following formula for $N_n^{(q)}(L)$ was first shown by Gustavson [43]; see also Rueppel [133, Chapter 4] for $q = 2$ and Niederreiter [93] for arbitrary q.

Theorem 7.1.6 *We have $N_n^{(q)}(0) = 1$ and*

$$N_n^{(q)}(L) = (q - 1)q^{\min(2L-1, 2n-2L)} \qquad \text{for } 1 \leq L \leq n.$$

Proof. Corollary 7.1.5 implies the recursions

$$N_{n+1}^{(q)}(L) = N_n^{(q)}(L) \qquad \text{for } 0 \leq L \leq n/2,$$

$$N_{n+1}^{(q)}(L) = qN_n^{(q)}(L) \qquad \text{for } L = (n+1)/2, \ n \text{ odd},$$

$$N_{n+1}^{(q)}(L) = qN_n^{(q)}(L) + (q-1)N_n^{(q)}(n+1-L) \qquad \text{for } (n+2)/2 \leq L \leq n+1,$$

and then the proof can be completed by straightforward induction on n.   □

Remark 7.1.7 The following alternative proof of Theorem 7.1.6 that is independent of Corollary 7.1.5 can be given for $1 \leq L \leq n/2$. Note first that in this case the linear recurrence relation over $\mathbf{F}_q$ of order $L_n(S) = L$ generating the first n terms of S is uniquely determined, for otherwise there would exist distinct rational functions g_n/f_n and k_n/h_n in $\mathbf{F}_q(x)$ with $\deg(f_n) = \deg(h_n) = L$ and

$$\nu_\infty\left(G - \frac{g_n}{f_n}\right) > n, \qquad \nu_\infty\left(G - \frac{k_n}{h_n}\right) > n,$$

where G is the generating function of S. But then

$$n < \nu_\infty\left(\frac{g_n}{f_n} - \frac{k_n}{h_n}\right) = \nu_\infty\left(\frac{g_n h_n - k_n f_n}{f_n h_n}\right) \leq -\nu_\infty(f_n h_n) = 2L \leq n,$$

a contradiction (in fact, the proof of Corollary 7.1.5 yields the formula $g_n/f_n = p_j/q_j$). Thus, by Theorem 7.1.4 there is a bijective correspondence between the objects counted by $N_n^{(q)}(L)$ and the continued fraction expansions $[A_1, \ldots, A_m]$ in $\mathbf{F}_q(x)$ with an arbitrary $m \geq 1$ and $\sum_{j=1}^m \deg(A_j) = L$. Now for any $d \geq 1$ the number of $A \in \mathbf{F}_q[x]$ with $\deg(A) = d$ is $(q-1)q^d$, and so

$$
\begin{aligned}
N_n^{(q)}(L) &= \sum_{m=1}^L \sum_{\substack{d_1,\ldots,d_m \geq 1 \\ d_1 + \cdots + d_m = L}} (q-1)q^{d_1} \cdots (q-1)q^{d_m} \\
&= q^L \sum_{m=1}^L (q-1)^m \sum_{\substack{d_1,\ldots,d_m \geq 1 \\ d_1 + \cdots + d_m = L}} 1 = q^L \sum_{m=1}^L (q-1)^m \binom{L-1}{m-1} \\
&= (q-1)q^L \sum_{m=0}^{L-1} \binom{L-1}{m}(q-1)^m = (q-1)q^{2L-1}.
\end{aligned}
$$

Definition 7.1.8 For a sequence S of elements of $\mathbf{F}_q$, the sequence $L_1(S), L_2(S), \ldots$ of linear complexities is called the **linear complexity profile** of S.

Remark 7.1.9 It follows from Theorem 7.1.4 that the linear complexity profile of S has the form

$$0, \ldots, 0, d_1, \ldots, d_1, d_1 + d_2, \ldots, d_1 + d_2, \ldots$$

with 0 occurring $d_1 - 1$ times and $\sum_{h=1}^{j} d_h$ occurring $d_j + d_{j+1}$ times for all $j \geq 1$, where $d_j = \deg(A_j)$ for all $j \geq 1$ and the A_j are the partial quotients in the continued fraction expansion (7.1) of the generating function G of S. In particular, the "jumps" in the linear complexity profile are just the degrees of the partial quotients.

The linear complexity profile is a useful tool for the assessment of keystreams. In order to establish benchmarks for statistical randomness tests based on the linear complexity profile, the behavior of the linear complexity profile of random sequences of elements of $\mathbf{F}_q$ has to be investigated. Such a study was initiated by Niederreiter [90]. The following natural stochastic model is used. Let μ_q be the uniform probability measure on $\mathbf{F}_q$ which assigns the measure $1/q$ to each element of $\mathbf{F}_q$. Let $\mathbf{F}_q^\infty$ be the sequence space over $\mathbf{F}_q$ and let μ_q^∞ be the complete product probability measure on $\mathbf{F}_q^\infty$ induced by μ_q. Then we are interested in properties of sequences $S \in \mathbf{F}_q^\infty$ that hold μ_q^∞-almost everywhere, i.e., for a set of sequences S of μ_q^∞-measure 1. Rather than proving the deeper theorems in Niederreiter [90] that need the theory of dynamical systems, we show some simpler results that can be obtained by the more elementary method found in Niederreiter [92].

Theorem 7.1.10 *Let ψ be a nonnegative function on the positive integers such that the series $\sum_{n=1}^{\infty} q^{-\psi(n)}$ converges. Then μ_q^∞-almost everywhere we have*

$$|L_n(S) - \frac{n}{2}| \leq \frac{1}{2}\psi(n) \qquad \text{for all sufficiently large } n.$$

Proof. For a fixed positive integer n let

$$D_n = \{S \in \mathbf{F}_q^\infty : L_n(S) > \frac{1}{2}(n + \psi(n))\}.$$

If $k(n)$ is the least integer $> (n + \psi(n))/2$, then for $S \in D_n$ we have $L_n(S) \geq k(n) \geq (n+1)/2$. Therefore from Theorem 7.1.6 we get, under the assumption that $k(n) \leq n$,

$$\begin{aligned}
\mu_q^\infty(D_n) &= q^{-n} \sum_{L=k(n)}^{n} N_n^{(q)}(L) = q^{-n} \sum_{L=k(n)}^{n} (q-1)q^{2n-2L} \\
&= (q-1)q^{n-2k(n)} \sum_{L=0}^{n-k(n)} q^{-2L} < \frac{1}{q+1}q^{2+n-2k(n)}.
\end{aligned}$$

Since $k(n) > (n + \psi(n))/2$, it follows that

$$\mu_q^\infty(D_n) < \frac{1}{q+1}q^{2-\psi(n)}.$$

If $k(n) > n$, then $\mu_q^\infty(D_n) = 0$, and so the above bound holds in all cases. From the hypothesis $\sum_{n=1}^{\infty} q^{-\psi(n)} < \infty$ we then obtain $\sum_{n=1}^{\infty} \mu_q^\infty(D_n) < \infty$. The Borel-Cantelli lemma now shows that the set of all $S \in \mathbf{F}_q^\infty$ for which $S \in D_n$ for infinitely many n has μ_q^∞-measure 0. In other words, μ_q^∞-almost everywhere we have $S \in D_n$ for at most finitely many n. From the definition of D_n it follows then that μ_q^∞-almost everywhere we have

$$L_n(S) \leq \frac{1}{2}(n + \psi(n)) \qquad \text{for all sufficiently large } n.$$

By a similar method we get an analogous lower bound.   □

Corollary 7.1.11 *We have*

$$\limsup_{n\to\infty} \frac{|L_n(S) - n/2|}{\log n} \leq \frac{1}{2\log q} \qquad \mu_q^\infty\text{-almost everywhere.}$$

In particular, we have

$$\lim_{n\to\infty} \frac{L_n(S)}{n} = \frac{1}{2} \qquad \mu_q^\infty\text{-almost everywhere.}$$

Proof. Apply Theorem 7.1.10 with the function

$$\psi(n) = \left(1 + \frac{1}{m}\right) \frac{\log n}{\log q},$$

where m is an arbitrary positive integer.   □

A survey of further results from the probabilistic theory of the linear complexity profile can be found in Niederreiter [95]. A briefer, but more recent account of facts on the linear complexity profile is available in Niederreiter [97].

7.2 Constructions of Almost Perfect Sequences

Corollary 7.1.11 suggests that it is relevant to study the deviations of $L_n(S)$ from $n/2$. There has been a strong interest in sequences S for which these deviations are bounded. Such sequences were first considered by Rueppel [133, Chapter 4] in the case $q = 2$.

Definition 7.2.1 If d is a positive integer, then a sequence S of elements of $\mathbf{F}_q$ is called **d-perfect** if

$$|2L_n(S) - n| \leq d \qquad \text{for all } n \geq 1.$$

A 1-perfect sequence is also called **perfect**. A sequence is called **almost perfect** if it is d-perfect for some d.

An alternative characterization of d-perfect sequences is based on continued fractions. The following theorem is implied by results in Niederreiter [89].

Theorem 7.2.2 *A sequence S of elements of $\mathbf{F}_q$ is d-perfect if and only if the generating function G of S is irrational and the partial quotients A_j in the continued fraction expansion of G satisfy $\deg(A_j) \leq d$ for all $j \geq 1$.*

Proof. Suppose first that G is irrational and that $\deg(A_j) \leq d$ for all $j \geq 1$. As in Theorem 7.1.4, for given $n \geq 1$ we determine $j = j(n) \geq 0$ by the condition

$$w_{j-1} + w_j \leq n < w_j + w_{j+1}.$$

Using (7.2) and the interpretation $\deg(A_0) = -1$, we can rewrite this condition in the form

$$2w_j - \deg(A_j) \leq n \leq 2w_j + \deg(A_{j+1}) - 1,$$

or equivalently,

$$1 - \deg(A_{j+1}) \leq 2w_j - n \leq \deg(A_j).$$

Since $L_n(S) = w_j$ by Theorem 7.1.4, we get

$$1 - d \leq 2L_n(S) - n \leq d,$$

and so S is d-perfect.

Conversely, suppose that S is d-perfect. Then G must be irrational and we have

$$2L_n(S) - n \leq d \qquad \text{for all } n \geq 1.$$

For given $j \geq 1$ we apply this bound with $n = 2w_j - \deg(A_j)$. Since then $L_n(S) = w_j$ by Theorem 7.1.4, we conclude that $\deg(A_j) \leq d$. $\qquad\qquad\square$

Remark 7.2.3 The proof of Theorem 7.2.2 shows that in order to establish that S is d-perfect, it suffices to prove that G is irrational and that

$$L_n(S) \leq \frac{n + d}{2} \qquad \text{for all } n \geq 1.$$

Similarly, it suffices to prove that

$$L_n(S) \geq \frac{n + 1 - d}{2} \qquad \text{for all } n \geq 1,$$

since then G is necessarily irrational and for given $j \geq 0$ we can apply this bound with $n = 2w_j + \deg(A_{j+1}) - 1$ to obtain $\deg(A_{j+1}) \leq d$ (this choice of n is not possible if $j = 0$ and $\deg(A_1) = 1$, but then the bound $\deg(A_1) \leq d$ is trivial). These arguments show also that for a perfect sequence S we have

$$L_n(S) = \left\lfloor \frac{n + 1}{2} \right\rfloor \qquad \text{for all } n \geq 1.$$

Example 7.2.4 Let $q = 2$ and let S be the sequence $s_1, s_2, \ldots$ of elements of $\mathbf{F}_2$ defined by $s_i = 1$ if $i = 2^h - 1$ for some integer $h \geq 1$, and $s_i = 0$ otherwise. Then the generating function G of S satisfies the identity $G^2 = xG + 1$ in $\mathbf{F}_2((x^{-1}))$, hence $G = x + G^{-1}$, which leads to the periodic continued fraction expansion

$$G = [x, x, x, \ldots].$$

Therefore, the sequence S is perfect by Theorem 7.2.2. This example is a special instance of a family of perfect binary sequences which were introduced in Niederreiter [88] and called

generalized Rueppel sequences. A generalized Rueppel sequence is determined by a sequence $m_1, m_2, \ldots$ of integers given by

$$m_1 = 1 \quad \text{and} \quad m_{h+1} = 2m_h + b_h \quad \text{for } h = 1, 2, \ldots,$$

where $b_h \in \{0, 1\}$ for all $h \geq 1$. Then a generalized Rueppel sequence $s_1, s_2, \ldots$ is defined by $s_i = 1$ if $i = m_h$ for some $h \geq 1$ and $s_i = 0$ otherwise. If $b_h = 0$ for all $h \geq 1$, then we get the **Rueppel sequence** which was introduced in Rueppel [133, Chapter 4].

Theorem 7.2.2 and the theorem of Lagrange for continued fractions in $\mathbf{F}_q((x^{-1}))$ show that a sequence of elements of $\mathbf{F}_q$ is almost perfect whenever its generating function is a quadratic irrational. However, on this general level we do not get an explicit bound on d for d-perfectness.

The set $\mathcal{P}_d^{(q)}$ of d-perfect sequences from $\mathbf{F}_q^\infty$ was investigated by Niederreiter and Vielhaber [100]. It is clear (e.g. from Theorem 7.2.2) that $\mathcal{P}_d^{(q)}$ has the cardinality of the continuum. On the other hand, it follows from a result of Niederreiter [90] that $\mu_q^\infty(\mathcal{P}_d^{(q)}) = 0$. In order to get more detailed information on the "size" of $\mathcal{P}_d^{(q)}$, one studies what is in a sense the Hausdorff dimension of $\mathcal{P}_d^{(q)}$. Namely, by mapping sequences from $\mathbf{F}_q^\infty$ to digit expansions of real numbers in base q, one can map $\mathcal{P}_d^{(q)}$ to a subset of the interval $[0, 1]$. Then according to a result in [100], this image of $\mathcal{P}_d^{(q)}$ in $[0, 1]$ has conventional Hausdorff dimension equal to

$$\frac{1}{2} + \frac{\log r_d^{(q)}}{2 \log q},$$

where $r_d^{(q)}$ is the largest real root of the polynomial $z^d - (q-1) \sum_{j=0}^{d-1} z^j$.

Now we present methods of constructing almost perfect sequences of elements of $\mathbf{F}_q$ by using global function fields over $\mathbf{F}_q$. All constructions work with the following ingredients:

$F/\mathbf{F}_q$: a global function field over $\mathbf{F}_q$;

P: a rational place of F;

t: a local parameter at P with $\deg((t)_\infty) = 2$;

f: a function in $F \backslash \mathbf{F}_q(t)$.

We first treat the case $\nu_P(f) \geq 0$ which was considered by Xing and Lam [176]. The local expansion of f at P has the form

$$f = \sum_{i=0}^\infty s_i t^i$$

with all $s_i \in \mathbf{F}_q$. From this expansion we read off the sequence S_1 of coefficients $s_1, s_2, \ldots$.

Theorem 7.2.5 *If $\nu_P(f) \geq 0$ and the integer d is such that $d \geq \deg((f)_\infty)$, then the sequence S_1 constructed above is d-perfect.*

Proof. In view of Remark 7.2.3 it suffices to prove that

$$L_n(S_1) \geq \frac{n + 1 - d}{2} \quad \text{for all } n \geq 1.$$

Fix $n \geq 1$, put $k = L_n(S_1)$, and write a linear recurrence relation over $\mathbf{F}_q$ of order k satisfied by $s_1, \ldots, s_n$ in the form

$$\sum_{h=0}^{k} a_h s_{i+h} = 0 \qquad \text{for } 1 \leq i \leq n-k, \tag{7.6}$$

where $a_h \in \mathbf{F}_q$ for $0 \leq h \leq k$ and $a_k = 1$. Consider the function

$$b = f \sum_{h=0}^{k} a_{k-h} t^h - \sum_{j=0}^{k} \left(\sum_{i=0}^{j} a_{k-j+i} s_i \right) t^j.$$

Note that b is a nonzero element of F since $a_k \neq 0$ and $f \notin \mathbf{F}_q(t)$. By applying the linear recurrence relation (7.6) and considering the local expansion of b at P, we obtain $\nu_P(b) \geq n+1$. On the other hand, the pole divisor of b satisfies

$$(b)_\infty \leq (f)_\infty + (t^k)_\infty.$$

Therefore

$$n + 1 \leq \nu_P(b) \leq \deg((b)_0) = \deg((b)_\infty) \leq d + 2k,$$

and so

$$k \geq \frac{n+1-d}{2}.$$

Thus, S_1 is d-perfect.   □

The case $\nu_P(f) < 0$ can be reduced to the situation considered above, as was pointed out in Xing et al. [185]. Indeed, if we put $v = -\nu_P(f) > 0$, then $\nu_P(t^v f) = 0$ and we have a local expansion at P of the form

$$t^v f = \sum_{i=0}^{\infty} s_i t^i$$

with all $s_i \in \mathbf{F}_q$. From this expansion we read off the sequence S_1' of coefficients $s_1, s_2, \ldots$. If $d \geq \deg((f)_\infty)$ as in Theorem 7.2.5, then

$$\deg((t^v f)_\infty) \leq \deg((f)_\infty) + \deg((t^v)_\infty) - \deg(vP) \leq d + v,$$

and so Theorem 7.2.5 implies the following result.

Corollary 7.2.6 *If $\nu_P(f) = -v < 0$ and the integer d is such that $d \geq \deg((f)_\infty)$, then the sequence S_1' constructed above is $(d+v)$-perfect.*

A systematic investigation of examples of almost perfect sequences that can be obtained from Theorem 7.2.5 was carried out by Kohel, Ling, and Xing [61]. This paper also discusses the efficient computation of local expansions by means of Hensel's lemma.

Example 7.2.7 Let $q = 2$, F be the rational function field $\mathbf{F}_2(x)$, and P be the zero of x. First we choose $t = x^2 + x$ and $f = x/(x + 1)$. Then we have the local expansion

$$f = \frac{x^2}{t} = \sum_{h=1}^{\infty} t^{2^h - 1}.$$

The sequence S_1 of coefficients $1, 0, 1, 0, 0, 0, 1, \ldots$ is perfect by Theorem 7.2.5. This is the same sequence as the generalized Rueppel sequence in Example 7.2.4. Next we choose $t = x^2 + x$ and $f = x$. Then we have the local expansion

$$f = \sum_{h=0}^{\infty} t^{2^h}.$$

The sequence S_1 of coefficients $1, 1, 0, 1, 0, 0, 0, 1, \ldots$ is perfect by Theorem 7.2.5. This is the Rueppel sequence mentioned in Example 7.2.4. Finally, we choose $t = x/(x^2 + x + 1)$ and $f = x$. Then we have the local expansion

$$f = t + t^2 + t^5 + t^6 + \cdots.$$

The sequence S_1 of coefficients $1, 1, 0, 0, 1, 1, \ldots$ is perfect by Theorem 7.2.5. This sequence is *not* a generalized Rueppel sequence.

The following result of Xing *et al.* [185] discusses the behavior of the sequences S_1 in Theorem 7.2.5 under the change of some initial terms.

Theorem 7.2.8 *If the sequence S_1 is as in Theorem 7.2.5 and the conditions of that theorem are satisfied, then any sequence S obtained by changing at most the first m terms of S_1 is $(d + 2m)$-perfect. Moreover, if the divisor inequality $m(t)_\infty \leq (f)_\infty$ holds, then S is still d-perfect.*

Proof. If the local expansion of f at P is

$$f = \sum_{i=0}^{\infty} s_i t^i,$$

then S_1 has the terms $s_1, s_2, \ldots$. Let $u_1, \ldots, u_m$ be m arbitrary elements of $\mathbf{F}_q$ and consider the new sequence S with the terms

$$u_1, \ldots, u_m, s_{m+1}, s_{m+2}, \ldots.$$

Put

$$b = f - \sum_{i=1}^{m} (s_i - u_i) t^i.$$

Then $\nu_P(b) \geq 0$ and $\deg((b)_\infty) \leq d + 2m$. Hence it follows from Theorem 7.2.5 that S is $(d + 2m)$-perfect. If $m(t)_\infty \leq (f)_\infty$, then $\deg((b)_\infty) \leq \deg((f)_\infty) \leq d$, and so S is d-perfect by Theorem 7.2.5. $\qquad \square$

Now we present two more constructions of almost perfect sequences due to Xing *et al.* [185]. Let $v = \nu_P(f)$ be arbitrary. Then the local expansion of f at P has the form

$$f = t^v \sum_{i=0}^{\infty} s_i t^i$$

with all $s_i \in \mathbf{F}_q$. From this expansion we read off the sequence S_2 of coefficients $s_0, s_1, \ldots$.

Theorem 7.2.9 *If $v = \nu_P(f)$ and the integer d is such that $d \geq \deg((f)_\infty)$, then the sequence S_2 constructed above is $(d + v - 1)$-perfect if $v > 0$ and $(d - v + 1)$-perfect if $v \leq 0$.*

Proof. First let $v > 0$. In view of Remark 7.2.3 it suffices to prove that

$$L_n(S_2) \geq \frac{n + 2 - d - v}{2} \qquad \text{for all } n \geq d + v - 1.$$

Fix $n \geq d + v - 1$, put $k = L_n(S_2)$, and write a linear recurrence relation over $\mathbf{F}_q$ of order k satisfied by $s_0, s_1, \ldots, s_{n-1}$ in the form

$$\sum_{h=0}^{k} a_h s_{i+h} = 0 \qquad \text{for } 0 \leq i \leq n - k - 1, \tag{7.7}$$

where $a_h \in \mathbf{F}_q$ for $0 \leq h \leq k$ and $a_k = 1$. Consider the function

$$b = f \sum_{h=0}^{k} a_{k-h} t^h - t^v \sum_{j=0}^{k-1} \left(\sum_{i=0}^{j} a_{k-j+i} s_i \right) t^j.$$

Note that b is a nonzero element of F since $a_k \neq 0$ and $f \notin \mathbf{F}_q(t)$. By applying the linear recurrence relation (7.7) and considering the local expansion of b at P, we obtain $\nu_P(b) \geq n + v$. On the other hand, the pole divisor of b satisfies

$$(b)_\infty \leq (f)_\infty + \left(t^{v+k-1} \right)_\infty.$$

Therefore

$$n + v \leq \nu_P(b) \leq \deg((b)_0) = \deg((b)_\infty) \leq d + 2(v + k - 1),$$

and so

$$k \geq \frac{n + 2 - d - v}{2}.$$

Thus, S_2 is $(d+v-1)$-perfect. In the case $v \leq 0$ the above argument is applied to ft^{-v+1}. $\square$

Example 7.2.10 Let $q = 3$, F be the rational function field $\mathbf{F}_3(x)$, and P be the zero of x. We choose $t = x^2 - x$ and $f = x$. Then we have the local expansion

$$f = -t + t^2 + t^3 - t^4 + t^5 + 0 \cdot t^6 + \cdots.$$

The sequence S_2 of coefficients $-1, 1, 1, -1, 1, 0, \ldots$ is perfect by Theorem 7.2.9.

In the last construction we suppose that $\nu_P(f) \leq 0$. Put $v = -\nu_P(f) \geq 0$. Then the local expansion of f at P can be written in the form

$$f = \sum_{j=1}^{v} r_j t^{j-v-1} + \sum_{i=0}^{\infty} s_i t^i$$

with all $r_j \in \mathbf{F}_q$ and $s_i \in \mathbf{F}_q$. From this expansion we read off the sequence S_3 of coefficients $s_1, s_2, \ldots$.

Theorem 7.2.11 *If $\nu_P(f) \leq 0$ and the integer d is such that $d \geq \deg((f)_\infty)$, then the sequence S_3 constructed above is d-perfect.*

Proof. In view of Remark 7.2.3 it suffices to prove that

$$L_n(S_3) \geq \frac{n+1-d}{2} \qquad \text{for all } n \geq 1.$$

Fix $n \geq 1$, put $k = L_n(S_3)$, and write a linear recurrence relation over $\mathbf{F}_q$ of order k satisfied by $s_1, \ldots, s_n$ in the form

$$\sum_{h=0}^{k} a_h s_{i+h} = 0 \qquad \text{for } 1 \leq i \leq n-k, \tag{7.8}$$

where $a_h \in \mathbf{F}_q$ for $0 \leq h \leq k$ and $a_k = 1$. Consider the function

$$b = \left(f - \sum_{j=1}^{v} r_j t^{j-v-1} \right) \sum_{h=0}^{k} a_{k-h} t^h - \sum_{j=0}^{k} \left(\sum_{i=0}^{j} a_{k-j+i} s_i \right) t^j.$$

Note that b is a nonzero element of F since $a_k \neq 0$ and $f \notin \mathbf{F}_q(t)$. By applying the linear recurrence relation (7.8) and considering the local expansion of b at P, we obtain $\nu_P(b) \geq n+1$. On the other hand, the pole divisor of b satisfies

$$(b)_\infty \leq (f)_\infty - vP + (t^v)_0 - vP + (t^k)_\infty.$$

Therefore

$$n+1 \leq \nu_P(b) \leq \deg((b)_0) = \deg((b)_\infty) \leq d + 2k,$$

and so

$$k \geq \frac{n+1-d}{2}.$$

Thus, S_3 is d-perfect. $\qquad\qquad\qquad\qquad\qquad\qquad\qquad\qquad\qquad\qquad\qquad\qquad$ $\square$

Example 7.2.12 Let $q = 3$, F be the rational function field $\mathbf{F}_3(x)$, and P be the zero of x. We choose $t = x^2 - x$ and $f = 1/x$. Then we have the local expansion

$$f = -t^{-1} - 1 + t + t^2 - t^3 + t^4 + 0 \cdot t^5 + 0 \cdot t^6 + \cdots.$$

The sequence S_3 of coefficients $1, 1, -1, 1, 0, 0, \ldots$ is perfect by Theorem 7.2.11.

7.3   A Construction of Perfect Hash Families

Perfect hash families are special families of maps between finite sets that have found various applications to cryptography, e.g. to broadcast encryption and secret sharing. They also have applications to other areas of computer science such as compiler design and information-retrieval systems. A detailed survey of perfect hash families and their applications was given by Czech, Havas, and Majewski [15].

Definition 7.3.1 Let n, m, and w be integers with $2 \leq w \leq m \leq n$ and let A and B be sets with $|A| = n$ and $|B| = m$. Then an (n, m, w)-**perfect hash family** is a family $\mathcal{H}$ of maps from A to B such that for any subset X of A with $|X| = w$ there exists at least one map in $\mathcal{H}$ that is injective when restricted to X.

We write $\mathrm{PHF}(H; n, m, w)$ for an (n, m, w)-perfect hash family $\mathcal{H}$ with $|\mathcal{H}| = H$. Let $H(n, m, w)$ denote the least positive H for which a $\mathrm{PHF}(H; n, m, w)$ exists. The interest lies in the asymptotic behavior of $H(n, m, w)$ as a function of n when m and w are fixed. According to a result of Fredman and Komlós [25], $H(n, m, w)$ is at least of the order of magnitude $\log n$ for fixed m and w. Elementary probabilistic and nonconstructive arguments show that $H(n, m, w) = O(\log n)$ for fixed m and w, with an implied constant depending only on m and w. Recently, Wang and Xing [170] introduced an explicit construction of perfect hash families based on global function fields which achieves $H(n, m, w) = O(\log n)$ for fixed m and w. In the following we describe this construction which will lead to this result in Corollary 7.3.7 below.

Let $F/\mathbf{F}_q$ be a global function field of genus g with $N(F) \geq 1$ and let G be a divisor of F. Consider the Riemann-Roch space $\mathcal{L}(G)$. To each rational place P of F with $P \notin \mathrm{supp}(G)$ we associate the map $h_P : \mathcal{L}(G) \longrightarrow \mathbf{F}_q$ defined by

$$h_P(f) = f(P) \qquad \text{for all } f \in \mathcal{L}(G).$$

Lemma 7.3.2 *If $\deg(G) \geq 2g + 1$ and P and Q are two different rational places of F that are not in $\mathrm{supp}(G)$, then the maps h_P and h_Q are different.*

Proof. Since $\deg(G - P - Q) \geq 2g - 1$, we obtain $\ell(G - P - Q) = \deg(G) - g - 1$ by the Riemann-Roch theorem. Similarly, we get $\ell(G - P) = \deg(G) - g$. Therefore there exists an $f \in \mathcal{L}(G - P) \backslash \mathcal{L}(G - P - Q)$. For this f it is easily seen that $f(P) = 0$ and $f(Q) \neq 0$, and so $h_P \neq h_Q$. $\square$

Theorem 7.3.3 *Let $F/\mathbf{F}_q$ be a global function field of genus g and let $P_1, \ldots, P_H$ be distinct rational places of F. Let G be a divisor of F with $\deg(G) \geq 2g + 1$ and $\mathrm{supp}(G) \cap \{P_1, \ldots, P_H\} = \emptyset$. Then there exists a perfect hash family $\mathrm{PHF}(H; q^{\deg(G) - g + 1}, q, w)$ whenever $2 \leq w \leq q$ and $H > \binom{w}{2} \deg(G)$.*

Proof. Put $\mathcal{P} = \{P_1, \ldots, P_H\}$ and let $\mathcal{H}$ be the set $\{h_P : P \in \mathcal{P}\}$ of maps from $\mathcal{L}(G)$ to $\mathbf{F}_q$. Then $|\mathcal{H}| = H$ by Lemma 7.3.2 and in the notation of perfect hash families we have $n = |\mathcal{L}(G)| = q^{\deg(G)-g+1}$ and $m = |\mathbf{F}_q| = q$. Let an integer w with $2 \leq w \leq q$ be given and consider an arbitrary subset X of $\mathcal{L}(G)$ with $|X| = w$. We have to find an $R \in \mathcal{P}$ such that h_R is injective when restricted to X. For $\mathcal{D}_X := \{(f-g)^2 : f, g \in X, f \neq g\}$ we have $|\mathcal{D}_X| \leq \binom{w}{2}$. Furthermore, the zeros of an element $(f-g)^2$ of $\mathcal{D}_X$ are exactly the zeros of $f - g$. Since $0 \neq f - g \in \mathcal{L}(G)$ and $\mathrm{supp}(G) \cap \mathcal{P} = \emptyset$, it follows by an argument in the proof of Theorem 6.1.4 that the number of zeros of $f - g$ from the set $\mathcal{P}$ is at most $\deg(G)$. Therefore, the number of places from $\mathcal{P}$ that are zeros for at least one element of $\mathcal{D}_X$ is at most $\binom{w}{2} \deg(G)$. By the condition $H > \binom{w}{2} \deg(G)$, there exists a place $R \in \mathcal{P}$ such that R is not a zero of any element of $\mathcal{D}_X$.

We claim that h_R is injective when restricted to X. Indeed, let $f, g \in X$ with $f \neq g$. Then $(f-g)^2 \in \mathcal{D}_X$, hence R is not a zero of $(f-g)^2$. This means that $f(R) \neq g(R)$, and so $h_R(f) \neq h_R(g)$. $\qquad\square$

Example 7.3.4 Let F be the rational function field over $\mathbf{F}_q$. Choose positive integers t, w, H with $2 \leq w \leq q$ and $\binom{w}{2}t < H \leq q+1$. Furthermore, choose distinct rational places $P_1, \ldots, P_H$ of F and a divisor G of F with $\deg(G) = t$ and $\mathrm{supp}(G) \cap \{P_1, \ldots, P_H\} = \emptyset$. Then Theorem 7.3.3 yields a PHF$(H; q^{t+1}, q, w)$.

Example 7.3.5 Let r be a prime power and put $q = r^2$. Let F be the Hermitian function field $\mathbf{F}_q(x, y)$ defined by $y^r + y = x^{r+1}$. Then $N(F) = r^3 + 1$ and $g = r(r-1)/2$ (see Example 3.1.12). Choose integers t, w, H with $t \geq r(r-1) + 1$, $2 \leq w \leq q$, and $\binom{w}{2}t < H \leq r^3 + 1$. Furthermore, choose distinct rational places $P_1, \ldots, P_H$ of F and a divisor G of F with $\deg(G) = t$ and $\mathrm{supp}(G) \cap \{P_1, \ldots, P_H\} = \emptyset$. Then Theorem 7.3.3 yields a PHF$(H; q^{t+1-r(r-1)/2}, q, w)$.

The following infinite class of examples of perfect hash families is obtained from the Garcia-Stichtenoth tower of global function fields.

Theorem 7.3.6 *Let $r \geq 4$ be a prime power and let c be a real number with $1 \leq c \leq (r-2)/2$. Then for every positive integer i there exists a perfect hash family*

$$\mathrm{PHF}\left((r-1)r^i; \lceil r^{2cr^i} \rceil, r^2, \left\lfloor \left(\frac{2r}{c+1}\right)^{1/2} \right\rfloor \right).$$

Proof. Put $q = r^2$ and consider the Garcia-Stichtenoth tower $K_1 \subseteq K_2 \subseteq \ldots$ of function fields over $\mathbf{F}_q$ defined by $K_1 = \mathbf{F}_q(x_1)$ and $K_i = K_{i-1}(x_i)$ for $i \geq 2$ with

$$x_i^r + x_i = \frac{x_{i-1}^r}{x_{i-1}^{r-1} + 1}.$$

Then $N(K_i) > (r-1)r^i$ and $g(K_i) < r^i$ for all $i \geq 1$ (see Example 5.4.1). Put

$$w = \left\lfloor \left(\frac{2r}{c+1}\right)^{1/2} \right\rfloor.$$

Then $2 \leq w \leq q$ by the definition of c. For each $i \geq 1$ put

$$H_i = (r - 1)r^i, \qquad t_i = \lfloor (c + 1)r^i \rfloor.$$

Then $t_i \geq 2g(K_i) + 1$ since $c \geq 1$, and it is straightforward to check that $H_i > \binom{w}{2}t_i$. Now we choose distinct rational places $P_1, \ldots, P_{H_i}$ of K_i and a divisor G_i of K_i with $\deg(G_i) = t_i$ and $\mathrm{supp}(G_i) \cap \{P_1, \ldots, P_{H_i}\} = \emptyset$. Then by Theorem 7.3.3 there exists a

$$\mathrm{PHF}\left((r - 1)r^i; r^{2(t_i - g(K_i) + 1)}, r^2, \left\lfloor \left(\frac{2r}{c + 1}\right)^{1/2} \right\rfloor\right).$$

The observation that

$$t_i - g(K_i) + 1 > (c + 1)r^i - r^i = cr^i$$

completes the proof.   □

Corollary 7.3.7 *For any integers $m \geq w \geq 2$, there exists a sequence of explicitly constructed perfect hash families* $\mathrm{PHF}(H_i; n_i, m, w)$, $i = 1, 2, \ldots$, *such that $n_i \to \infty$ as $i \to \infty$ and $H_i \leq C \log n_i$ for all $i \geq 1$ with a number C depending only on m and w.*

Proof. For given $m \geq w \geq 2$, let r be the least prime power with $r^2 \geq m$ and $r \geq w^2$. Choose c such that $c + 1 = 2rw^{-2}$. Then $1 \leq c \leq (r - 2)/2$ as required in Theorem 7.3.6. Now we fix a positive integer i. Then Theorem 7.3.6 implies the existence of an explicitly constructed

$$\mathrm{PHF}((r - 1)r^i; n_i, r^2, w) \qquad \text{with } n_i = \lceil r^{2cr^i} \rceil.$$

Since $r^2 \geq m$, it is obvious that we can explicitly construct a $\mathrm{PHF}(H_0; r^2, m, w)$ with H_0 depending only on m and w, e.g. by choosing a bijection on each w-subset of the given r^2-set A and extending it to A in an arbitrary manner. By forming all compositions $h_2 \circ h_1$ of maps

$$h_1 \in \mathrm{PHF}((r - 1)r^i; n_i, r^2, w), \qquad h_2 \in \mathrm{PHF}(H_0; r^2, m, w),$$

we get an explicit construction of a

$$\mathrm{PHF}((r - 1)r^i H_0; n_i, m, w).$$

Finally, it is clear that

$$H_i := (r - 1)r^i H_0 \leq C \log n_i$$

with C depending only on m and w.   □

7.4   Hash Families and Authentication Schemes

We consider now hash families of cryptographic significance that are of a somewhat different type than those in the previous section. For integers $n \geq m \geq 2$ and sets A and B with $|A| = n$ and $|B| = m$, it will be convenient to call a family $\mathcal{H}$ of maps from A to B with $|\mathcal{H}| = H \geq 1$ an $(H; n, m)$ **hash family.**

Definition 7.4.1 Let $\varepsilon > 0$ be a real number. Then an $(H; n, m)$ hash family $\mathcal{H}$ is called ε-**almost strongly universal** if the following two conditions are satisfied:

(i) for any element $a \in A$ and any element $b \in B$, there exist exactly H/m maps $h \in \mathcal{H}$ such that $h(a) = b$;

(ii) for any two distinct elements $a_1, a_2 \in A$ and any two (not necessarily distinct) elements $b_1, b_2 \in B$, there exist at most $\varepsilon H/m$ maps $h \in \mathcal{H}$ such that $h(a_1) = b_1$ and $h(a_2) = b_2$.

Remark 7.4.2 It follows from part (ii) of Definition 7.4.1 that an ε-almost strongly universal $(H; n, m)$ hash family $\mathcal{H}$ enjoys the following property: for any two distinct elements $a_1, a_2 \in A$, there are at most εH maps $h \in \mathcal{H}$ such that $h(a_1) = h(a_2)$. An $(H; n, m)$ hash family with the latter property is called ε-**almost universal**.

The following construction is due to Xing, Wang, and Lam [187]. Let $F/\mathbf{F}_q$ be a global function field of genus g with $N(F) \geq 1$ and let $\mathcal{P}$ be a nonempty set of rational places of F. Furthermore, let D be a positive divisor of F with $\mathrm{supp}(D) \cap \mathcal{P} = \emptyset$. Put $G = D - R$ with a fixed rational place $R \in \mathcal{P}$. Consider the Riemann-Roch space $\mathcal{L}(G)$. To each ordered pair $(P, \alpha) \in \mathcal{P} \times \mathbf{F}_q$ we associate the map $h_{(P,\alpha)} : \mathcal{L}(G) \longrightarrow \mathbf{F}_q$ defined by

$$h_{(P,\alpha)}(f) = f(P) + \alpha \qquad \text{for all } f \in \mathcal{L}(G).$$

Lemma 7.4.3 Let $\mathcal{H} = \{h_{(P,\alpha)} : (P, \alpha) \in \mathcal{P} \times \mathbf{F}_q\}$. If $\deg(D) \geq 2g + 1$, then the cardinality of $\mathcal{H}$ is equal to $|\mathcal{P}|q$.

Proof. If suffices to prove that if $h_{(P,\alpha)} = h_{(Q,\beta)}$ for (P, α) and (Q, β) in $\mathcal{P} \times \mathbf{F}_q$, then $(P, \alpha) = (Q, \beta)$. Thus, assume that

$$h_{(P,\alpha)}(f) = h_{(Q,\beta)}(f) \qquad \text{for all } f \in \mathcal{L}(G).$$

Then, in particular, we get

$$\alpha = h_{(P,\alpha)}(0) = h_{(Q,\beta)}(0) = \beta.$$

This implies that

$$f(P) = f(Q) \qquad \text{for all } f \in \mathcal{L}(G). \tag{7.9}$$

Note that $\mathbf{F}_q \cap \mathcal{L}(G) = \{0\}$ since G is not positive, and $\mathbf{F}_q \subseteq \mathcal{L}(D)$ since D is positive. Furthermore,

$$\ell(D) = \deg(D) + 1 - g = \deg(G) + 2 - g = \ell(G) + 1 \tag{7.10}$$

by the Riemann-Roch theorem, and so $\mathcal{L}(D) = \mathcal{L}(G) \bigoplus \mathbf{F}_q$. Together with (7.9) this yields

$$f(P) = f(Q) \qquad \text{for all } f \in \mathcal{L}(D),$$

and then Lemma 7.3.2 shows that $P = Q$. $\qquad \square$

Theorem 7.4.4 *Let $F/\mathbf{F}_q$ be a global function field of genus g and let $\mathcal{P}$ be a nonempty set of rational places of F. Suppose that D is a positive divisor of F with $\deg(D) \geq 2g + 1$ and $\mathrm{supp}(D) \cap \mathcal{P} = \emptyset$. Then there exists an ε-almost strongly universal $(H; n, m)$ hash family with*

$$H = |\mathcal{P}|q, \quad n = q^{\deg(D)-g}, \quad m = q, \quad \varepsilon = \frac{\deg(D)}{|\mathcal{P}|}.$$

Proof. Let $\mathcal{H} = \{h_{(P,\alpha)} : (P, \alpha) \in \mathcal{P} \times \mathbf{F}_q\}$ be as in Lemma 7.4.3, then $H = |\mathcal{H}| = |\mathcal{P}|q$ by that lemma. We have $n = |A| = |\mathcal{L}(G)| = q^{\deg(D)-g}$ by (7.10) and $m = |B| = |\mathbf{F}_q| = q$. Given $a \in A = \mathcal{L}(G)$ and $b \in B = \mathbf{F}_q$, there exist exactly $|\mathcal{P}| = H/m$ pairs $(P, \alpha) \in \mathcal{P} \times \mathbf{F}_q$ such that

$$h_{(P,\alpha)}(a) = a(P) + \alpha = b,$$

since each choice of P determines α uniquely. Thus, the condition (i) in Definition 7.4.1 is satisfied.

In order to verify condition (ii) in Definition 7.4.1, we choose two distinct elements $a_1, a_2 \in A$ and any two elements $b_1, b_2 \in B$ and consider

$$
\begin{aligned}
k : \quad &= \quad |\{h_{(P,\alpha)} \in \mathcal{H} : h_{(P,\alpha)}(a_1) = b_1,\ h_{(P,\alpha)}(a_2) = b_2\}| \\
&= \quad |\{(P, \alpha) \in \mathcal{P} \times \mathbf{F}_q : a_1(P) + \alpha = b_1,\ a_2(P) + \alpha = b_2\}| \\
&= \quad |\{(P, \alpha) \in \mathcal{P} \times \mathbf{F}_q : (a_1 - a_2 - b_1 + b_2)(P) = 0,\ a_2(P) + \alpha = b_2\}|.
\end{aligned}
$$

As $a_1 - a_2 \in \mathcal{L}(G) \setminus \{0\}$ and $b_1 - b_2 \in \mathbf{F}_q$, we get $0 \neq a_1 - a_2 - b_1 + b_2 \in \mathcal{L}(D)$; compare with the proof of Lemma 7.4.3. Then by an argument used in the proof of Theorem 6.1.4, the number of zeros of $a_1 - a_2 - b_1 + b_2$ from the set $\mathcal{P}$ is at most $\deg(D)$. Since each choice of P determines a unique α from the identity $a_2(P) + \alpha = b_2$, we have

$$k \leq \deg(D) = \frac{\deg(D)}{|\mathcal{P}|} \cdot \frac{H}{m}.$$

Hence we can take $\varepsilon = \deg(D)/|\mathcal{P}|$. $\square$

Example 7.4.5 Let F be the rational function field over $\mathbf{F}_q$ and let $\mathcal{P}$ be the set of all rational places of F. For an integer d with $2 \leq d \leq q$, choose a positive divisor D of F with $\deg(D) = d$ and $\mathrm{supp}(D) \cap \mathcal{P} = \emptyset$. Then Theorem 7.4.4 yields an ε-almost strongly universal $(H; n, m)$ hash family with

$$H = (q + 1)q, \quad n = q^d, \quad m = q, \quad \varepsilon = \frac{d}{q + 1} < 1.$$

Example 7.4.6 Let r be a prime power and put $q = r^2$. Let F be the Hermitian function field $\mathbf{F}_q(x, y)$ defined by $y^r + y = x^{r+1}$. Then $N(F) = r^3 + 1$ and $g(F) = r(r-1)/2$ (see Example 3.1.12). Fix a rational place P of F and put $D = dP$ for an integer d with $r(r-1) + 1 \leq d < r^3$. Let $\mathcal{P}$ be the set of all rational places of F except P. Then Theorem 7.4.4 yields an ε-almost strongly universal $(H; n, m)$ hash family with

$$H = r^5, \quad n = r^{2d - r(r-1)}, \quad m = r^2, \quad \varepsilon = \frac{d}{r^3} < 1.$$

Example 7.4.7 Let $r \geq 3$ be a prime power and put $q = r^2$. Consider the Garcia-Stichtenoth tower $K_1 \subseteq K_2 \subseteq \ldots$ of function fields over $\mathbf{F}_q$ as in the proof of Theorem 7.3.6. For each positive integer i fix a rational place P_i of K_i and put $D_i = d_i P_i$ for an integer d_i with $2r^i \leq d_i \leq (r-1)r^i$. Let $\mathcal{P}_i$ be a set of rational places of K_i with $P_i \notin \mathcal{P}_i$ and $|\mathcal{P}_i| = (r-1)r^i$. Then Theorem 7.4.4 yields an ε_i-almost strongly universal $(H_i; n_i, m)$ hash family with

$$H_i = (r-1)r^{i+2}, \quad n_i = r^{2(d_i - g(K_i))} > r^{2(d_i - r^i)}, \quad m = r^2, \quad \varepsilon_i = \frac{d_i}{(r-1)r^i} \leq 1.$$

If we choose $d_i = cr^i$ for all $i \geq 1$ with a fixed integer c satisfying $2 \leq c \leq r-1$, then we get a fixed value $\varepsilon = c/(r-1)$.

An **authentication scheme** (without secrecy) consists of a finite set A of plaintexts (or source states), a finite set B of authenticators, and a finite set K of keys, together with an authentication map $f : A \times K \longrightarrow B$. When the transmitter wants to send the plaintext $a \in A$ using a key $k \in K$, which is secretly shared with the receiver, then the authenticator $b = f(a, k) \in B$ is formed and the (signed) message (a, b) is transmitted, i.e., the authenticator b is appended to the plaintext a. Upon receipt of the message (a, b), its authenticity is examined by checking whether $b = f(a, k)$ or not, using the secret key k. Two types of attacks by an opponent are considered:

(i) an **impersonation attack** in which the opponent places a (new) message (a', b') into the channel;

(ii) a **substitution attack** in which the opponent observes a message (a, b) in the channel and changes it to a message (a', b') with $a' \neq a$.

We say that a message (a, b) is **valid** if there exists a key $k \in K$ such that $b = f(a, k)$.

We assume that the plaintexts and keys are uniformly distributed. The **deception probabilities** p_I and p_S are the probabilities of success for the opponent when trying an impersonation attack and a substitution attack, respectively. More precisely, p_I is the maximum probability for a message (a, b) to be valid, and p_S is the maximum conditional probability for a message (a', b') to be valid, given that the valid message (a, b) with $a' \neq a$ has been observed. It is known that both p_I and p_S are at least $1/|B|$ (see [154, Chapter 10]). Naturally, an important aim in the construction of authentication schemes is to keep the deception probabilities small.

It was pointed out by Stinson [153] that ε-almost strongly universal $(H; n, m)$ hash families can be interpreted as authentication schemes with minimum deception probability $p_I = 1/|B|$.

Proposition 7.4.8 *If there exists an ε-almost strongly universal $(H; n, m)$ hash family, then there exists an authentication scheme for n plaintexts, having m authenticators and H keys, such that $p_I = 1/m$ and $p_S \leq \varepsilon$.*

Proof. Let $\mathcal{H}$ be an ε-almost strongly universal $(H; n, m)$ hash family of maps from A to B, so that $|\mathcal{H}| = H$, $|A| = n$, and $|B| = m$. We can index the elements of $\mathcal{H}$ by keys k

from a set K, i.e., $\mathcal{H} = \{h_k : k \in K\}$ with $|K| = H$. Now we define the authentication map $f : A \times K \longrightarrow B$ by

$$f(a, k) = h_k(a) \qquad \text{for all } (a, k) \in A \times K.$$

Then $p_I = 1/m$ follows from condition (i) in Definition 7.4.1 and $p_S \leq \varepsilon$ follows from condition (ii) in Definition 7.4.1. $\square$

Therefore, the construction and the examples of ε-almost strongly universal $(H; n, m)$ hash families in this section also lead to useful authentication schemes. For instance, Example 7.4.7 yields a sequence of authentication schemes which are better, in a sense that can be made precise, than the authentication schemes constructed by Bierbrauer [9] and Helleseth and Johansson [50]; see Xing, Wang, and Lam [187] for the detailed comparison. We refer to Simmons [147] and Stinson [154, Chapter 10] for further background on authentication schemes.

Chapter 8

Applications to Low-Discrepancy Sequences

In the theory of uniform distribution of sequences in number theory and also in quasi-Monte Carlo methods in numerical analysis, one of the key problems is to distribute points as uniformly as possible over a unit cube of arbitrary dimension. The precise formulation of this problem leads to the concept of star discrepancy and to the requirement of constructing low-discrepancy point sets and sequences. There are several classical constructions of low-discrepancy sequences, and consequently of low-discrepancy point sets, but by far the most powerful construction methods were developed recently and they are based on global function fields, in particular on global function fields with many rational places. These methods improve on all previous methods of obtaining low-discrepancy sequences and yield, in fact, low-discrepancy sequences with parameters which are best possible in terms of their order of magnitude. In this chapter we first present the relevant background on star discrepancy and on the theory of (t, m, s)-nets and (t, s)-sequences. Then we describe in detail two construction principles for low-discrepancy sequences that depend in an essential way on the choice of suitable places of global function fields.

8.1   Background on (t, m, s)-Nets and (t, s)-Sequences

For any integer $s \geq 1$ we denote by I^s the s-dimensional unit cube $[0, 1]^s$. We consider (finite) point sets and (infinite) sequences of points in I^s, where the term "point set" is used in the sense of the combinatorial notion of "multiset", that is, a set in which multiplicity of elements is allowed and taken into account. We introduce the star discrepancy which is a measure for the deviation from uniformity for point sets and sequences. For a subinterval J of I^s and for a point set P consisting of N points $\mathbf{x}_0, \mathbf{x}_1, \ldots, \mathbf{x}_{N-1} \in I^s$ we write $A(J; P)$ for the number of integers n with $0 \leq n \leq N - 1$ for which $\mathbf{x}_n \in J$. We put

$$R(J; P) = \frac{A(J; P)}{N} - \lambda_s(J), \qquad (8.1)$$

where λ_s is the s-dimensional Lebesgue measure.

Definition 8.1.1 The **star discrepancy** $D_N^*(P)$ of the point set P is defined by

$$D_N^*(P) = \sup_J |R(J; P)|,$$

where the supremum is extended over all subintervals J of I^s with one vertex at the origin. For a sequence S of points in I^s, the **star discrepancy** $D_N^*(S)$ is meant to be the star discrepancy of the first N terms of S.

The star discrepancy is a crucial quantity in the quasi-Monte Carlo method for numerical integration. This method is usually applied in a multidimensional setting and, like any scheme for numerical integration, it is invoked in situations where an integral is very complicated or impossible to evaluate analytically. For a given continuous integrand f on I^s we approximate its integral by

$$\int_{I^s} f \, d\lambda_s \approx \frac{1}{N} \sum_{n=0}^{N-1} f(\mathbf{x}_n),$$

where $\mathbf{x}_0, \mathbf{x}_1, \ldots, \mathbf{x}_{N-1}$ are suitably chosen points from I^s. For any choice of these points we have the error bound

$$\left| \int_{I^s} f \, d\lambda_s - \frac{1}{N} \sum_{n=0}^{N-1} f(\mathbf{x}_n) \right| \le V(f) D_N^*(P), \tag{8.2}$$

where $D_N^*(P)$ is the star discrepancy of the point set P formed by $\mathbf{x}_0, \mathbf{x}_1, \ldots, \mathbf{x}_{N-1}$. The number $V(f)$ depends only on the integrand f and measures the oscillation of f. Concretely, $V(f)$ is the variation of f on I^s in the sense of Hardy and Krause and the error bound (8.2) is meaningful only in the case where $V(f)$ is finite. It is clear from this error bound that the desirable point sets P are those for which $D_N^*(P)$ is small. Furthermore, if S is a sequence of points $\mathbf{x}_0, \mathbf{x}_1, \ldots$ in I^s, then we get a convergent numerical integration scheme, i.e.,

$$\lim_{N \to \infty} \frac{1}{N} \sum_{n=0}^{N-1} f(\mathbf{x}_n) = \int_{I^s} f \, d\lambda_s, \tag{8.3}$$

whenever $V(f)$ is finite and $\lim_{N \to \infty} D_N^*(S) = 0$. In fact, the theory of uniform distribution of sequences shows that the conditions on f for the validity of (8.3) can be relaxed and that all we need is that f be Riemann integrable on I^s. For the proof of this statement and of (8.2) we refer to the book of Kuipers and Niederreiter [62]. Detailed information on the quasi-Monte Carlo method and discrepancy theory can be found in the books of Drmota and Tichy [20], Hua and Wang [54], and Niederreiter [94] and in the survey articles of Niederreiter [86] and Spanier and Maize [149].

As we have seen above, an important motivation for the construction of point sets P with small star discrepancy $D_N^*(P)$ stems from the quasi-Monte Carlo method, but this construction problem is also of great theoretical interest. It is widely believed that for any dimension $s \ge 1$ and any $N \ge 2$ the least possible star discrepancy that can be achieved for a point set P of N points in I^s has order of magnitude $N^{-1}(\log N)^{s-1}$, and this is known to be true for $s = 1$ and $s = 2$. We call a point set P of N points in I^s a **low-discrepancy point set** if

$$D_N^*(P) = O(N^{-1}(\log N)^{s-1}).$$ (8.4)

Similarly, a sequence S of points in I^s is called a **low-discrepancy sequence** if

$$D_N^*(S) = O(N^{-1}(\log N)^s) \qquad \text{for all } N \geq 2.$$ (8.5)

In both (8.4) and (8.5), the implied constant is assumed to be independent of N. The order of magnitude in (8.5) is conjectured to be the least that can be obtained in dimension s, and this has been proved for $s = 1$.

The most powerful current methods for the construction of low-discrepancy point sets and sequences are based on the theory of (t, m, s)-nets and (t, s)-sequences, which are point sets, respectively sequences, satisfying strong uniformity properties with respect to their distribution in I^s. The idea of the notion of a (t, m, s)-net P is to try to make the star discrepancy $D_N^*(P)$ small by requiring that the quantity $R(J; P)$ in (8.1) be equal to 0 for a large, but finite family of subintervals J of I^s. Given an integer $b \geq 2$, it will be convenient to call an interval of the form

$$J = \prod_{i=1}^{s} [a_i b^{-d_i}, (a_i + 1)b^{-d_i}) \subseteq I^s$$

with integers $d_i \geq 0$ and $0 \leq a_i < b^{d_i}$ for $1 \leq i \leq s$ an **elementary interval in base** b.

Definition 8.1.2 For integers $b \geq 2$ and $0 \leq t \leq m$, a (t, m, s)-**net in base** b is a point set P consisting of b^m points in I^s such that $R(J; P) = 0$ for every elementary interval $J \subseteq I^s$ in base b with $\lambda_s(J) = b^{t-m}$.

Note that every elementary interval $J \subseteq I^s$ in base b with $\lambda_s(J) \geq b^{t-m}$ can be written as the disjoint union of elementary intervals in base b with λ_s-measure b^{t-m}. This implies that a (t, m, s)-net P in base b satisfies $R(J; P) = 0$ for every elementary interval $J \subseteq I^s$ in base b with $\lambda_s(J) \geq b^{t-m}$. A consequence of this fact is that a (t, m, s)-net in base b is also a (u, m, s)-net in base b for all integers u with $t \leq u \leq m$. Therefore it is clear that smaller values of t mean stronger uniformity properties of a (t, m, s)-net in base b. The number t is often called the **quality parameter** of a (t, m, s)-net in base b. These considerations also suggest the following terminology.

Definition 8.1.3 A (t, m, s)-net P in base b is a **strict** (t, m, s)-**net in base** b if t is the least value u such that P is a (u, m, s)-net in base b.

Explicit upper bounds on the star discrepancy of (t, m, s)-nets in base b can be found in Niederreiter [87, Section 3], [94, Chapter 4]. We mention here only the following simplified bound: for any (t, m, s)-net P in base b with $m \geq 1$ we have

$$D_N^*(P) \leq B_b(s, t)N^{-1}(\log N)^{s-1} + O(b^t N^{-1}(\log N)^{s-2}),$$

where $N = b^m$ and the implied constant in the Landau symbol depends only on b and s. Here

$$B_b(s,t) = b^t \left(\frac{b-1}{2 \log b} \right)^{s-1}$$

if either (i) $s = 2$; or (ii) $b = 2, s = 3, 4$; in all other cases

$$B_b(s,t) = \frac{b^t}{(s-1)!} \left(\frac{\lfloor b/2 \rfloor}{\log b} \right)^{s-1}.$$

Because of the exponential dependence of $B_b(s,t)$ on the quality parameter t, even a small decrease in the value of t leads to a considerable amelioration of the discrepancy bound.

Remark 8.1.4 For $m = t$ and $m = t + 1$, a (t,m,s)-net in base b always exists. If $m = t$, then any point set of b^t points in $[0,1)^s$ forms a (t,m,s)-net in base b. If $m = t + 1$, then the point set consisting of the points $(n/b, \ldots, n/b) \in I^s, n = 0, 1, \ldots, b-1$, each taken with multiplicity b^t, forms a (t,m,s)-net in base b. For $m \geq t + 2$ there are combinatorial obstructions to the general existence of (t,m,s)-nets in base b. This was first pointed out in Niederreiter [87, Section 5] where it was shown, for instance, that for $m \geq 2$ a $(0,m,s)$-net in base b can exist only if $s \leq b + 1$ (by the way, if b is a prime power, then we will see in Example 8.3.2 that "only if" can be replaced by "if and only if"). Later, a combinatorial equivalence between (t,m,s)-nets in base b and so-called generalized orthogonal arrays was established, and this led to lower bounds on the quality parameter t in a (t,m,s)-net in base b. We refer to Clayman et al. [14] for a recent account and tables of such lower bounds.

Now we introduce the sequence analogs of (t,m,s)-nets, namely (t,s)-sequences. For a base $b \geq 2$ we write $Z_b = \{0, 1, \ldots, b-1\}$ for the set of digits in base b. Given a real number $x \in [0,1]$, let

$$x = \sum_{j=1}^{\infty} y_j b^{-j} \qquad \text{with all } y_j \in Z_b \tag{8.6}$$

be a b-adic expansion of x, where the case in which $y_j = b - 1$ for all but finitely many j is allowed. For an integer $m \geq 1$ we define the truncation

$$[x]_{b,m} = \sum_{j=1}^{m} y_j b^{-j}.$$

It should be emphasized that this truncation operates on the *expansion* of x and not on x itself, since it may yield different results depending on which b-adic expansion of x is used. If $\mathbf{x} = (x^{(1)}, \ldots, x^{(s)}) \in I^s$ and the $x^{(i)}, 1 \leq i \leq s$, are given by prescribed b-adic expansions, then we define

$$[\mathbf{x}]_{b,m} = \left([x^{(1)}]_{b,m}, \ldots, [x^{(s)}]_{b,m} \right).$$

Definition 8.1.5 Let $b \geq 2$ and $t \geq 0$ be integers. A sequence $\mathbf{x}_0, \mathbf{x}_1, \ldots$ of points in I^s is a (t, s)-**sequence in base** b if for all integers $k \geq 0$ and $m > t$ the points $[\mathbf{x}_n]_{b,m}$ with $kb^m \leq n < (k+1)b^m$ form a (t, m, s)-net in base b. Here the coordinates of all points of the sequence are given by prescribed b-adic expansions of the form (8.6).

By the remarks following Definition 8.1.2, it is clear that a (t, s)-sequence in base b is also a (u, s)-sequence in base b for all integers $u \geq t$. This suggests the next definition.

Definition 8.1.6 A (t, s)-sequence S in base b is a **strict** (t, s)-**sequence in base** b if t is the least value u such that S is a (u, s)-sequence in base b.

Detailed information on upper bounds for the star discrepancy of (t, s)-sequences is provided in Niederreiter [87, Section 4], [94, Chapter 4]; compare also with [102, Remark 2]. We state here only the following simplified bound: for any (t, s)-sequence S in base b we have

$$D_N^*(S) \leq C_b(s, t) N^{-1} (\log N)^s + O(b^t N^{-1} (\log N)^{s-1}) \qquad \text{for all } N \geq 2, \qquad (8.7)$$

where the implied constant in the Landau symbol depends only on b and s. Here

$$C_b(s, t) = \frac{b^t}{s} \left(\frac{b-1}{2 \log b} \right)^s$$

if either (i) $s = 2$; or (ii) $b = 2, s = 3, 4$; in all other cases

$$C_b(s, t) = \frac{b^t}{s!} \cdot \frac{b-1}{2 \lfloor b/2 \rfloor} \left(\frac{\lfloor b/2 \rfloor}{\log b} \right)^s.$$

In particular, any (t, s)-sequence in base b is a low-discrepancy sequence in the sense of (8.5). As in the case of (t, m, s)-nets, smaller values of t mean stronger uniformity properties of a (t, s)-sequence in base b, and we call t the **quality parameter** of a (t, s)-sequence in base b. Even small improvements in the value of t lead to considerably better discrepancy bounds for (t, s)-sequences.

It is trivial that a (t, s)-sequence in base b yields a (t, m, s)-net in base b for every integer $m \geq t$. The following simple result shows that we can even gain one dimension in the net.

Lemma 8.1.7 *If there exists a* (t, s)-*sequence in base* b, *then for every integer* $m \geq t$ *there exists a* $(t, m, s+1)$-*net in base* b.

Proof. Let $\mathbf{x}_0, \mathbf{x}_1, \ldots$ be a (t, s)-sequence in base b. Given $m \geq t$, we show that the points

$$\mathbf{y}_n = \left([\mathbf{x}_n]_{b,m}, \frac{n}{b^m} \right) \in I^{s+1}, \quad n = 0, 1, \ldots, b^m - 1,$$

form a $(t, m, s+1)$-net P in base b. Let

$$J = \prod_{i=1}^{s+1} [a_i b^{-d_i}, (a_i + 1) b^{-d_i}) \subseteq I^{s+1}$$

be an elementary interval in base b with $\lambda_{s+1}(J) = b^{t-m}$; hence

$$\sum_{i=1}^{s+1} d_i = m - t.$$

Then $\mathbf{y}_n \in J$ if and only if the following two conditions are satisfied:

(i) $a_{s+1}b^{m-d_{s+1}} \leq n < (a_{s+1}+1)b^{m-d_{s+1}}$;

(ii) $[\mathbf{x}_n]_{b,m} \in J' := \prod_{i=1}^{s}[a_ib^{-d_i}, (a_i+1)b^{-d_i})$.

Since we have $m - d_{s+1} \geq t$ and $\mathbf{x}_0, \mathbf{x}_1, \ldots$ is a (t, s)-sequence in base b, the points

$$[\mathbf{x}_n]_{b,m-d_{s+1}} \quad \text{with } a_{s+1}b^{m-d_{s+1}} \leq n < (a_{s+1}+1)b^{m-d_{s+1}}$$

form a $(t, m - d_{s+1}, s)$-net in base b. Consequently, the point set P' consisting of

$$[\mathbf{x}_n]_{b,m} \quad \text{with } a_{s+1}b^{m-d_{s+1}} \leq n < (a_{s+1}+1)b^{m-d_{s+1}}$$

is a $(t, m - d_{s+1}, s)$-net in base b. Thus $A(J'; P') = b^t$, hence $A(J; P) = b^t$, and so $R(J; P) = 0$. $\qquad\square$

Remark 8.1.8 If we combine results on the combinatorial obstructions to the existence of (t, m, s)-nets in base b (see Remark 8.1.4) with Lemma 8.1.7, then we get necessary conditions for the existence of (t, s)-sequences in base b. For instance, a $(0, s)$-sequence in base b can exist only if $s \leq b$ (by the way, if b is a prime power, then we will see in Example 8.3.2 that "only if" can be replaced by "if and only if"). The currently best general lower bound on the quality parameter t in a (t, s)-sequence in base b was shown by Niederreiter and Xing [104], and it says that for every base $b \geq 2$ and every dimension $s \geq 1$, a necessary condition for the existence of a (t, s)-sequence in base b is

$$t \geq \frac{s}{b} - \log_b \frac{(b-1)s + b + 1}{2}, \tag{8.8}$$

where $\log_b$ denotes the logarithm to the base b. A special variant of this bound will be proved in Theorem 8.2.16.

Expository accounts of the theory of (t, m, s)-nets and (t, s)-sequences are given in the books of Laywine and Mullen [73, Chapter 15] and Niederreiter [94, Chapter 4]. Surveys of various constructions of (t, m, s)-nets and (t, s)-sequences can be found in the papers of Niederreiter [96], [98] and Niederreiter and Xing [104], [109], [113]. A chronological listing of constructions of (t, m, s)-nets and (t, s)-sequences is available when combining the articles of Mullen, Mahalanabis, and Niederreiter [83] and Clayman et al. [14], and these two papers contain also tables of the quality parameter t. Recent improvements on these tables are discussed in Schmid [135].

8.2 The Digital Method

We describe a general principle for the construction of (t, m, s)-nets and (t, s)-sequences which was introduced in Niederreiter [87, Section 6] and is called the **digital method**. Most of the known constructions of (t, m, s)-nets and (t, s)-sequences are based on the digital method. We present this method in a simplified version which shows, however, all the essential features of the method.

First we explain the digital method for the construction of (t, m, s)-nets. We fix a base $b \geq 2$ and integers $m \geq 1$ and $s \geq 1$. From Section 8.1 we recall the notation $Z_b = \{0, 1, \ldots, b-1\}$ for the set of digits in base b. Now we choose a finite commutative ring R with identity and of order b. We set up a map $\phi_m : R^m \longrightarrow [0, 1]$ by selecting a bijection $\eta : R \longrightarrow Z_b$ and putting

$$\phi_m(r_1, \ldots, r_m) = \sum_{j=1}^{m} \eta(r_j) b^{-j} \qquad \text{for } (r_1, \ldots, r_m) \in R^m.$$

Furthermore, we choose $m \times m$ matrices $C_1, \ldots, C_s$ over R, the so-called **generating matrices**. Finally, we define the point set consisting of the b^m points

$$(\phi_m(\mathbf{n}C_1), \ldots, \phi_m(\mathbf{n}C_s)) \in I^s \tag{8.9}$$

with the row vector $\mathbf{n}$ running through R^m (the order of the points is irrelevant).

Definition 8.2.1 If the point set (8.9) forms a (strict) (t, m, s)-net in base b, then it is called a **digital (strict) (t, m, s)-net constructed over R**. A point set which is a digital (strict) (t, m, s)-net constructed over R for some finite commutative ring R with identity and of order b is also called a **digital (strict) (t, m, s)-net in base b**.

Any point set (8.9) is a (t, m, s)-net in base b for some t, just take $t = m$. A more useful result for the determination of a suitable quality parameter t is provided by the following theorem. For $1 \leq i \leq s$ we write $\mathbf{c}_1^{(i)}, \ldots, \mathbf{c}_m^{(i)}$ for the column vectors of the generating matrix C_i. Furthermore, for any s-tuple $\mathbf{d} = (d_1, \ldots, d_s)$ of integers with $0 \leq d_i \leq m$ for $1 \leq i \leq s$ and $d := \sum_{i=1}^{s} d_i > 0$ we introduce the $m \times d$ matrix

$$C_{\mathbf{d}} = \left(\mathbf{c}_1^{(1)} \ldots \mathbf{c}_{d_1}^{(1)} \ldots \mathbf{c}_1^{(s)} \ldots \mathbf{c}_{d_s}^{(s)} \right).$$

Theorem 8.2.2 Let t be an integer with $0 \leq t < m$. Then the point set (8.9) is a digital (t, m, s)-net constructed over R if and only if for any s-tuple $\mathbf{d} = (d_1, \ldots, d_s)$ of nonnegative integers with $\sum_{i=1}^{s} d_i = m - t$ the system of homogeneous linear equations

$$\mathbf{n}C_{\mathbf{d}} = \mathbf{0} \in R^{m-t} \tag{8.10}$$

has exactly b^t solutions $\mathbf{n} \in R^m$.

Proof. Let

$$J = \prod_{i=1}^{s} \left[a_i b^{-d_i}, (a_i + 1) b^{-d_i} \right) \subseteq I^s$$

be an elementary interval in base b with $\lambda_s(J) = b^{t-m}$, that is, with $\sum_{i=1}^{s} d_i = m - t$. For a given $\mathbf{n} \in R^m$, the corresponding point in (8.9) lies in J if and only if

$$\phi_m(\mathbf{n}C_i) \in \left[a_i b^{-d_i}, (a_i + 1) b^{-d_i} \right) \qquad \text{for } 1 \le i \le s.$$

This condition means that for $1 \le i \le s$ the first d_i b-adic digits of $\phi_m(\mathbf{n}C_i)$ and $a_i b^{-d_i}$ agree, and this is easily seen to be equivalent to $\mathbf{n}C_{\mathbf{d}} = \mathbf{a}$ for some $\mathbf{a} \in R^{m-t}$ depending only on J. Furthermore, if J runs through all possibilities for fixed $\mathbf{d}$, then $\mathbf{a}$ runs through R^{m-t}. Note that the property that the point set P in (8.9) forms a (t, m, s)-net in base b means that $A(J; P) = b^t$ for all J with $\lambda_s(J) = b^{t-m}$, and this is now seen to be equivalent to $\mathbf{n}C_{\mathbf{d}} = \mathbf{a}$ having exactly b^t solutions $\mathbf{n} \in R^m$ for all $\mathbf{a}$ and $\mathbf{d}$. By considering the group homomorphism $\mathbf{n} \in R^m \mapsto \mathbf{n}C_{\mathbf{d}} \in R^{m-t}$, we conclude that it suffices to impose the last condition on the homogeneous case $\mathbf{a} = \mathbf{0}$ only. □

We arrive at a more convenient formulation of Theorem 8.2.2 in the case where R is a finite field. In this case we write q instead of b, so that we have $R = \mathbf{F}_q$. As above we consider the generating matrices $C_1, \ldots, C_s$ of the point set (8.9), and for $1 \le i \le s$ let $\mathbf{c}_1^{(i)}, \ldots, \mathbf{c}_m^{(i)} \in \mathbf{F}_q^m$ denote the column vectors of C_i. The following notion is crucial.

Definition 8.2.3 Let d be an integer with $0 \le d \le m$. The system $C = \{\mathbf{c}_j^{(i)} \in \mathbf{F}_q^m : 1 \le i \le s, 1 \le j \le m\}$ of vectors is a (d, m, s)-**system over** $\mathbf{F}_q$ if for any nonnegative integers $d_1, \ldots, d_s$ with $\sum_{i=1}^{s} d_i = d$ the vectors $\mathbf{c}_j^{(i)}, 1 \le j \le d_i, 1 \le i \le s$, are linearly independent over $\mathbf{F}_q$ (this property is assumed to be vacuously satisfied for $d = 0$). The largest value of d such that C is a (d, m, s)-system over $\mathbf{F}_q$ is denoted by $\varrho(C)$.

Theorem 8.2.4 *The point set* (8.9) *obtained by the digital method with $R = \mathbf{F}_q$ is a digital (t, m, s)-net constructed over $\mathbf{F}_q$ if and only if the system $C = \{\mathbf{c}_j^{(i)} \in \mathbf{F}_q^m : 1 \le i \le s, 1 \le j \le m\}$ of column vectors of the generating matrices is a (d, m, s)-system over $\mathbf{F}_q$ with $d = m - t$. In particular, the point set* (8.9) *is a digital strict (t, m, s)-net constructed over $\mathbf{F}_q$ with $t = m - \varrho(C)$.*

Proof. This follows from Theorem 8.2.2, Definition 8.2.3, and elementary linear algebra. □

Example 8.2.5 Let q and m be arbitrary and let $s = 2$. Choose a basis $\{\mathbf{b}_1, \ldots, \mathbf{b}_m\}$ of $\mathbf{F}_q^m$ and define the generating matrix C_1 with column vectors $\mathbf{c}_j^{(1)} = \mathbf{b}_j$ for $1 \le j \le m$ and the generating matrix C_2 with column vectors $\mathbf{c}_j^{(2)} = \mathbf{b}_{m-j+1}$ for $1 \le j \le m$. Then it is clear that the system $C = \{\mathbf{c}_j^{(i)} : 1 \le i \le 2, 1 \le j \le m\}$ satisfies $\varrho(C) = m$, and so the corresponding point set is a digital $(0, m, 2)$-net constructed over $\mathbf{F}_q$ by Theorem 8.2.4.

Remark 8.2.6 It is easy to prove that for given q and $m \geq 2$, there can be no digital $(0, m, s)$-net constructed over $\mathbf{F}_q$ for $s \geq q+2$. For otherwise there would exist an (m, m, s)-system $C = \{\mathbf{c}_j^{(i)} \in \mathbf{F}_q^m : 1 \leq i \leq s, 1 \leq j \leq m\}$ over $\mathbf{F}_q$ by Theorem 8.2.4. Then $\{\mathbf{c}_1^{(1)}, \ldots, \mathbf{c}_m^{(1)}\}$ is a basis of $\mathbf{F}_q^m$. In the representation of each $\mathbf{c}_1^{(i)}, 2 \leq i \leq s$, in terms of this basis, the coefficient of $\mathbf{c}_m^{(1)}$ must be nonzero by the definition of an (m, m, s)-system over $\mathbf{F}_q$. Thus, there exist nonzero $\beta_i \in \mathbf{F}_q, 2 \leq i \leq s$, such that $\beta_i \mathbf{c}_1^{(i)} - \mathbf{c}_m^{(1)}$ is a linear combination of $\mathbf{c}_1^{(1)}, \ldots, \mathbf{c}_{m-1}^{(1)}$. Let γ_i be the coefficient of $\mathbf{c}_{m-1}^{(1)}$ in the last representation. Since $s \geq q + 2$, two of the elements $\gamma_2, \ldots, \gamma_s$ must be identical, say $\gamma_h = \gamma_k$ with $2 \leq h < k \leq s$. Then by subtraction we see that $\beta_h \mathbf{c}_1^{(h)} - \beta_k \mathbf{c}_1^{(k)}$ is a linear combination of $\mathbf{c}_1^{(1)}, \ldots, \mathbf{c}_{m-2}^{(1)}$, a contradiction to C being an (m, m, s)-system over $\mathbf{F}_q$. Note that by Remark 8.1.4, for any given b and $m \geq 2$ there can be no $(0, m, s)$-net in base b for $s \geq b+2$, but this is a bit more difficult to prove.

If we know how to obtain digital nets constructed over finite fields, then we can get digital nets in an arbitrary base b by a straightforward direct product construction.

Theorem 8.2.7 Let $b = q_1 \cdots q_h$ be a product of prime powers. Suppose that for $1 \leq r \leq h$ there exists a digital (t_r, m, s)-net constructed over $\mathbf{F}_{q_r}$. Then there exists a digital (t, m, s)-net in base b with

$$t = \max_{1 \leq r \leq h} t_r.$$

Proof. By Theorem 8.2.4, for $1 \leq r \leq h$ there exist generating matrices $C_1^{(r)}, \ldots, C_s^{(r)} \in \mathbf{F}_{q_r}^{m \times m}$ for which the system of column vectors is an $(m - t_r, m, s)$-system over $\mathbf{F}_{q_r}$. Now let R be the ring direct product $\prod_{r=1}^{h} \mathbf{F}_{q_r}$. We define generating matrices $C_1, \ldots, C_s \in R^{m \times m}$ in the following way. For $1 \leq i \leq s$ each entry of C_i is the element of R that is given as the h-tuple of the corresponding entries of the $C_i^{(r)}, 1 \leq r \leq h$. Now we check the criterion in Theorem 8.2.2 with $t = \max(t_1, \ldots, t_h)$, where we can assume that $t < m$. For a given $\mathbf{d} = (d_1, \ldots, d_s) \in \mathbf{N}_0^s$ with $\sum_{i=1}^{s} d_i = m - t$ we study the number of solutions $\mathbf{n} \in R^m$ of the system (8.10). Because of the direct product structure of R, this is equivalent to considering, for $1 \leq r \leq h$, the system

$$\mathbf{n}_r C_\mathbf{d}^{(r)} = 0 \in \mathbf{F}_{q_r}^{m-t} \tag{8.11}$$

with unknown $\mathbf{n}_r \in \mathbf{F}_{q_r}^m$. Since $m - t \leq m - t_r$ for $1 \leq r \leq h$, the matrix $C_\mathbf{d}^{(r)}$ has rank $m - t$ for each r. Thus, the system (8.11) has exactly q_r^t solutions for each r. Consequently, the system (8.10) has exactly $\prod_{r=1}^{h} q_r^t = b^t$ solutions. $\square$

Now we describe the digital method for the construction of sequences. We fix a base $b \geq 2$ and a dimension $s \geq 1$. As in the construction of digital nets, we choose a finite commutative ring R with identity and of order b. We set up a map $\phi_\infty : R^\infty \longrightarrow [0, 1]$ by selecting a bijection $\eta : R \longrightarrow Z_b$ and putting

$$\phi_\infty(r_1, r_2, \ldots) = \sum_{j=1}^{\infty} \eta(r_j) b^{-j} \qquad \text{for } (r_1, r_2, \ldots) \in R^\infty.$$

Furthermore, we choose **generating matrices** $C_1, \ldots, C_s$ over R with infinitely many rows and columns. At this stage we have to proceed in a somewhat more subtle manner than before since the order of terms in a sequence matters. For $n = 0, 1, \ldots$ let

$$n = \sum_{l=0}^{\infty} a_l(n) b^l$$

be the digit expansion of n in base b, where $a_l(n) \in Z_b$ for $l \geq 0$ and $a_l(n) = 0$ for all sufficiently large l. Choose a bijection $\psi : Z_b \longrightarrow R$ with $\psi(0) = 0$ and associate with n the sequence

$$\mathbf{n} = (\psi(a_0(n)), \psi(a_1(n)), \ldots) \in R^{\infty}.$$

Now we define the sequence $\mathbf{x}_0, \mathbf{x}_1, \ldots$ of points in I^s by

$$\mathbf{x}_n = (\phi_{\infty}(\mathbf{n}C_1), \ldots, \phi_{\infty}(\mathbf{n}C_s)) \qquad \text{for } n = 0, 1, \ldots. \tag{8.12}$$

Note that the products $\mathbf{n}C_i$ are well defined since $\mathbf{n}$ contains only finitely many nonzero terms.

Definition 8.2.8 If the sequence (8.12) forms a (strict) (t, s)-sequence in base b, then it is called a **digital (strict) (t, s)-sequence constructed over** R. Here the truncations are required to operate on the expansions provided by ϕ_{∞}. A sequence which is a digital (strict) (t, s)-sequence constructed over R for some finite commutative ring R with identity and of order b is also called a **digital (strict) (t, s)-sequence in base** b.

The following result provides an analog of Theorem 8.2.2. For a matrix M over R with infinitely many rows and columns and an integer $m \geq 1$ we let $M^{(m \times m)}$ be the $m \times m$ submatrix of M obtained from the first m rows and columns of M. Given the generating matrices $C_1, \ldots, C_s$ above, we can then form the matrices $C_1^{(m \times m)}, \ldots, C_s^{(m \times m)}$ and so the $m \times d$ matrix $C_{\mathbf{d}}^{(m \times m)}$ as in (8.10).

Theorem 8.2.9 *Let $t \geq 0$ be an integer. Then the sequence (8.12) is a digital (t, s)-sequence constructed over R if and only if for any integer $m > t$ and any s-tuple $\mathbf{d} = (d_1, \ldots, d_s)$ of nonnegative integers with $\sum_{i=1}^{s} d_i = m - t$ the system of homogeneous linear equations*

$$\mathbf{n}^{(m)} C_{\mathbf{d}}^{(m \times m)} = \mathbf{0} \in R^{m-t} \tag{8.13}$$

has exactly b^t solutions $\mathbf{n}^{(m)} \in R^m$.

Proof. We proceed by Definition 8.1.5 and consider, for integers $k \geq 0$ and $m > t$, the points $[\mathbf{x}_n]_{b,m}$ with $kb^m \leq n < (k+1)b^m$. From (8.12) we get

$$[\mathbf{x}_n]_{b,m} = (\sigma_m(\mathbf{n}C_1), \ldots, \sigma_m(\mathbf{n}C_s)),$$

where $\sigma_m = \phi_m \circ \pi_m$ and $\pi_m : R^{\infty} \longrightarrow R^m$ is the projection to the first m terms of a sequence from R^{∞}. Thus, $\sigma_m(\mathbf{n}C_i)$ depends only on the first m terms of $\mathbf{n}C_i \in R^{\infty}$. Next we note that in the range $kb^m \leq n < (k+1)b^m$, the digits $a_l(n)$ of n are prescribed for $l \geq m$, whereas the $a_l(n)$ with $0 \leq l \leq m - 1$ can vary freely over Z_b. Then a brief calculation shows that the point set consisting of the $[\mathbf{x}_n]_{b,m}$ with $kb^m \leq n < (k + 1)b^m$ is given by

$$\left(\phi_m(\mathbf{n}^{(m)} C_1^{(m \times m)} + \mathbf{r}_1), \ldots, \phi_m(\mathbf{n}^{(m)} C_s^{(m \times m)} + \mathbf{r}_s) \right)$$

with $\mathbf{n}^{(m)}$ running through R^m and fixed $\mathbf{r}_1, \ldots, \mathbf{r}_s \in R^m$. Now the same argument as that used in the proof of Theorem 8.2.2 demonstrates that the above point set is a (t, m, s)-net in base b if and only if the system (8.13) has exactly b^t solutions $\mathbf{n}^{(m)} \in R^m$ for all possible d. $\qquad\square$

We consider again the special case $R = \mathbf{F}_q$. Let $C_1, \ldots, C_s$ be the generating matrices of the sequence (8.12), and for $1 \leq i \leq s$ let $\mathbf{c}_j^{(i)} \in \mathbf{F}_q^\infty, j = 1, 2 \ldots$, denote the columns of C_i. Then we set up the system

$$C^{(\infty)} = \{\mathbf{c}_j^{(i)} \in \mathbf{F}_q^\infty : 1 \leq i \leq s \text{ and } j \geq 1\}. \tag{8.14}$$

For $m \geq 1$ let $\pi_m : \mathbf{F}_q^\infty \longrightarrow \mathbf{F}_q^m$ be the projection as above and put

$$C^{(m)} = \{\pi_m(\mathbf{c}_j^{(i)}) \in \mathbf{F}_q^m : 1 \leq i \leq s, 1 \leq j \leq m\}.$$

Now we define

$$\tau(C^{(\infty)}) = \sup_{m \geq 1} \left(m - \varrho(C^{(m)}) \right), \tag{8.15}$$

where $\varrho(C^{(m)})$ is given by Definition 8.2.3.

Theorem 8.2.10 *If the system $C^{(\infty)}$ in (8.14) satisfies $\tau(C^{(\infty)}) < \infty$, then the sequence (8.12) obtained by the digital method with $R = \mathbf{F}_q$ is a digital strict (t, s)-sequence constructed over $\mathbf{F}_q$ with $t = \tau(C^{(\infty)})$.*

Proof. Note that the system $C^{(m)}$ is just the system of column vectors of the matrices $C_1^{(m \times m)}, \ldots, C_s^{(m \times m)}$. Then the result follows from Theorem 8.2.9 and elementary linear algebra. $\qquad\square$

Example 8.2.11 On the basis of Theorem 8.2.10 it is easy to obtain a digital $(0, 1)$-sequence constructed over $\mathbf{F}_q$: just choose the generating matrix C_1 in such a way that each matrix $C_1^{(m \times m)}$, $m = 1, 2, \ldots$, has rank m. In the special case where each $C_1^{(m \times m)}$ is an identity matrix over $\mathbf{F}_q$, we get a well-known sequence, namely a generalized van der Corput sequence in base q (compare with [94, p. 25]).

The following result is an analog of Theorem 8.2.7 and it is proved in a similar way using Theorem 8.2.9.

Theorem 8.2.12 *Let $b = q_1 \cdots q_h$ be a product of prime powers. Suppose that for $1 \leq r \leq h$ there exists a digital (t_r, s)-sequence constructed over $\mathbf{F}_{q_r}$. Then there exists a digital (t, s)-sequence in base b with*

$$t = \max_{1 \leq r \leq h} t_r.$$

In Lemma 8.1.7 we have shown that a (t, s)-sequence yields nets in dimension $s + 1$. The following lemma establishes an analog for digital (t, s)-sequences and digital nets.

Lemma 8.2.13 *If there exists a digital (t, s)-sequence constructed over R, then for every integer $m \geq \max(t, 1)$ there exists a digital $(t, m, s + 1)$-net constructed over R.*

Proof. Let $\mathbf{x}_0, \mathbf{x}_1, \ldots$ be a digital (t, s)-sequence constructed over R, where we choose the bijections $\eta : R \longrightarrow Z_b$ and $\psi : Z_b \longrightarrow R$ in the construction in such a way that they are inverses of each other. As in the proof of Lemma 8.1.7 we consider the point set P consisting of

$$\mathbf{y}_n = \left([\mathbf{x}_n]_{b,m}, \frac{n}{b^m}\right) \in I^{s+1}, \quad n = 0, 1, \ldots, b^m - 1,$$

where $b = |R|$ as usual. We have already verified in that proof that P is a $(t, m, s + 1)$-net in base b. Thus, it remains to show that P is obtained by the digital method. If $C_1, \ldots, C_s$ are the generating matrices in (8.12), then it is easy to see that for the first s generating matrices of P we can take $C_1^{(m \times m)}, \ldots, C_s^{(m \times m)}$; compare with the proof of Theorem 8.2.9. For the last generating matrix of P we take the $m \times m$ matrix over R with 1's on the secondary diagonal and 0's elsewhere. $\square$

Remark 8.2.14 It follows from Remark 8.2.6 and Lemma 8.2.13 that for any given q, there can be no digital $(0, s)$-sequence constructed over $\mathbf{F}_q$ for $s \geq q + 1$. Note that by Remark 8.1.8, for any given b, there can be no $(0, s)$-sequence in base b with $s \geq b + 1$, but this is a bit more difficult to prove.

Further information on the digital method for general R can be found in Larcher, Niederreiter, and Schmid [64] and Niederreiter [94, Chapter 4]. In the remainder of this section we concentrate on the case $R = \mathbf{F}_q$ in the digital method. The following quantity $d_q(s)$ is relevant. The fact that the definition of $d_q(s)$ is meaningful, i.e., that for every q and s there exists a digital (t, s)-sequence constructed over $\mathbf{F}_q$ for some t, was first proved by Niederreiter [91] and will also become apparent from the results in Section 8.3.

Definition 8.2.15 For any prime power q and any dimension $s \geq 1$ let $d_q(s)$ be the least value of t such that there exists a digital (t, s)-sequence constructed over $\mathbf{F}_q$.

The observation in Remark 8.2.14 says that $d_q(s) \geq 1$ for $s \geq q+1$. Obviously, the lower bound in (8.8) is also valid for $d_q(s)$. This lower bound on $d_q(s)$ can be proved directly by following an argument due to Schmid and Wolf [136] which shows an interesting connection with the theory of linear codes.

Theorem 8.2.16 *For all prime powers q and all dimensions $s \geq 1$ we have*

$$d_q(s) \geq \frac{s}{q} - \log_q \frac{(q - 1)s + q + 1}{2}.$$

Proof. Let t be such that there exists a digital (t,s)-sequence constructed over $\mathbf{F}_q$. Then by Lemma 8.2.13, for every integer $m > t$ there exists a digital $(t,m,s+1)$-net constructed over $\mathbf{F}_q$. We put

$$h = \left\lfloor \frac{(q-1)s}{q} \right\rfloor + 1$$

and consider $m = t + h$. If $s + 1 \leq m$, then $t \geq s/q$, and we are done. Thus, we can assume $s + 1 > m$. By Theorem 8.2.4 we get an $(h,m,s+1)$-system $\{\mathbf{c}_j^{(i)} \in \mathbf{F}_q^m : 1 \leq i \leq s+1, 1 \leq j \leq m\}$ over $\mathbf{F}_q$. Note that by the definition of an $(h,m,s+1)$-system over $\mathbf{F}_q$, any h of the vectors $\mathbf{c}_1^{(i)}, 1 \leq i \leq s+1$, are linearly independent over $\mathbf{F}_q$. Set up the $m \times (s+1)$ matrix H with the column vectors $\mathbf{c}_1^{(1)}, \ldots, \mathbf{c}_1^{(s+1)}$. The null space of H is a linear code over $\mathbf{F}_q$ of length $s+1$ and dimension at least $s+1-m$, and by Lemma 6.1.2 the minimum distance of this linear code is at least $h+1$. By passing to a linear subspace, we get a linear $[s+1, s+1-m]$ code over $\mathbf{F}_q$ with minimum distance at least $h+1$. The Plotkin bound in Proposition 6.2.2 then yields

$$h + 1 \leq \frac{(s+1)q^{s-m}(q-1)}{q^{s+1-m} - 1}$$

and hence

$$q^{s-t-h} \leq \frac{h+1}{(h+1)q - (s+1)(q-1)}.$$

This implies

$$
\begin{aligned}
t &\geq s - h + \log_q \left(q - \frac{(q-1)(s+1)}{h+1} \right) \\
&\geq \frac{s}{q} - 1 + \log_q \left(q - \frac{(q-1)(s+1)}{h+1} \right).
\end{aligned}
$$

From

$$h \geq s - \frac{s}{q} + \frac{1}{q}$$

it follows that

$$q - \frac{(q-1)(s+1)}{h+1} \geq \frac{2q}{(q-1)s + q + 1},$$

and so the desired lower bound is shown in all cases. $\qquad\square$

A useful survey of digital nets and digital (t,s)-sequences constructed over $\mathbf{F}_q$ was given by Larcher [63]. The recent work of Niederreiter and Pirsic [99] establishes a different approach to the determination of the quality parameter of a digital net constructed over $\mathbf{F}_q$ and develops further analogies with the theory of linear codes. The method of Lawrence et al. [72] for obtaining good digital nets constructed over $\mathbf{F}_q$ from good linear codes over $\mathbf{F}_q$ should also be mentioned in this context.

8.3 A Construction Using Rational Places

It is a useful consequence of the theory developed in Section 8.2 that, in order to construct digital nets and (t,s)-sequences in an arbitrary base, it suffices to know how to obtain digital

(t, s)-sequences constructed over a finite field. In this and the next section, we show that global function fields provide powerful tools for the generation of digital (t, s)-sequences constructed over $\mathbf{F}_q$. In the present section we describe the construction in Niederreiter and Xing [102] which uses global function fields with many rational places.

Let a dimension $s \geq 1$ and a prime power q be given. According to the principles of the digital method, in order to obtain a digital (t, s)-sequence constructed over $\mathbf{F}_q$, we need to define suitable generating matrices over $\mathbf{F}_q$ or, equivalently, a suitable system $C^{(\infty)}$ as in (8.14). The ingredients of the construction are the following. We choose a global function field $F/\mathbf{F}_q$ satisfying $N(F/\mathbf{F}_q) \geq s + 1$ and $s + 1$ distinct rational places $P_\infty, P_1, \ldots, P_s$ of F. Furthermore, we choose a positive nonspecial divisor D of F with $\ell(D) = 1$. The existence of such a divisor was discussed in Remark 6.1.11, with the result that such a divisor is available in all cases of practical interest.

The details of the construction are as follows. Since D is nonspecial, we have $\ell(D + jP_i) = j + 1$ for $1 \leq i \leq s$ and $j \geq 1$ by Corollary 1.1.4, and so for each $1 \leq i \leq s$ and $j \geq 1$ we can choose a function

$$f_j^{(i)} \in \mathcal{L}(D + jP_i) \backslash \mathcal{L}(D + (j-1)P_i). \qquad (8.16)$$

Let z be a local parameter at P_∞ and consider the local expansions of the $f_j^{(i)}$ at P_∞ which are of the form

$$f_j^{(i)} = z^{-v} \sum_{r=0}^{\infty} b_{r,j}^{(i)} z^r \qquad \text{for } 1 \leq i \leq s \text{ and } j \geq 1,$$

where $v := \nu_{P_\infty}(D) \geq 0$ and all coefficients $b_{r,j}^{(i)} \in \mathbf{F}_q$. For $1 \leq i \leq s$ and $j \geq 1$ we now define

$$c_{r,j}^{(i)} = \begin{cases} b_{r,j}^{(i)} & \text{for } 0 \leq r \leq v - 1, \\ b_{r+1,j}^{(i)} & \text{for } r \geq v. \end{cases}$$

Then we set up the system

$$C^{(\infty)} = \{ \mathbf{c}_j^{(i)} = \left(c_{0,j}^{(i)}, c_{1,j}^{(i)}, \ldots \right) \in \mathbf{F}_q^{\infty} : 1 \leq i \leq s \text{ and } j \geq 1 \}.$$

We write $S(P_\infty, P_1, \ldots, P_s; D)$ for a sequence obtained from this system by the digital method. Note that $\mathbf{c}_j^{(i)}$ is the jth column of the ith generating matrix of such a sequence $S(P_\infty, P_1, \ldots, P_s; D)$.

Theorem 8.3.1 *Let $F/\mathbf{F}_q$ be a global function field of genus g with $N(F/\mathbf{F}_q) \geq s + 1$, let $P_\infty, P_1, \ldots, P_s$ be distinct rational places of F, and let D be a positive nonspecial divisor of F with $\ell(D) = 1$. Then $S(P_\infty, P_1, \ldots, P_s; D)$ is a digital (g, s)-sequence constructed over $\mathbf{F}_q$.*

Proof. In view of Theorem 8.2.10 it suffices to show that

$$\tau(C^{(\infty)}) \leq g,$$

or on account of (8.15),

$$\varrho(C^{(m)}) \geq m - g \qquad \text{for all } m > g.$$

This means that we have to verify the following property: for any integer $m > g$ and any nonnegative integers $d_1, \ldots, d_s$ with $\sum_{i=1}^{s} d_i = m - g$, the vectors

$$\pi_m(\mathbf{c}_j^{(i)}) = \left(c_{0,j}^{(i)}, \ldots, c_{m-1,j}^{(i)} \right) \in \mathbf{F}_q^m \qquad \text{for } 1 \leq j \leq d_i, 1 \leq i \leq s,$$

are linearly independent over $\mathbf{F}_q$. Let H be the set of i with $1 \leq i \leq s$ for which $d_i \geq 1$, and suppose that we have

$$\sum_{i \in H} \sum_{j=1}^{d_i} a_j^{(i)} \pi_m(\mathbf{c}_j^{(i)}) = \mathbf{0} \in \mathbf{F}_q^m$$

for some $a_j^{(i)} \in \mathbf{F}_q$. Note that $\deg(D) = g$, and so $v \leq g < m$. Therefore we get

$$\sum_{i \in H} \sum_{j=1}^{d_i} a_j^{(i)} b_{r,j}^{(i)} = 0 \qquad \text{for } 0 \leq r \leq m \text{ with } r \neq v. \tag{8.17}$$

Now we consider the element $f \in F$ given by

$$f = \sum_{i \in H} \sum_{j=1}^{d_i} a_j^{(i)} (f_j^{(i)} - b_{v,j}^{(i)}) = z^{-v} \sum_{\substack{r=0 \\ r \neq v}}^{\infty} \left(\sum_{i \in H} \sum_{j=1}^{d_i} a_j^{(i)} b_{r,j}^{(i)} \right) z^r.$$

From (8.17) we obtain $\nu_{P_\infty}(f) \geq m + 1 - v$. Together with $\mathcal{L}(D) = \mathbf{F}_q$ and the choice of the $f_j^{(i)}$ in (8.16) this shows that

$$f \in \mathcal{L}(D + \sum_{i=1}^{s} d_i P_i - (m+1)P_\infty).$$

But $\deg(D + \sum_{i=1}^{s} d_i P_i - (m+1)P_\infty) = -1$, and so we must have $f = 0$, that is,

$$\sum_{i \in H} \sum_{j=1}^{d_i} a_j^{(i)} \left(f_j^{(i)} - b_{v,j}^{(i)} \right) = 0.$$

For fixed $h \in H$ we can write

$$e_h := \sum_{j=1}^{d_h} a_j^{(h)} \left(f_j^{(h)} - b_{v,j}^{(h)} \right) = - \sum_{i \in H \setminus \{h\}} \sum_{j=1}^{d_i} a_j^{(i)} \left(f_j^{(i)} - b_{v,j}^{(i)} \right). \tag{8.18}$$

By the choice of the $f_j^{(h)}$ in (8.16) we have

$$\nu_{P_h} \left(f_j^{(h)} - b_{v,j}^{(h)} \right) = -\nu_{P_h}(D) - j \qquad \text{for all } j \geq 1.$$

If not all $a_j^{(h)} = 0$, then this would imply

$$\nu_{P_h}(e_h) \leq -\nu_{P_h}(D) - 1.$$

By considering the right-hand side of (8.18) and recalling (8.16), we get

$$\nu_{P_h}(e_h) \geq -\nu_{P_h}(D),$$

a contradiction. Thus $a_j^{(h)} = 0$ for $1 \leq j \leq d_h$. $\qquad\square$

Example 8.3.2 Let q be arbitrary and choose F to be the rational function field over $\mathbf{F}_q$. Then Theorem 8.3.1 yields a digital $(0, s)$-sequence constructed over $\mathbf{F}_q$ for $1 \le s \le q$. The restriction on s is best possible by Remark 8.2.14. Using Definition 8.2.15, we can say that $d_q(s) = 0$ for $1 \le s \le q$. This fact was first proved by Faure [22] for primes q and later by Niederreiter [87] for general q. In combination with Lemma 8.2.13 we see that there exists a digital $(0, m, s)$-net constructed over $\mathbf{F}_q$ whenever $m \ge 1$ and $1 \le s \le q + 1$. For $m \ge 2$ the restriction on s is best possible by Remark 8.2.6.

The desire to optimize the construction in this section leads to the definition of the quantity $V_q(s)$ below. Since $A(q) > 0$ by Lemma 5.2.4, this definition is meaningful.

Definition 8.3.3 For any prime power q and any dimension $s \ge 1$ let

$$V_q(s) = \min\{g \ge 0 : N_q(g) \ge s + 1\}.$$

Corollary 8.3.4 *For every prime power q and every dimension $s \ge 1$ there exists a digital $(V_q(s), s)$-sequence constructed over $\mathbf{F}_q$.*

Proof. For $s \le q$ we have $V_q(s) = 0$ and the result follows from Example 8.3.2. If $s \ge q+1$, then let $F/\mathbf{F}_q$ be a global function field of genus $g = V_q(s)$ with $N(F/\mathbf{F}_q) \ge s + 1$. Then $N(F/\mathbf{F}_q) \ge q + 2 \ge 4$, and so by Remark 6.1.11 there exists a positive nonspecial divisor D of F with $\ell(D) = 1$. The rest follows from Theorem 8.3.1. $\Box$

In view of Corollary 8.3.4, the numbers $d_q(s)$ and $V_q(s)$ introduced in Definitions 8.2.15 and 8.3.3, respectively, are related by

$$d_q(s) \le V_q(s) \qquad \text{for all } s \ge 1. \tag{8.19}$$

Thus, $d_q(s)$ can be bounded from above by establishing upper bounds on $V_q(s)$. The following method of bounding $V_q(s)$ which is based on the class field theory of global function fields was developed by Niederreiter and Xing [102].

Let K be a global function field with $N(K) \ge 1$ and fix an abelian closure K^{ab} of K (see Remark 2.5.2). We recall some definitions from Section 2.7 in a special case. For a prime number ℓ and a nonempty set $\mathcal{P}$ of rational places of K, the $(\ell, \mathcal{P})$-Hilbert class field $H(\ell, \mathcal{P})$ of K is defined to be the maximal unramified abelian ℓ-extension of K inside K^{ab} in which all places in $\mathcal{P}$ split completely. We can iterate this construction: with $K_1 = H(\ell, \mathcal{P})$ and $\mathcal{P}_1$ being the set of all places of K_1 that lie over the places in $\mathcal{P}$, let K_2 be the $(\ell, \mathcal{P}_1)$-Hilbert class field of K_1, and so on. In this way we get a tower

$$K = K_0 \subseteq K_1 \subseteq K_2 \subseteq \ldots,$$

which is called the $(\ell, \mathcal{P})$-Hilbert class field tower of K. This tower is called infinite if $K_n \ne K_{n+1}$ for all $n \ge 0$.

Proposition 8.3.5 *For a given prime power q, suppose that there exists a global function field $K/\mathbf{F}_q$ of genus $g > 1$ with $N(K/\mathbf{F}_q) \geq 1$ which has an infinite $(\ell, \mathcal{P})$-Hilbert class field tower for some prime number ℓ and some nonempty set $\mathcal{P}$ of rational places of K. Then we have*

$$V_q(s) \leq \frac{g-1}{|\mathcal{P}|} Ms + 1 \qquad \text{for all } s \geq 1,$$

where

$$M = \max\left(\frac{|\mathcal{P}|}{q+1}, \ell\right).$$

Proof. Let K be as in the proposition and let

$$K = K_0 \subset K_1 \subset K_2 \subset \ldots$$

be the infinite $(\ell, \mathcal{P})$-Hilbert class field tower of K. We have $[K_n : K_{n-1}] = \ell^{m_n}$ with $m_n \geq 1$ for $n = 1, 2, \ldots$ By Galois theory, for every $n \geq 1$ there is a chain of fields

$$K_{n-1} = L_{n,0} \subset L_{n,1} \subset \ldots \subset L_{n,m_n} = K_n$$

such that $[L_{n,j} : L_{n,j-1}] = \ell$ for $1 \leq j \leq m_n$. If $\mathcal{P}_{n,j}$ is the set of all places of $L_{n,j}$ that lie over the places in $\mathcal{P}$, then

$$|\mathcal{P}_{n,j}| = [L_{n,j} : K]|\mathcal{P}| \qquad \text{for } 0 \leq j \leq m_n \text{ and } n \geq 1.$$

Since the extension $L_{n,j}/K$ is unramified, we have

$$g(L_{n,j}) = (g-1)[L_{n,j} : K] + 1$$

by the Hurwitz genus formula. For $s \leq q$ the result of the proposition is trivial. Now let $s \geq q + 1$. If $s \leq |\mathcal{P}| - 1$, then

$$V_q(s) \leq g \leq \frac{g-1}{|\mathcal{P}|} \cdot \frac{|\mathcal{P}|}{q+1} s + 1 \leq \frac{g-1}{|\mathcal{P}|} Ms + 1.$$

If $s \geq |\mathcal{P}|$, then we can find some $n \geq 1$ and $1 \leq j \leq m_n$ such that

$$|\mathcal{P}_{n,j-1}| \leq s \leq |\mathcal{P}_{n,j}| - 1 \leq N_q(g(L_{n,j})) - 1.$$

Hence

$$
\begin{aligned}
V_q(s) &\leq g(L_{n,j}) = (g-1)[L_{n,j} : K] + 1 \\
&= \ell(g-1)[L_{n,j-1} : K] + 1 = \ell(g-1)\frac{|\mathcal{P}_{n,j-1}|}{|\mathcal{P}|} + 1 \\
&\leq \frac{\ell(g-1)}{|\mathcal{P}|} s + 1 \leq \frac{g-1}{|\mathcal{P}|} Ms + 1,
\end{aligned}
$$

and so the proposition is shown in all cases. $\qquad\square$

Theorem 8.3.6 *For all prime powers q we have*

$$d_q(s) \le \frac{c}{\log q}\, s + 1 \qquad \text{for all } s \ge 1,$$

where $c > 0$ is an absolute constant.

Proof. First we consider the case where q is sufficiently large, say $q \ge 2^{24}$ as in the proof of Theorem 5.2.9. We proceed in analogy with that proof. We can choose n of the order of magnitude $\log q$ such that the conditions in Theorem 5.2.8 are satisfied. By the proof of Theorem 5.2.8, with F being the rational function field over $\mathbf{F}_q$, we can construct a global function field $K/\mathbf{F}_q$ and a set $\mathcal{P}$ of rational places of K such that $|\mathcal{P}|$ is of the order of magnitude n^2 and the Golod-Shafarevich condition is satisfied for $\ell = 2$. Thus, the $(2, \mathcal{P})$-Hilbert class field tower of K is infinite. Furthermore, the genus of K has the order of magnitude n. An application of Proposition 8.3.5 then shows that

$$V_q(s) \le \frac{c_1 n}{n^2}\, s + 1 \le \frac{c_2}{\log q}\, s + 1 \qquad \text{for all } s \ge 1,$$

with absolute constants $c_1 > 0$ and $c_2 > 0$. Because of (8.19) we get the same bound for $d_q(s)$. For small q, say for $q < 2^{24}$, we can refer to the bound for $d_q(s)$ in Theorem 8.4.4 which will be proved in an independent manner. $\qquad\square$

Theorem 8.3.7 *Let $q = p^e$ with a prime p and an odd integer $e \ge 3$ and let m be the least prime factor of e. If p is odd and $q > 27$, then*

$$d_q(s) \le \frac{\lceil 2(2q^{1/m} + 3)^{1/2}\rceil + 1}{q^{1/m} + 1}\, s + 1 \qquad \text{for all } s \ge 1.$$

If $p = 2$ and e is composite, then

$$d_q(s) \le \frac{2\lceil 2(2q^{1/m} + 2)^{1/2}\rceil + 4}{q^{1/m} + 1}\, s + 1 \qquad \text{for all } s \ge 1.$$

Proof. Because of (8.19) it suffices to prove the same bounds for $V_q(s)$. In both cases we apply Proposition 8.3.5 with $\ell = 2$. In the first case, the proof of Theorem 5.3.1 yields a global function field $K/\mathbf{F}_q$ of genus

$$g = \lceil 2(2q^{1/m} + 3)^{1/2}\rceil + 2$$

and a set $\mathcal{P}$ of rational places of K with $|\mathcal{P}| = 2q^{1/m} + 2$ such that the $(2, \mathcal{P})$-Hilbert class field tower of K is infinite (since the Golod-Shafarevich condition is satisfied for $\ell = 2$). In the second case, the proof of Theorem 5.3.3 yields a global function field $K/\mathbf{F}_q$ of genus

$$g = 2\lceil 2(2q^{1/m} + 2)^{1/2}\rceil + 5$$

and a set $\mathcal{P}$ of rational places of K with $|\mathcal{P}| = 2q^{1/m} + 2$ such that the $(2, \mathcal{P})$-Hilbert class field tower of K is infinite. $\qquad\square$

Remark 8.3.8 The case $q = 27$ is not covered in Theorem 8.3.7, but here we can again go back to the proof of Theorem 5.3.1 and observe that if we use this proof with $q = m = 3$ and $n = 8$, then we get a global function field $K/\mathbf{F}_{27}$ of genus $g = 7$. If we choose a set $\mathcal{P}$ of rational places of K with $|\mathcal{P}| = 5$, then the $(2, \mathcal{P})$-Hilbert class field tower of K is infinite since the Golod-Shafarevich condition is satisfied for $\ell = 2$. Therefore, Proposition 8.3.5 yields

$$V_{27}(s) \le \frac{12}{5} s + 1 \qquad \text{for all } s \ge 1,$$

and in view of (8.19) we get the same bound for $d_{27}(s)$.

Remark 8.3.9 An analogy with Theorem 8.3.7 would suggest that if q is a square, then

$$d_q(s) \le \frac{c}{q^{1/4}} s + 1 \qquad \text{for all } s \ge 1,$$

where $c > 0$ is an absolute constant. This is indeed true and can be proved as follows. Put $q = r^2$ and note that we can take r sufficiently large, since for small r we can refer to the bound in Theorem 8.4.4 which will be proved in an independent manner. For odd q we use the proof of Theorem 5.3.1 with $m = 2$. We put $n = \lceil 2(2r + 3)^{1/2} \rceil + 3$, then for sufficiently large r we have $n \le (r^2 - r)/2$, the number of monic irreducible polynomials of degree 2 in $\mathbf{F}_r[x]$. We construct the global function field $K/\mathbf{F}_q$ of genus $g = n - 1$ as in the proof of Theorem 5.3.1 and also the set $\mathcal{P}$ of rational places of K with $|\mathcal{P}| = 2r + 2$. Then the $(2, \mathcal{P})$-Hilbert class field tower of K is infinite since the Golod-Shafarevich condition is satisfied for $\ell = 2$. The desired result follows then from (8.19) and Proposition 8.3.5. For even q we use the proof of Theorem 5.3.3 with $m = 2$. We put $n = \lceil 2(2r + 2)^{1/2} \rceil + 3$, but now we let $f_1, \ldots, f_n$ be n distinct monic irreducible polynomials of degree 3 in $\mathbf{F}_r[x]$. This choice is possible since $n \le (r^3 - r)/3$ for sufficiently large r. The genus of $K/\mathbf{F}_q$ is now $g = 3n - 1$. The set $\mathcal{P}$ of rational places of K with $|\mathcal{P}| = 2r + 2$ is constructed as in the proof of Theorem 5.3.3. Then everything goes through as in the case of odd q.

For squares q there is an alternative method of bounding $d_q(s)$ which is based on an explicit tower of global function fields. The resulting bound due to Xing and Niederreiter [179] is quite simple and actually better than that in Remark 8.3.9, except in the case where q is the square of a prime. First we need an auxiliary result on different exponents in special extensions.

Lemma 8.3.10 *For a prime power r, let $F/\mathbf{F}_{r^2}$ be a global function field, let $w \in F$, and let P be a place of F satisfying $\nu_P(w) = -m < 0$ with $\gcd(m, r) = 1$. Let $E = F(y)$ with $y^r + y = w$ and let K be a field with $F \subseteq K \subseteq E$. Let P' and P'' be places of K, respectively E, lying over P. Then the different exponent $d(P''|P')$ is given by*

$$d(P''|P') = ([E : K] - 1)(m + 1).$$

Proof. Since the polynomial $x^r + x \in \mathbf{F}_r[x]$ has all its roots in $\mathbf{F}_{r^2}$, it is clear that E/F is a Galois extension. If $e(P''|P')$ is the ramification index, then

$$\nu_{P''}(w) = e(P''|P)\nu_P(w) = -e(P''|P)m,$$

but also

$$\nu_{P''}(w) = \nu_{P''}(y^r + y) = r\nu_{P''}(y).$$

From $\gcd(m,r) = 1$ it follows that r divides $e(P''|P)$. On the other hand, $e(P''|P) \leq [E : F] \leq r$, hence $e(P''|P) = [E : F] = r$. Thus, P is totally ramified in the extension E/F and $\nu_{P''}(y) = -m$.

Let $z \in F$ with $\nu_P(z) = 1$ and let the integers $h \geq 1$ and k be such that $kr - hm = 1$. Then $t = y^h z^k$ satisfies $\nu_{P''}(t) = 1$. For any $\sigma \in \mathrm{Gal}(E/F)\backslash\{\mathrm{id}\}$ we have $\sigma(y) = y + c$ for some nonzero $c \in \mathbf{F}_{r^2}$, thus

$$t - \sigma(t) = t(1 - (1 + cy^{-1})^h) = -t\sum_{j=1}^{h}\binom{h}{j}c^j y^{-j}.$$

Note that $h \neq 0$ in $\mathbf{F}_{r^2}$, hence $\nu_{P''}(t - \sigma(t)) = 1 + m$. Now with $G = \mathrm{Gal}(E/K)$ we get by a standard argument (compare with the proof of [152, Proposition III.7.8(c)]),

$$d(P''|P') = \sum_{\sigma\in G\backslash\{\mathrm{id}\}} \nu_{P''}(t - \sigma(t)),$$

and then the lemma follows immediately. $\square$

Theorem 8.3.11 *If the prime power q is a square, then*

$$d_q(s) \leq \frac{ps}{q^{1/2} - 1} \qquad \text{for all } s \geq 1,$$

where p is the unique prime factor of q.

Proof. Put $q = r^2$ and $r = p^e$. Let $E_1 \subseteq E_2 \subseteq \ldots$ be the tower of global function fields over $\mathbf{F}_q$ constructed by Garcia and Stichtenoth [30] (see also Example 5.4.2), that is, $E_1 = \mathbf{F}_q(x_1)$ is a rational function field and $E_{n+1} = E_n(z_{n+1})$ for $n = 1, 2, \ldots$ with

$$z_{n+1}^r + z_{n+1} = x_n^{r+1} \qquad \text{and} \qquad x_{n+1} = \frac{z_{n+1}}{x_n}.$$

Then E_{n+1}/E_n is a Galois extension of degree r for each $n \geq 1$. Hence there exists a chain of fields

$$E_n = K_{n,0} \subset K_{n,1} \subset \ldots \subset K_{n,e} = E_{n+1}$$

such that $[K_{n,i+1} : K_{n,i}] = p$ for $0 \leq i \leq e - 1$. From results in [30] we know that for all $n \geq 1$ we have

$$g(E_n) \leq r^n + r^{n-1}, \qquad N(E_n) \geq (q - 1)r^{n-1}.$$

The last inequality implies

$$N(K_{n,i}) \geq \frac{N(E_{n+1})}{[E_{n+1} : K_{n,i}]} \geq p^i(q-1)r^{n-1} \qquad \text{for } 0 \leq i \leq e. \qquad (8.20)$$

Next we establish an upper bound for $g(K_{n,i})$. From [30] we know that for each place P of E_n that is ramified in the extension E_{n+1}/E_n we have $\nu_P(x_n) = -1$, and therefore we obtain $\nu_P(x_n^{r+1}) = -r - 1$. It follows then from the first part of the proof of Lemma 8.3.10 that P is totally ramified in E_{n+1}/E_n. According to [30], the sum of the degrees of these places P is equal to $r^{\lfloor n/2 \rfloor}$, and so the same holds for the sum of the degrees of the places P' of $K_{n,i}$ that are ramified in $E_{n+1}/K_{n,i}$, where $0 \le i \le e - 1$. For any such P' and the unique place P'' of E_{n+1} lying over it we have

$$d(P''|P') = (p^{e-i} - 1)(r + 2)$$

by Lemma 8.3.10. By combining these facts with the Hurwitz genus formula, we obtain

$$2g(E_{n+1}) - 2 = p^{e-i}(2g(K_{n,i}) - 2) + r^{\lfloor n/2 \rfloor}(r + 2)(p^{e-i} - 1)$$

for $0 \le i \le e$, and so

$$
\begin{aligned}
g(K_{n,i}) &\le \frac{p^i}{r}(g(E_{n+1}) - 1) - \frac{1}{2}r^{\lfloor n/2 \rfloor - 1}(r + 2)(r - p^i) + 1 \\
&\le p^i\left(r^n + r^{n-1}\right).
\end{aligned}
$$

The result of the theorem is trivial for $1 \le s \le q$ (compare with Example 8.3.2). If $s \ge q + 1$, then there are integers $n \ge 1$ and $1 \le i \le e$ such that

$$p^{i-1}(q - 1)r^{n-1} \le s \le p^i(q - 1)r^{n-1} - 1.$$

In view of (8.19) and (8.20) we get

$$d_q(s) \le V_q(s) \le g(K_{n,i}) \le \frac{p}{r-1}p^{i-1}(q-1)r^{n-1} \le \frac{ps}{r-1},$$

which is the desired bound. □

All the bounds on $d_q(s)$ given above are of the form $d_q(s) = O(s)$. In the light of Theorem 8.2.16, this order of magnitude in s is best possible. In fact, the lower bound (8.8) shows that even for general (t, s)-sequences in base q, the quality parameter t must be at least of the order of magnitude s as a function of s. Constructions of (t, s)-sequences in base q prior to that by Niederreiter and Xing [102] did not achieve this optimal order of magnitude. Sobol' [148] was the first to construct (t, s)-sequences, but his construction is for the base $q = 2$ only and it yields $d_2(s) = O(s \log s)$. This construction was improved and generalized to arbitrary q by Niederreiter [91]. The construction by Niederreiter works with rational function fields and it still achieves only $d_q(s) = O(s \log s)$.

The first construction using general global function fields is due to Niederreiter and Xing [101]. It yields many smaller quality parameters t than previous constructions, but no improvement on the bound $d_q(s) = O(s \log s)$. Further remarks on this construction can be found in Niederreiter and Xing [104]. The latter paper also introduces a new construction of (t, s)-sequences in base q using global function fields with many rational places. This construction is not as effective as that of the sequences $S(P_\infty, P_1, \ldots, P_s; D)$ in this section, but quite a bit simpler. It works with a global function field $F/\mathbf{F}_q$ of genus g with $N(F/\mathbf{F}_q) \ge s + 1$ and it yields a digital (t, s)-sequence constructed over $\mathbf{F}_q$ with

$t = 0$ if $g = 0$ and $t = 2g - 1$ if $g > 0$. Recent tables of upper bounds on $d_q(s)$ for $q = 2, 3, 4, 5, 8, 16$ and $1 \leq s \leq 50$ are available in Niederreiter [98].

If we combine Theorems 8.3.6 and 8.2.12, then we see that for any integer $b \geq 2$ we can construct digital (t, s)-sequences in base b with

$$t \leq \frac{c}{\log q_1} s + 1 \qquad \text{for all } s \geq 1,$$

where q_1 is the least prime power in the canonical factorization of b as a product of pairwise coprime prime powers. The bound $t = O(s)$ is of the best possible order of magnitude in s in view of (8.8). By Lemma 8.2.13 we get an immediate consequence for the construction of digital nets in base b.

We briefly consider the implications for the discrepancy bound (8.7), in particular for the coefficient $C_b(s, t)$ of the leading term. Since for any fixed $b \geq 2$ we can get a value $t = t(b, s)$ of the quality parameter with $t(b, s) = O(s)$, we obtain

$$\log C_b(s, t(b, s)) \leq -s \log s + O(s),$$

where the implied constant depends only on b. Thus, $C_b(s, t(b, s))$ tends to 0 at a super-exponential rate as $s \to \infty$. This rate of convergence is faster than the convergence rates achieved by all previous methods.

8.4   A Construction Using Arbitrary Places

In the previous section we have either described or mentioned three constructions of digital (t, s)-sequences based on global function fields, namely the constructions by Niederreiter and Xing [101], [102], [104]. Besides these three constructions, there is only one more known construction of digital (t, s)-sequences using general global function fields, namely that by Xing and Niederreiter [179] which we will describe in the present section. This construction is more flexible than the other three constructions since it allows us to work with places of arbitrary degree. If only rational places are employed in this construction, then it is of the same quality as the construction of the sequences $S(P_\infty, P_1, \ldots, P_s; D)$ in Section 8.3, in the sense that in this case the two constructions yield the same value of the quality parameter t. Another aspect is that, in cases of practical interest, it is trivial to find the auxiliary divisor D that is needed in the construction to be described here, whereas the divisor D for the construction in Section 8.3 is sometimes harder to find.

Let a dimension $s \geq 1$ and a prime power q be given. As before, we intend to generate a digital (t, s)-sequence constructed over $\mathbf{F}_q$, and for this purpose we have to define suitable generating matrices over $\mathbf{F}_q$ or, equivalently, a suitable system $C^{(\infty)}$ as in (8.14). The ingredients of the construction are the following. Let $F/\mathbf{F}_q$ be a global function field of genus g which contains at least one rational place P_∞, and let D be a divisor of F with $\deg(D) = 2g$ and $P_\infty \notin \text{supp}(D)$. Furthermore, we choose s distinct places $P_1, \ldots, P_s$ of F with $P_i \neq P_\infty$ for $1 \leq i \leq s$, and we put $e_i = \deg(P_i)$ for $1 \leq i \leq s$.

We now describe the construction in detail. Note that $\ell(D) = g + 1$ by the Riemann-Roch theorem. We choose a basis of $\mathcal{L}(D)$ in the following way. Observe that $\ell(D - P_\infty) = g$

by the Riemann-Roch theorem and $\ell(D - (2g + 1)P_\infty) = 0$, hence there exist integers $0 = n_0 < n_1 < \cdots < n_g \leq 2g$ such that

$$\ell(D - n_u P_\infty) = \ell(D - (n_u + 1)P_\infty) + 1 \qquad \text{for } 0 \leq u \leq g.$$

Now we choose

$$w_u \in \mathcal{L}(D - n_u P_\infty) \backslash \mathcal{L}(D - (n_u + 1)P_\infty) \qquad \text{for } 0 \leq u \leq g.$$

Then it is easily seen that $\{w_0, w_1, \ldots, w_g\}$ is a basis of $\mathcal{L}(D)$. For each $i = 1, \ldots, s$ we consider the chain

$$\mathcal{L}(D) \subset \mathcal{L}(D + P_i) \subset \mathcal{L}(D + 2P_i) \subset \cdots$$

of vector spaces over $\mathbf{F}_q$. By starting from the basis $\{w_0, w_1, \ldots, w_g\}$ of $\mathcal{L}(D)$ and successively adding basis vectors at each step of the chain, we obtain for each $n \geq 1$ a basis

$$\{w_0, w_1, \ldots, w_g, f_1^{(i)}, f_2^{(i)} \ldots, f_{ne_i}^{(i)}\}$$

of $\mathcal{L}(D + nP_i)$. Now let z be a local parameter at P_∞. For $r = 0, 1, \ldots$ we put

$$z_r = \begin{cases} z^r & \text{if } r \notin \{n_0, n_1, \ldots, n_g\}, \\ w_u & \text{if } r = n_u \text{ for some } u \in \{0, 1, \ldots, g\}. \end{cases}$$

Note that $\nu_{P_\infty}(z_r) = r$ for all $r \geq 0$. For $1 \leq i \leq s$ and $j \geq 1$ we have $f_j^{(i)} \in \mathcal{L}(D + kP_i)$ for some $k \geq 1$ and also $P_\infty \notin \mathrm{supp}(D + kP_i)$, hence $\nu_{P_\infty}(f_j^{(i)}) \geq 0$. Thus, we have local expansions at P_∞ of the form

$$f_j^{(i)} = \sum_{r=0}^{\infty} a_{r,j}^{(i)} z_r \qquad \text{for } 1 \leq i \leq s \text{ and } j \geq 1,$$

where all coefficients $a_{r,j}^{(i)} \in \mathbf{F}_q$. For $1 \leq i \leq s$ and $j \geq 1$ we now define the sequence of elements $c_{r,j}^{(i)} \in \mathbf{F}_q, r = 0, 1, \ldots$, as follows: first we consider the sequence of elements $a_{r,j}^{(i)}, r = 0, 1, \ldots$, and then we delete those terms with $r = n_u$ for some $u \in \{0, 1, \ldots, g\}$. Then we set up the system

$$C^{(\infty)} = \{\mathbf{c}_j^{(i)} = \left(c_{0,j}^{(i)}, c_{1,j}^{(i)}, \ldots\right) \in \mathbf{F}_q^\infty : 1 \leq i \leq s \text{ and } j \geq 1\}.$$

We write $S_1(P_\infty, P_1, \ldots, P_s; D)$ for a sequence obtained from this system by the digital method. Note that $\mathbf{c}_j^{(i)}$ is the jth column of the ith generating matrix of such a sequence $S_1(P_\infty, P_1, \ldots, P_s; D)$.

Theorem 8.4.1 *Let $F/\mathbf{F}_q$ be a global function field of genus g which contains at least one rational place P_∞, let D be a divisor of F with $\deg(D) = 2g$ and $P_\infty \notin \mathrm{supp}(D)$, and let $P_1, \ldots, P_s$ be distinct places of F with $P_i \neq P_\infty$ for $1 \leq i \leq s$. Then $S_1(P_\infty, P_1, \ldots, P_s; D)$ is a digital (t, s)-sequence constructed over $\mathbf{F}_q$ with*

$$t = g + \sum_{i=1}^{s} (e_i - 1),$$

where $e_i = \deg(P_i)$ for $1 \leq i \leq s$.

Proof. As in the proof of Theorem 8.3.1, it suffices to show the following property: for any integer $m > t$ and any nonnegative integers $d_1, \ldots, d_s$ with $\sum_{i=1}^{s} d_i = m - t$, the vectors

$$\pi_m(\mathbf{c}_j^{(i)}) = \left(c_{0,j}^{(i)}, \ldots, c_{m-1,j}^{(i)}\right) \in \mathbf{F}_q^m \qquad \text{for } 1 \le j \le d_i, 1 \le i \le s,$$

are linearly independent over $\mathbf{F}_q$. Let H be the set of i with $1 \le i \le s$ for which $d_i \ge 1$, and suppose that we have

$$\sum_{i \in H} \sum_{j=1}^{d_i} b_j^{(i)} \pi_m(\mathbf{c}_j^{(i)}) = 0 \in \mathbf{F}_q^m$$

for some $b_j^{(i)} \in \mathbf{F}_q$. With $R = \{n_0, n_1, \ldots, n_g\}$ this means that

$$\sum_{i \in H} \sum_{j=1}^{d_i} b_j^{(i)} a_{r,j}^{(i)} = 0 \tag{8.21}$$

for the first m nonnegative integers r that are not in R. Now we consider the element $f \in F$ given by

$$f = \sum_{i \in H} \sum_{j=1}^{d_i} b_j^{(i)} \left(f_j^{(i)} - \sum_{u=0}^{g} a_{n_u,j}^{(i)} w_u \right) = \sum_{\substack{r=0 \\ r \notin R}}^{\infty} \left(\sum_{i \in H} \sum_{j=1}^{d_i} b_j^{(i)} a_{r,j}^{(i)} \right) z_r.$$

If we use (8.21) as well as $m > g$ and $n_g \le 2g$, then we arrive at $\nu_{P_\infty}(f) \ge m + g + 1$. Together with the choice of the $f_j^{(i)}$ this shows that

$$f \in \mathcal{L}\left(D + \sum_{i=1}^{s} \left(\left\lfloor \frac{d_i - 1}{e_i} \right\rfloor + 1 \right) P_i - (m + g + 1) P_\infty \right).$$

But

$$\deg \left(D + \sum_{i=1}^{s} \left(\left\lfloor \frac{d_i - 1}{e_i} \right\rfloor + 1 \right) P_i - (m + g + 1) P_\infty \right)$$

$$= 2g + \sum_{i=1}^{s} \left(\left\lfloor \frac{d_i - 1}{e_i} \right\rfloor + 1 \right) e_i - (m + g + 1) \le t + \sum_{i=1}^{s} d_i - m - 1 = -1,$$

and so we must have $f = 0$, that is,

$$\sum_{i \in H} \sum_{j=1}^{d_i} b_j^{(i)} f_j^{(i)} =: w \in \mathcal{L}(D).$$

For fixed $h \in H$ we can write

$$\sum_{j=1}^{d_h} b_j^{(h)} f_j^{(h)} = w - \sum_{i \in H \setminus \{h\}} \sum_{j=1}^{d_i} b_j^{(i)} f_j^{(i)}. \tag{8.22}$$

If not all $b_j^{(h)} = 0$, then by choice of the $f_j^{(h)}$ we have

$$\sum_{j=1}^{d_h} b_j^{(h)} f_j^{(h)} \in \mathcal{L}(D + k P_h) \setminus \mathcal{L}(D) \qquad \text{for some } k \ge 1,$$

and so

$$\nu_{P_h}\left(\sum_{j=1}^{d_h} b_j^{(h)} f_j^{(h)}\right) \leq -\nu_{P_h}(D) - 1.$$

By considering the right-hand side of (8.22), we get

$$\nu_{P_h}\left(\sum_{j=1}^{d_h} b_j^{(h)} f_j^{(h)}\right) \geq -\nu_{P_h}(D),$$

a contradiction. Thus $b_j^{(h)} = 0$ for $1 \leq j \leq d_h$. □

Remark 8.4.2 If we choose all places $P_1, \ldots, P_s$ in Theorem 8.4.1 to be rational, then we get the same quality parameter $t = g$ as in Theorem 8.3.1. If we let F be the rational function field $\mathbf{F}_q(x)$, choose P_∞ to be the pole of x, and let $P_1, \ldots, P_s$ be the zeros of the distinct monic irreducible polynomials $f_1, \ldots, f_s$ in $\mathbf{F}_q[x]$, respectively, then Theorem 8.4.1 yields the same quality parameter

$$t = \sum_{i=1}^{s}(\deg(f_i) - 1)$$

as the construction of Niederreiter [91].

Example 8.4.3 Put $q = 3$ and let $F = \mathbf{F}_3(x, y)$ be the Artin-Schreier extension of the rational function field $\mathbf{F}_3(x)$ with

$$y^3 - y = \frac{x^3 - x}{(x^2 + x - 1)^2}.$$

Then F has genus $g = 4$. The four rational places of $\mathbf{F}_3(x)$ split completely in the extension $F/\mathbf{F}_3(x)$, and so $N(F) = 12$. Furthermore, the zero of $x^2 + x - 1$ is totally ramified in the extension $F/\mathbf{F}_3(x)$, hence F has at least one place of degree 2. Let $P_\infty, P_1, \ldots, P_{11}$ be the 12 rational places of F and let P_{12} be a place of F of degree 2. If $d_3(12)$ is given by Definition 8.2.15, then Theorem 8.4.1 yields $d_3(12) \leq g + 1 = 5$, whereas the construction in Section 8.3 produces only the bound $d_3(12) \leq V_3(12) \leq 6$.

We consider now an implication of the construction in this section for the quantity $d_q(s)$ in Definition 8.2.15. The following upper bound on $d_q(s)$ has the form $d_q(s) = O(s)$ like the bound in Theorem 8.3.6, but the implied constant is asymptotically weaker as $q \to \infty$. On the other hand, the bound is more explicit than that in Theorem 8.3.6 and its proof has the advantage that it is based on explicitly given global function fields.

Theorem 8.4.4 *For every prime power q and every dimension $s \geq 1$ we have*

$$d_q(s) \leq \frac{3q - 1}{q - 1}(s - 1) - \frac{(2q + 4)(s - 1)^{1/2}}{(q^2 - 1)^{1/2}} + 2.$$

Proof. Consider the tower $F_1 \subseteq F_2 \subseteq \ldots$ of function fields over $\mathbf{F}_q$, where $F_1 = \mathbf{F}_q(x_1)$ is a rational function field and $F_{n+1} = F_n(z_{n+1})$ for $n = 1, 2, \ldots$ with

$$z_{n+1}^q + z_{n+1} = x_n^{q+1} \quad \text{and} \quad x_{n+1} = \frac{z_{n+1}}{x_n}.$$

If for each $n \geq 1$ we let $E_n = F_n \cdot \mathbf{F}_{q^2}$ be a constant field extension of F_n, then we obtain the tower $E_1 \subseteq E_2 \subseteq \ldots$ of function fields over $\mathbf{F}_{q^2}$ constructed by Garcia and Stichtenoth [30] (see also Example 5.4.2). It follows from the invariance of the genus under constant field extensions and from the genus bound in [30] that

$$g(F_n) = g(E_n) \leq q^n + q^{n-1} - q^{(n+1)/2} - 2q^{(n-1)/2} + 1 \qquad \text{for all } n \geq 1.$$

By another result from [30] we have

$$N(E_n/\mathbf{F}_{q^2}) \geq (q^2 - 1)q^{n-1} + 1 \qquad \text{for all } n \geq 1.$$

Let Q_1 be the pole of x_1 in F_1. We will prove by induction on n that Q_1 is totally ramified in F_n/F_1 and that the unique place Q_n of F_n lying over Q_1 satisfies $\nu_{Q_n}(x_n) = -1$. This being trivial for $n = 1$, we assume that it has been shown for some $n \geq 1$. If Q_{n+1} is a place of F_{n+1} lying over Q_n, then with $e(Q_{n+1}|Q_n)$ denoting the ramification index we get

$$\nu_{Q_{n+1}}(x_n^{q+1}) = e(Q_{n+1}|Q_n)\nu_{Q_n}(x_n^{q+1}) = -(q+1)e(Q_{n+1}|Q_n),$$

but also

$$\nu_{Q_{n+1}}(x_n^{q+1}) = \nu_{Q_{n+1}}\left(z_{n+1}^q + z_{n+1}\right) = q\nu_{Q_{n+1}}(z_{n+1}).$$

Thus, q divides $e(Q_{n+1}|Q_n)$. On the other hand, $e(Q_{n+1}|Q_n) \leq [F_{n+1} : F_n] \leq q$, hence $e(Q_{n+1}|Q_n) = [F_{n+1} : F_n] = q$, and so Q_1 is totally ramified in F_{n+1}/F_1. Moreover, $\nu_{Q_{n+1}}(z_{n+1}) = -q - 1$, thus

$$\nu_{Q_{n+1}}(x_{n+1}) = \nu_{Q_{n+1}}\left(\frac{z_{n+1}}{x_n}\right) = -q - 1 - q\nu_{Q_n}(x_n) = -1,$$

and the induction is complete. Note that we have shown in particular that $N(F_n/\mathbf{F}_q) \geq 1$ for all $n \geq 1$.

Since E_n/F_n is an unramified extension of degree 2, we have

$$N(F_n/\mathbf{F}_q) + 2B_2(F_n/\mathbf{F}_q) = N(E_n/\mathbf{F}_{q^2}) \geq (q^2 - 1)q^{n-1} + 1 \quad \text{for all } n \geq 1,$$

where $B_2(F_n/\mathbf{F}_q)$ denotes the number of places of F_n of degree 2. Together with $N(F_n/\mathbf{F}_q) \geq 1$ this yields

$$N(F_n/\mathbf{F}_q) + B_2(F_n/\mathbf{F}_q) \geq \frac{1}{2}(q^2 - 1)q^{n-1} + 1 \qquad \text{for all } n \geq 1. \qquad (8.23)$$

Now we are ready to prove the bound for $d_q(s)$ in the theorem. For all s we will get this bound by applying the construction of the sequences $S_1(P_\infty, P_1, \ldots, P_s; D)$ in this section with a suitable global function field F_n from the tower described above. First let $1 \leq s \leq q$. Then an obvious application of Theorem 8.4.1 to $F_1 = \mathbf{F}_q(x_1)$ yields $d_q(s) = 0$, and the bound holds. Next let $q + 1 \le s < \frac{1}{2}(q^2 - 1) + 1$. Since $N(F_1/\mathbf{F}_q) = q + 1$ and $B_2(F_1/\mathbf{F}_q) = \frac{1}{2}(q^2 - q)$, it follows from Theorem 8.4.1, applied to F_1 with q rational places (the remaining rational place serving as P_∞) and $s - q$ places of degree 2, that

$$d_q(s) \le s - q \le \frac{3q - 1}{q - 1}(s - 1) - \frac{(2q + 4)(s - 1)^{1/2}}{(q^2 - 1)^{1/2}} + 2.$$

Finally, let $s \ge \frac{1}{2}(q^2 - 1) + 1$. Then there exists some $n \ge 1$ such that

$$\frac{1}{2}(q^2 - 1)q^{n-1} + 1 \le s \le \frac{1}{2}(q^2 - 1)q^n.$$

Recall that we have $N(F_{n+1}/\mathbf{F}_q) \ge 1$ and (8.23), so that F_{n+1} has at least one rational place which can serve as P_∞ and s additional places of degree ≤ 2. Thus, Theorem 8.4.1 yields

$$
\begin{aligned}
d_q(s) &\le g(F_{n+1}) + s \le q^{n+1} + q^n - q^{(n+2)/2} - 2q^{n/2} + 1 + s \\
&\le \frac{2q}{q - 1}(s - 1) - (q + 2)\left(\frac{2q(s - 1)}{q^2 - 1}\right)^{1/2} + 1 + s \\
&\le \frac{3q - 1}{q - 1}(s - 1) - \frac{(2q + 4)(s - 1)^{1/2}}{(q^2 - 1)^{1/2}} + 2,
\end{aligned}
$$

and so the desired bound is shown in all cases. □

Remark 8.4.5 Theorem 8.4.4 gives decent bounds on $d_q(s)$ for small q. But even for small q if may be possible to get better bounds from class field theory. For instance, in the case $q = 5$ consider the construction in the proof of the lower bound for $A(5)$ in Theorem 5.5.3. In that proof we obtained a global function field $K/\mathbf{F}_5$ of genus $g = 12$ and a set $\mathcal{P}$ of rational places of K with $|\mathcal{P}| = 8$ such that the Golod-Shafarevich condition is satisfied for $\ell = 2$. Thus, the $(2, \mathcal{P})$-Hilbert class field tower of K is infinite. Then Proposition 8.3.5 yields

$$V_5(s) \le \frac{11}{4}s + 1 \qquad \text{for all } s \ge 1,$$

and in view of (8.19) we have the same bound for $d_5(s)$. This bound is better than that in Theorem 8.4.4 for sufficiently large s.

If we want to optimize the construction of the sequences $S_1(P_\infty, P_1, \ldots, P_s; D)$ in this section for a given dimension $s \ge 1$ and a given prime power q, then we are led to the following considerations. For a global function field $F/\mathbf{F}_q$ with $N(F/\mathbf{F}_q) \ge 1$, define $\delta_s(F)$ as follows. Exclude one rational place of F and list all other places of F according to nondecreasing degrees. If $P_1, \ldots, P_s$ are the first s places in this list, then put

$$\delta_s(F) = \sum_{i=1}^{s}(\deg(P_i) - 1).$$

Now define

$$X_q(s) = \min_F(g(F) + \delta_s(F)),$$

where the minimum is extended over all global function fields $F/\mathbf{F}_q$ with $N(F/\mathbf{F}_q) \geq 1$. Then Theorem 8.4.1 yields

$$d_q(s) \leq X_q(s) \qquad \text{for all } s \geq 1,$$

which is analogous to (8.19). Except for the consequences of the obvious fact that

$$X_q(s) \leq V_q(s) \qquad \text{for all } s \geq 1,$$

not much is known about the quantity $X_q(s)$.

Appendix A

Curves and Their Function Fields

We discuss the connection between algebraic curves over finite fields and global function fields. For facts on algebraic curves and varieties, we refer to the books of Hartshorne [45] and Silverman [146].

A.1 Transcendence Degree

For a given field k, a **function field** over k is an extension field F of k such that there is at least one element $x \in F$ that is transcendental over k. The field k is called the constant field of F. Furthermore, as in Section 1.1, the field k is called the **full constant field** of F if k is algebraically closed in F, that is, if each element of F that is algebraic over k belongs to k.

Let F be a function field that is finitely generated over k. The **transcendence degree** of F over k is defined to be the least positive integer t such that for some $x_1, \ldots, x_t \in F$, the field F is algebraic over $k(x_1, \ldots, x_t)$. We say then that F is an **algebraic function field of t variables** over k.

A.2 Affine Spaces

If $\mathbf{F}_q$ is the finite field of order q and n a positive integer, then **affine n-space** over $\mathbf{F}_q$, denoted by $\mathbf{A}^n$, is the Cartesian n-space

$$\mathbf{A}^n := \mathbf{A}^n(\bar{\mathbf{F}}_q) = \{(a_1, \ldots, a_n) : a_i \in \bar{\mathbf{F}}_q \text{ for } i = 1, \ldots, n\},$$

where $\bar{\mathbf{F}}_q$ is a fixed algebraic closure of $\mathbf{F}_q$. For a positive integer m, the $\mathbf{F}_{q^m}$-**rational point set** is the subset

$$\mathbf{A}^n(\mathbf{F}_{q^m}) = \{(a_1, \ldots, a_n) : a_i \in \mathbf{F}_{q^m} \text{ for } i = 1, \ldots, n\}.$$

An element P of $\mathbf{A}^n$ is called a **point**, and each a_i is called a **coordinate** of P. A point $P = (a_1, \ldots, a_n) \in \mathbf{A}^n(\mathbf{F}_{q^m})$ is called $\mathbf{F}_{q^m}$-**rational**. We also call P **rational** if P is $\mathbf{F}_q$-rational. It is clear that the number of $\mathbf{F}_{q^m}$-rational points of $\mathbf{A}^n$ is q^{mn}. Define the Galois action of $\mathrm{Gal}(\bar{\mathbf{F}}_q/\mathbf{F}_q)$ on $\mathbf{A}^n$ by

$$\sigma(P) = (\sigma(a_1), \ldots, \sigma(a_n))$$

for any $\sigma \in \mathrm{Gal}(\bar{\mathbf{F}}_q/\mathbf{F}_q)$ and $P = (a_1, \ldots, a_n) \in \mathbf{A}^n$. It is clear that for any two automorphisms σ and τ of $\mathrm{Gal}(\bar{\mathbf{F}}_q/\mathbf{F}_q)$ we have $(\sigma\tau)(P) = \sigma(\tau(P))$. In particular, we define the Frobenius action as follows:

$$\pi(a_1, \ldots, a_n) = (a_1^q, \ldots, a_n^q).$$

Using the above action, we can characterize the $\mathbf{F}_q$-rational points by

$$\mathbf{A}^n(\mathbf{F}_q) = \{P \in \mathbf{A}^n : \sigma(P) = P \text{ for all } \sigma \in \mathrm{Gal}(\bar{\mathbf{F}}_q/\mathbf{F}_q)\} = \{P \in \mathbf{A}^n : \pi(P) = P\}.$$

For a point $P \in \mathbf{A}^n$, we call the set

$$\{\sigma(P) : \sigma \in \mathrm{Gal}(\bar{\mathbf{F}}_q/\mathbf{F}_q)\}$$

a **closed point** over $\mathbf{F}_q$ or an $\mathbf{F}_q$-**closed point** (we just say closed point if there is no confusion). Two points in a closed point over $\mathbf{F}_q$ are called $\mathbf{F}_q$-**conjugate**.

Let $P = (a_1, \ldots, a_n)$ be a point in a closed point P. Then

$$\mathsf{P} = \{\sigma(P) : \sigma \in \mathrm{Gal}(\bar{\mathbf{F}}_q/\mathbf{F}_q)\} = \{\sigma(P) : \sigma \in \mathrm{Gal}(\mathbf{F}_q(a_1, \ldots, a_n)/\mathbf{F}_q)\}.$$

The cardinality of P is called the **degree** of P, denoted by $\deg(\mathsf{P})$. It is equal to the degree of the extension $\mathbf{F}_q(a_1, \ldots, a_n)/\mathbf{F}_q$.

For a positive integer r, an $\mathbf{F}_q$-closed point P of degree m splits into $\gcd(r, m)$ $\mathbf{F}_{q^r}$-closed points of degree $m/\gcd(r, m)$, i.e., $\mathsf{P} = \cup_{j=1}^{\gcd(r,m)} \mathsf{P}_j$ and each P_j is an $\mathbf{F}_{q^r}$-closed point of degree $m/\gcd(r, m)$. In particular, an $\mathbf{F}_q$-closed point of degree m splits into m $\mathbf{F}_{q^r}$-rational points if and only if m divides r.

A.3   Projective Spaces

Roughly speaking, projective spaces are obtained by adding "points at infinity" to affine spaces. In order to do so, we have to go to affine spaces of higher dimension.

Definition A.3.1 A **projective n-space** over $\mathbf{F}_q$, denoted by $\mathbf{P}^n(\bar{\mathbf{F}}_q)$ or $\mathbf{P}^n$, is the set of equivalence classes of nonzero $(n+1)$-tuples $(a_0, a_1, \ldots, a_n)$ of elements of $\bar{\mathbf{F}}_q$ under the equivalence relation given by

$$(a_0, a_1, \ldots, a_n) \sim (b_0, b_1, \ldots, b_n)$$

if there exists a nonzero element λ of $\bar{\mathbf{F}}_q$ such that $a_i = \lambda b_i$ for all $i = 0, 1, \ldots, n$. An element of $\mathbf{P}^n$ is called a **point**. We denote the equivalence class containing the $(n+1)$-tuple $(a_0, a_1, \ldots, a_n)$ by $[a_0, a_1, \ldots, a_n]$. Thus, $[a_0, a_1, \ldots, a_n]$ and $[\lambda a_0, \lambda a_1, \ldots, \lambda a_n]$ stand for the same point if $\lambda \neq 0$. For a point $P = [a_0, a_1, \ldots, a_n]$, each a_i is called a **homogeneous coordinate** for P.

A point in the set

$$\mathbf{P}^n(\mathbf{F}_{q^m}) := \{[a_0, a_1, \ldots, a_n] \in \mathbf{P}^n : a_i \in \mathbf{F}_{q^m} \text{ for } i = 0, 1, \ldots, n\}$$

is called $\mathbf{F}_{q^m}$**-rational**. Since a point in $\mathbf{P}^n$ is an equivalence class, the point $[a_0, a_1, \ldots, a_n]$ of $\mathbf{P}^n$ is $\mathbf{F}_{q^m}$-rational if and only if there exists a nonzero element $\lambda \in \bar{\mathbf{F}}_q$ such that $\lambda a_i \in \mathbf{F}_{q^m}$ for all $i = 0, 1, \ldots, n$. This is equivalent to saying that

$$(a_0/a_i, \ldots, a_{i-1}/a_i, a_{i+1}/a_i, \ldots, a_n/a_i) \in \mathbf{A}^n(\mathbf{F}_{q^m})$$

provided that $a_i \neq 0$. The number of $\mathbf{F}_q$-rational points in $\mathbf{P}^n(\bar{\mathbf{F}}_q)$ is $(q^{n+1} - 1)/(q - 1)$.

Consider the Galois action of $\mathrm{Gal}(\bar{\mathbf{F}}_q/\mathbf{F}_q)$ on $\mathbf{P}^n$ given by

$$\sigma([a_0, a_1, \ldots, a_n]) = [\sigma(a_0), \sigma(a_1), \ldots, \sigma(a_n)]$$

for any $\sigma \in \mathrm{Gal}(\bar{\mathbf{F}}_q/\mathbf{F}_q)$ and $[a_0, a_1, \ldots, a_n] \in \mathbf{P}^n$. Since

$$\sigma[\lambda a_0, \lambda a_1, \ldots, \lambda a_n] = [\sigma(\lambda)\sigma(a_0), \sigma(\lambda)\sigma(a_1), \ldots, \sigma(\lambda)\sigma(a_n)],$$

the above action is well defined. In particular, we define the Frobenius action by

$$\pi([a_0, a_1, \ldots, a_n]) = [a_0^q, a_1^q, \ldots, a_n^q].$$

Using the above action, we can characterize the $\mathbf{F}_q$-rational points by

$$\mathbf{P}^n(\mathbf{F}_q) = \{P \in \mathbf{P}^n : \sigma(P) = P \text{ for all } \sigma \in \mathrm{Gal}(\bar{\mathbf{F}}_q/\mathbf{F}_q)\} = \{P \in \mathbf{P}^n : \pi(P) = P\}.$$

Similarly, we can define closed points in $\mathbf{P}^n$. For a point $P \in \mathbf{P}^n$, we call the set

$$\{\sigma(P) : \sigma \in \mathrm{Gal}(\bar{\mathbf{F}}_q/\mathbf{F}_q)\}$$

a **closed point** over $\mathbf{F}_q$ or an $\mathbf{F}_q$**-closed point** (we just say closed point if there is no confusion). Two points in a closed point over $\mathbf{F}_q$ are called $\mathbf{F}_q$**-conjugate**.

Let $P = [a_0, a_1, \ldots, a_n]$ with $a_i \neq 0$ be a point in a closed point P. Then

$$\mathsf{P} = \{\sigma(P) : \sigma \in \mathrm{Gal}(\bar{\mathbf{F}}_q/\mathbf{F}_q)\} = \{\sigma(P) : \sigma \in \mathrm{Gal}(\mathbf{F}_q(a_0/a_i, \ldots, a_n/a_i)/\mathbf{F}_q)\}.$$

The cardinality of P is called the **degree** of P, denoted by deg(P). It is equal to the degree of the extension $\mathbf{F}_q(a_0/a_i, \ldots, a_n/a_i)/\mathbf{F}_q$.

For a positive integer r, an $\mathbf{F}_q$-closed point P of degree m splits into $\gcd(r, m)$ $\mathbf{F}_{q^r}$-closed points of degree $m/\gcd(r, m)$, i.e., $\mathsf{P} = \cup_{j=1}^{\gcd(r,m)} \mathsf{P}_j$ and each P_j is an $\mathbf{F}_{q^r}$-closed point of degree $m/\gcd(r, m)$. In particular, an $\mathbf{F}_q$-closed point of degree m splits into m $\mathbf{F}_{q^r}$-rational points if and only if m divides r.

We consider the following map for each $i = 1, \ldots, n+1$:

$$\phi_i : \mathbf{A}^n \longrightarrow \mathbf{P}^n, \qquad (a_1, \ldots, a_n) \mapsto [a_1, \ldots, a_{i-1}, 1, a_i, \ldots, a_n]. \tag{A.1}$$

Then ϕ_i is injective. Thus, $\mathbf{A}^n$ can be embedded into $\mathbf{P}^n$. Moreover, for a point $P \in \mathbf{A}^n$, it is easy to see that $\{\sigma(\phi_i(P)) : \sigma \in \mathrm{Gal}(\bar{\mathbf{F}}_q/\mathbf{F}_q)\} = \{\phi_i(\sigma(P)) : \sigma \in \mathrm{Gal}(\bar{\mathbf{F}}_q/\mathbf{F}_q)\}$. This means that ϕ_i induces a bijective map between the closed points of $\mathbf{A}^n$ and those closed points of $\mathbf{P}^n$ with the ith homogeneous coordinates not equal to zero. Furthermore, the degree of a closed point P of $\mathbf{A}^n$ is equal to the degree of $\phi_i(\mathsf{P})$.

A.4 Affine Varieties

Let $\mathbf{F}_q[X] := \mathbf{F}_q[x_1, \ldots, x_n]$ and $\bar{\mathbf{F}}_q[X] := \bar{\mathbf{F}}_q[x_1, \ldots, x_n]$ be polynomial rings over $\mathbf{F}_q$ and $\bar{\mathbf{F}}_q$, respectively, in the same number n of variables. For a polynomial $f(X) \in \bar{\mathbf{F}}_q[X]$, we denote the set of **zeros** of $f(X)$ by

$$V(f) = \{P \in \mathbf{A}^n = \mathbf{A}^n(\bar{\mathbf{F}}_q) : f(P) = 0\}.$$

More generally, for a subset S of $\bar{\mathbf{F}}_q[X]$ we define the **zero set** $V(S)$ of S to be the set of common zeros of all polynomials in S, i.e.,

$$V(S) = \{P \in \mathbf{A}^n : f(P) = 0 \text{ for all } f \in S\}.$$

Definition A.4.1 An **affine algebraic set** is any set of the form $V(S)$ for some subset S of $\bar{\mathbf{F}}_q[X]$, and $V(S)$ is said to be **defined over $\mathbf{F}_q$** if S is a subset of $\mathbf{F}_q[X]$.

We denote by $V/\mathbf{F}_q$ an affine algebraic set V defined over $\mathbf{F}_q$. There always exists a finite subset T of $\mathbf{F}_q[X]$ such that $V/\mathbf{F}_q$ is the zero set $V(T)$ of T. The set of $\mathbf{F}_q$-rational points of $V/\mathbf{F}_q$ is

$$V(\mathbf{F}_q) = V \cap \mathbf{A}^n(\mathbf{F}_q),$$

i.e, a point $P \in V$ is $\mathbf{F}_q$-rational if and only if $\pi(P) = P$.

For an n-tuple $J = (j_1, \ldots, j_n)$ of nonnegative integers, we abbreviate $x_1^{j_1} \cdots x_n^{j_n}$ by X^J. For a polynomial $f(X) = \sum c_J X^J \in \bar{\mathbf{F}}_q[X]$ and an automorphism $\sigma \in \mathrm{Gal}(\bar{\mathbf{F}}_q/\mathbf{F}_q)$, we define the action $\sigma(f(X)) = \sum \sigma(c_J) X^J$. Thus, $f(X)$ belongs to $\mathbf{F}_q[X]$ if and only if $\sigma(f) = f$ for all $\sigma \in \mathrm{Gal}(\bar{\mathbf{F}}_q/\mathbf{F}_q)$. This is equivalent to $\pi(f) = f$.

For a point $P \in \mathbf{A}^n$, a polynomial $f(X) \in \bar{\mathbf{F}}_q[X]$, and an automorphism $\sigma \in \mathrm{Gal}(\bar{\mathbf{F}}_q/\mathbf{F}_q)$, we have

$$\sigma(f(P)) = \sigma(f)(\sigma(P)).$$

From the above identity we see that a closed point P is a subset of $V/\mathbf{F}_q$ as long as one of the points in P belongs to V. Thus, it makes sense to speak of a **closed point** of an affine algebraic set defined over $\mathbf{F}_q$.

For an affine algebraic set $V/\mathbf{F}_q$, we put

$$I(V) = \{f \in \bar{\mathbf{F}}_q[X] : f(P) = 0 \text{ for all } P \in V\}$$

and

$$I(V/\mathbf{F}_q) = I(V) \cap \mathbf{F}_q[X] = \{f \in \mathbf{F}_q[X] : f(P) = 0 \text{ for all } P \in V\}.$$

Then it is easy to verify that $I(V)$, respectively $I(V/\mathbf{F}_q)$, is an ideal of $\bar{\mathbf{F}}_q[X]$, respectively $\mathbf{F}_q[X]$.

An affine algebraic set $V \subseteq \mathbf{A}^n$ is called **reducible** if there exist two affine algebraic sets $V_1, V_2 \subseteq \mathbf{A}^n$ such that $V_i \neq V$ for $i = 1, 2$ and $V = V_1 \cup V_2$. If V is nonempty and not reducible, then V is said to be **irreducible**. An affine algebraic set V is irreducible if and only if $I(V)$ is a prime ideal of $\bar{\mathbf{F}}_q[X]$. For an affine algebraic set $V/\mathbf{F}_q$, we say also that

$V/\mathbf{F}_q$ is **absolutely irreducible** if V is irreducible. Note that it is not enough to check that $I(V/\mathbf{F}_q)$ is a prime ideal of $\mathbf{F}_q[X]$. For example, $V = V(x^2 + y^2)$ is not irreducible in $\mathbf{A}^2(\bar{\mathbf{F}}_3)$, but $I(V/\mathbf{F}_3) = (x^2 + y^2)$ is a prime ideal of $\mathbf{F}_3[x, y]$. Conversely, if $V/\mathbf{F}_q$ is (absolutely) irreducible, then it is easy to verify that $I(V/\mathbf{F}_q)$ is a prime ideal of $\mathbf{F}_q[X]$.

Definition A.4.2 An affine algebraic set $V/\mathbf{F}_q$ is called an **affine variety** if V is irreducible.

Definition A.4.3 Let $V/\mathbf{F}_q$ be an affine variety. The **coordinate ring** of $V/\mathbf{F}_q$ is defined by

$$\mathbf{F}_q[V] = \mathbf{F}_q[X]/I(V/\mathbf{F}_q).$$

This is an integral domain since $I(V/\mathbf{F}_q)$ is a prime ideal. The quotient field of $\mathbf{F}_q[V]$ is called the **function field** of V, denoted by $\mathbf{F}_q(V)$. We also have the **absolute coordinate ring** of V defined by

$$\bar{\mathbf{F}}_q[V] = \bar{\mathbf{F}}_q[X]/I(V)$$

and the **absolute function field** $\bar{\mathbf{F}}_q(V)$ of V, which is the quotient field of $\bar{\mathbf{F}}_q[V]$.

For two polynomials $a, b \in \bar{\mathbf{F}}_q[X]$ with $a - b \in I(V)$, we have $a(P) = b(P)$ for all $P \in V$. This implies that each element f of $\bar{\mathbf{F}}_q[V]$ induces a well-defined function from V to $\bar{\mathbf{F}}_q$. Consequently, each element f/g of $\bar{\mathbf{F}}_q(V)$ with $f, g \in \bar{\mathbf{F}}_q[V]$ defines a function from $V \setminus V(g)$ to $\bar{\mathbf{F}}_q$.

A point $P \in V$ is called a **zero**, respectively **pole**, of a function $h \in \bar{\mathbf{F}}_q(V)$ if there exist two elements $f, g \in \bar{\mathbf{F}}_q[V]$ such that $h = f/g$ and $f(P) = 0$ and $g(P) \neq 0$, respectively $f(P) \neq 0$ and $g(P) = 0$. If P is not a pole of h, then $h(P)$ is well defined.

We have defined the Galois action of $\mathrm{Gal}(\bar{\mathbf{F}}_q/\mathbf{F}_q)$ on the polynomial ring $\bar{\mathbf{F}}_q[X]$. This action can be defined on $\bar{\mathbf{F}}_q[V]$ for an affine variety $V/\mathbf{F}_q$ because of the following result: if $V/\mathbf{F}_q$ is an affine variety, then for any two polynomials $a, b \in \bar{\mathbf{F}}_q[X]$ with $a - b \in I(V)$ and any automorphism $\sigma \in \mathrm{Gal}(\bar{\mathbf{F}}_q/\mathbf{F}_q)$ we have $\sigma(a)(P) = \sigma(b)(P)$ for all $P \in V$. Furthermore, we can extend the action to the quotient field $\bar{\mathbf{F}}_q(V)$ by $\sigma(f/g) = \sigma(f)/\sigma(g)$.

We now consider the relationship between $\mathbf{F}_q[V]$, respectively $\mathbf{F}_q(V)$, and $\bar{\mathbf{F}}_q[V]$, respectively $\bar{\mathbf{F}}_q(V)$, for an affine variety $V/\mathbf{F}_q$. For two polynomials $f, g \in \mathbf{F}_q[X]$, we note that $f - g \in I(V)$ if and only if $f - g \in I(V/\mathbf{F}_q)$. Thus, $\mathbf{F}_q[V]$ is naturally embedded into $\bar{\mathbf{F}}_q[V]$, and therefore $\mathbf{F}_q(V)$ is a subfield of $\bar{\mathbf{F}}_q(V)$. Furthermore, we have

$$\bar{\mathbf{F}}_q[V] = \bar{\mathbf{F}}_q \cdot \mathbf{F}_q[V], \qquad \bar{\mathbf{F}}_q(V) = \bar{\mathbf{F}}_q \cdot \mathbf{F}_q(V),$$

$$\mathbf{F}_q[V] = \{f \in \bar{\mathbf{F}}_q[V] : \sigma(f) = f \text{ for all } \sigma \in \mathrm{Gal}(\bar{\mathbf{F}}_q/\mathbf{F}_q)\},$$

$$\mathbf{F}_q(V) = \{h \in \bar{\mathbf{F}}_q(V) : \sigma(h) = h \text{ for all } \sigma \in \mathrm{Gal}(\bar{\mathbf{F}}_q/\mathbf{F}_q)\}.$$

The field $\mathbf{F}_q$ is the full constant field of the function field $\mathbf{F}_q(V)$.

Definition A.4.4 Let $V/\mathbf{F}_q$ be an affine variety and $P \in V$. The **local ring** of V at P is defined by

$$\bar{\mathbf{F}}_q(V)_P = \{h \in \bar{\mathbf{F}}_q(V) : h(P) \text{ is well defined}\}.$$

Its maximal ideal is

$$\overline{M}_P := \overline{M}_P(V) = \{h \in \bar{\mathbf{F}}_q(V)_P : h(P) = 0\}.$$

Furthermore, if P is an $\mathbf{F}_q$-rational point of V, then the **local ring** of $V/\mathbf{F}_q$ at P is

$$\mathbf{F}_q(V)_P = \{h \in \mathbf{F}_q(V) : h(P) \text{ is well defined}\} = \bar{\mathbf{F}}_q(V)_P \cap \mathbf{F}_q(V)$$

and its maximal ideal is

$$M_P := M_P(V) = \{h \in \mathbf{F}_q(V)_P : h(P) = 0\} = \overline{M}_P \cap \mathbf{F}_q(V).$$

Definition A.4.5 The **dimension** of the affine variety $V/\mathbf{F}_q$ is defined to be the transcendence degree of $\bar{\mathbf{F}}_q(V)$ over $\bar{\mathbf{F}}_q$, denoted by $\dim(V)$.

Definition A.4.6 Let $V/\mathbf{F}_q \subseteq \mathbf{A}^n$ be an affine variety and $f_1, \ldots, f_m \in \mathbf{F}_q[X]$ a set of generators for $I(V)$. Then V is **smooth** at a point $P \in V$ if the $m \times n$ matrix

$$\left(\frac{\partial f_i}{\partial x_j}(P) \right)_{1 \le i \le m, 1 \le j \le n}$$

has rank $n - \dim(V)$. If V is smooth at every point of V, then we say that V is **smooth**.

A.5 Projective Varieties

Definition A.5.1 A polynomial $f \in \bar{\mathbf{F}}_q[X] = \bar{\mathbf{F}}_q[x_0, x_1, \ldots, x_n]$ (we denote $(x_0, x_1, \ldots, x_n)$ by X in this section) is called **homogeneous** of degree d if

$$f(\lambda x_0, \lambda x_1, \ldots, \lambda x_n) = \lambda^d f(x_0, x_1, \ldots, x_n)$$

for all $\lambda \in \bar{\mathbf{F}}_q$. An ideal of $\bar{\mathbf{F}}_q[X]$ is **homogeneous** if it is generated by homogeneous polynomials.

Definition A.5.2 A **projective algebraic set** is any set V of the form

$$Z(I) = \{P \in \mathbf{P}^n : f(P) = 0 \text{ for all homogeneous } f \in I\}$$

for some homogeneous ideal I of $\bar{\mathbf{F}}_q[X]$, and V is said to be **defined over $\mathbf{F}_q$** if I is generated by homogeneous polynomials from $\mathbf{F}_q[X]$.

We denote by $V/\mathbf{F}_q$ a projective algebraic set V defined over $\mathbf{F}_q$. The set of $\mathbf{F}_q$-rational points of a projective algebraic set $V = Z(I)$ over $\mathbf{F}_q$ is

$$V(\mathbf{F}_q) = V \cap \mathbf{P}^n(\mathbf{F}_q),$$

i.e, a point $P \in V$ is $\mathbf{F}_q$-rational if and only if $\pi(P) = P$.

For a point $P \in \mathbf{P}^n$, a homogeneous polynomial $f(X) \in \bar{\mathbf{F}}[X]$, and an automorphism $\sigma \in \text{Gal}(\bar{\mathbf{F}}_q/\mathbf{F}_q)$, we have

$$\sigma(f(P)) = \sigma(f)(\sigma(P)).$$

From the above identity we see that a closed point P is a subset of V as long as one of the points in P belongs to V. Thus, it makes sense to speak of a **closed point** of a projective algebraic set over $\mathbf{F}_q$.

For a projective algebraic set $V/\mathbf{F}_q$, we let $I(V)$ be the ideal of $\bar{\mathbf{F}}_q[X]$ generated by the set

$$\{f \in \bar{\mathbf{F}}_q[X] : f \text{ is homogeneous and } f(P) = 0 \text{ for all } P \in V\}.$$

A projective algebraic set $V \subseteq \mathbf{P}^n$ is called **reducible** if there exist two projective algebraic sets $V_1, V_2 \subseteq \mathbf{P}^n$ such that $V_i \neq V$ for $i = 1, 2$ and $V = V_1 \cup V_2$. If V is nonempty and not reducible, then V is said to be **irreducible**. A projective algebraic set V is irreducible if and only if $I(V)$ is a prime ideal of $\bar{\mathbf{F}}_q[X]$. For a projective algebraic set $V/\mathbf{F}_q$, we say also that $V/\mathbf{F}_q$ is **absolutely irreducible** if V is irreducible.

Definition A.5.3 A projective algebraic set $V/\mathbf{F}_q$ is called a **projective variety** if V is irreducible.

Lemma A.5.4 *Let ϕ_i be defined as in (A.1) and put $U_i = \phi_i(\mathbf{A}^n)$. If V is a projective variety, then $\phi_i^{-1}(V \cap U_i)$ is either an affine variety or the empty set for any $1 \leq i \leq n+1$. Moreover, there exists $1 \leq j \leq n+1$ such that $\phi_j^{-1}(V \cap U_j) \neq \emptyset$.*

Definition A.5.5 Let $V/\mathbf{F}_q$ be a projective variety and choose i such that $\phi_i^{-1}(V \cap U_i) \neq \emptyset$. The **dimension** of V is defined to be the dimension of $\phi_i^{-1}(V \cap U_i)$, denoted by $\dim(V)$. The **function field** $\mathbf{F}_q(V)$ of V is the function field $\mathbf{F}_q(\phi_i^{-1}(V \cap U_i))$ and the **absolute function field** $\bar{\mathbf{F}}_q(V)$ of V is the absolute function field $\bar{\mathbf{F}}_q(\phi_i^{-1}(V \cap U_i))$.

Let $V/\mathbf{F}_q$ be a projective variety. Then $\mathbf{F}_q$ is the full constant field of the function field $\mathbf{F}_q(V)$. Local rings of $V/\mathbf{F}_q$ are defined in analogy with Definition A.4.4.

Definition A.5.6 Let $V/\mathbf{F}_q \subseteq \mathbf{P}^n$ be a projective variety and $P = [a_0, a_1, \ldots, a_n] \in V$. Suppose that $a_i \neq 0$. Then V is **smooth** at P if $\phi_{i+1}^{-1}(V \cap U_{i+1})$ is smooth at $(a_0/a_i, \ldots, a_{i-1}/a_i, a_{i+1}/a_i, \ldots, a_n/a_i)$. If V is smooth at every point of V, then we say that V is **smooth**.

A.6 Projective Curves

A projective variety of dimension 1 defined over $\mathbf{F}_q$ is called a **projective (algebraic) curve** over $\mathbf{F}_q$. For a smooth projective curve $\mathcal{X}/\mathbf{F}_q$ and a point P of $\mathcal{X}$, the local ring $\bar{\mathbf{F}}_q(\mathcal{X})_P$ is a discrete valuation ring of $\bar{\mathbf{F}}_q(\mathcal{X})$ and its unique maximal ideal $\overline{M}_P$ is a place of $\bar{\mathbf{F}}_q(\mathcal{X})$. The intersection $\bar{\mathbf{F}}_q(\mathcal{X})_P \cap \mathbf{F}_q(\mathcal{X})$ is also a discrete valuation ring of $\mathbf{F}_q(\mathcal{X})$ and the intersection $\overline{M}_P \cap \mathbf{F}_q(\mathcal{X})$ is a place of degree m of $\mathbf{F}_q(\mathcal{X})$, where m is the degree of the closed point of $\mathcal{X}$ containing P.

Definition A.6.1 Let $\mathcal{X}_1/\mathbf{F}_q \subseteq \mathbf{P}^l$ and $\mathcal{X}_2/\mathbf{F}_q \subseteq \mathbf{P}^n$ be two smooth projective curves. A **morphism over $\mathbf{F}_q$** (or **$\mathbf{F}_q$-morphism**) from $\mathcal{X}_1$ to $\mathcal{X}_2$ is a map of the form

$$\eta : \mathcal{X}_1 \longrightarrow \mathcal{X}_2, \quad P \in \mathcal{X}_1 \mapsto [f_0(P), f_1(P), \ldots, f_n(P)],$$

where $f_0, f_1, \ldots, f_n \in \mathbf{F}_q(\mathcal{X}_1)$ have the property that for every point $P \in \mathcal{X}_1$, there exists a function $h_P \in \bar{\mathbf{F}}_q(\mathcal{X}_1)$ such that $(f_i h_P)(P)$ is well defined for $0 \leq i \leq n$ and $(f_j h_P)(P) \neq 0$ for some j with $0 \leq j \leq n$. Then we put specifically

$$\eta(P) = [(f_0 h_P)(P), (f_1 h_P)(P), \ldots, (f_n h_P)(P)] \in \mathcal{X}_2.$$

Furthermore, η is said to be an $\mathbf{F}_q$-**isomorphism** if there exists a morphism ψ over $\mathbf{F}_q$ from $\mathcal{X}_2/\mathbf{F}_q$ to $\mathcal{X}_1/\mathbf{F}_q$ such that $\eta \circ \psi$ and $\psi \circ \eta$ are identity maps on $\mathcal{X}_2$ and $\mathcal{X}_1$, respectively. In this case, we say that $\mathcal{X}_1/\mathbf{F}_q$ is $\mathbf{F}_q$-**isomorphic** to $\mathcal{X}_2/\mathbf{F}_q$.

The conditions on η in Definition A.6.1 may be relaxed somewhat since it is a well-known theorem that on a smooth projective curve, any rational map defined over $\mathbf{F}_q$ is already a morphism over $\mathbf{F}_q$. The following results describe the relationship between curves and function fields.

Theorem A.6.2 *The map $\delta : \mathcal{X}/\mathbf{F}_q \mapsto \mathbf{F}_q(\mathcal{X})$ yields a natural correspondence between smooth projective curves over $\mathbf{F}_q$ and global function fields (of one variable) with full constant field $\mathbf{F}_q$, up to isomorphisms.*

The phrase "up to isomorphisms" in the above theorem refers to the result that two smooth projective curves over $\mathbf{F}_q$ are $\mathbf{F}_q$-isomorphic if and only if their function fields are $\mathbf{F}_q$-isomorphic. The map δ in Theorem A.6.2 also induces a correspondence between closed points of $\mathcal{X}$ and places of $\mathbf{F}_q(\mathcal{X})$ as follows.

Theorem A.6.3 *Let $\mathcal{X}/\mathbf{F}_q$ be a smooth projective curve. Then:*

(i) *δ induces a correspondence $P \in \mathcal{X} \mapsto \overline{\mathsf{M}}_P$ between points of $\mathcal{X}$ and places of $\bar{\mathbf{F}}_q(\mathcal{X})$.*

(ii) *If P is a closed point of $\mathcal{X}$ and P_1, P_2 are two points in P, then $\overline{\mathsf{M}}_{P_1} \cap \mathbf{F}_q(\mathcal{X}) = \overline{\mathsf{M}}_{P_2} \cap \mathbf{F}_q(\mathcal{X})$ is a place of degree $|\mathsf{P}|$ of $\mathbf{F}_q(\mathcal{X})$. Thus, δ induces a bijective correspondence between closed points of degree m of $\mathcal{X}/\mathbf{F}_q$ and places of degree m of $\mathbf{F}_q(\mathcal{X})$ for all $m \geq 1$. In particular, δ induces a bijective correspondence between $\mathbf{F}_q$-rational points of $\mathcal{X}/\mathbf{F}_q$ and rational places of $\mathbf{F}_q(\mathcal{X})$.*

To every smooth projective curve we can associate a uniquely determined genus which stems from the Riemann-Roch theorem for curves and is again a nonnegative integer. It is an important fact that the correspondence δ in Theorem A.6.2 preserves the genus, i.e., the curve $\mathcal{X}/\mathbf{F}_q$ and its function field $\mathbf{F}_q(\mathcal{X})$ have the same genus.

Bibliography

[1] M. Abdón and F. Torres, On maximal curves in characteristic two, *Manuscripta Math.* **99**, 39–53 (1999).

[2] I. Aleshnikov, V. Deolalikar, P.V. Kumar, and H. Stichtenoth, Towards a basis for the space of regular functions in a tower of function fields meeting the Drinfeld-Vladut bound, *Finite Fields and Applications* (D. Jungnickel and H. Niederreiter, eds.), Springer, Berlin, to appear.

[3] B. Anglès and C. Maire, A note on tamely ramified towers of global function fields, *Finite Fields Appl.*, to appear.

[4] E. Artin and J. Tate, *Class Field Theory*, 2nd ed., Addison-Wesley, Redwood City, CA, 1990.

[5] R. Auer, Ray class fields of global function fields with many rational places, Dissertation, University of Oldenburg, 1999.

[6] R. Auer, Curves over finite fields with many rational points obtained by ray class field extensions, *Algorithmic Number Theory* (W. Bosma, ed.), Lecture Notes in Computer Science, Vol. **1838**, pp. 127–134, Springer, Berlin, 2000.

[7] R. Auer, Ray class fields of global function fields with many rational places, *Acta Arith.* **95**, 97–122 (2000).

[8] P. Beelen and R. Pellikaan, The Newton polygon of plane curves with many rational points, *Designs, Codes and Cryptography* **21**, 41–68 (2000).

[9] J. Bierbrauer, Universal hashing and geometric codes, *Designs, Codes and Cryptography* **11**, 207–221 (1997).

[10] I.F. Blake, Curves with many points and their applications, *Applied Algebra, Algebraic Algorithms and Error-Correcting Codes* (M. Fossorier *et al.*, eds.), Lecture Notes in Computer Science, Vol. **1719**, pp. 55–64, Springer, Berlin, 1999.

[11] I.F. Blake, G. Seroussi, and N.P. Smart, *Elliptic Curves in Cryptography*, London Math. Soc. Lecture Note Series, Vol. **265**, Cambridge University Press, Cambridge, 1999.

[12] L. Carlitz, A class of polynomials, *Trans. Amer. Math. Soc.* **43**, 167–182 (1938).

[13] J.W.S. Cassels and A. Fröhlich (eds.), *Algebraic Number Theory*, Academic Press, London, 1967.

[14] A.T. Clayman, K.M. Lawrence, G.L. Mullen, H. Niederreiter, and N.J.A. Sloane, Updated tables of parameters of (t, m, s)-nets, *J. Combinatorial Designs* **7**, 381–393 (1999).

[15] Z.J. Czech, G. Havas, and B.S. Majewski, Perfect hashing, *Theoretical Computer Science* **182**, 1–143 (1997).

[16] B. de Mathan, Approximations diophantiennes dans un corps local, *Bull. Soc. Math. France Suppl. Mém.* **21** (1970).

[17] M. Deuring, Die Typen der Multiplikatorenringe elliptischer Funktionenkörper, *Abh. Math. Sem. Univ. Hamburg* **14**, 197–272 (1941).

[18] M. Deuring, *Lectures on the Theory of Algebraic Functions of One Variable*, Lecture Notes in Math., Vol. **314**, Springer, Berlin, 1973.

[19] C.S. Ding, H. Niederreiter, and C.P. Xing, Some new codes from algebraic curves, *IEEE Trans. Inform. Theory* **46**, 2638–2642 (2000).

[20] M. Drmota and R.F. Tichy, *Sequences, Discrepancies and Applications*, Lecture Notes in Math., Vol. **1651**, Springer, Berlin, 1997.

[21] A. Enge, *Elliptic Curves and Their Applications to Cryptography: An Introduction*, Kluwer, Boston, 1999.

[22] H. Faure, Discrépance de suites associées à un système de numération (en dimension s), *Acta Arith.* **41**, 337–351 (1982).

[23] G.-L. Feng and T.R.N. Rao, A simple approach for construction of algebraic-geometric codes from affine plane curves, *IEEE Trans. Inform. Theory* **40**, 1003–1012 (1994).

[24] G.-L. Feng and T.R.N. Rao, Improved geometric Goppa codes. Part I: Basic theory, *IEEE Trans. Inform. Theory* **41**, 1678–1693 (1995).

[25] M.L. Fredman and J. Komlós, On the size of separating systems and families of perfect hash functions, *SIAM J. Algebraic Discrete Methods* **5**, 61–68 (1984).

[26] G. Frey, M. Perret, and H. Stichtenoth, On the different of abelian extensions of global fields, *Coding Theory and Algebraic Geometry* (H. Stichtenoth and M.A. Tsfasman, eds.), Lecture Notes in Math., Vol. **1518**, pp. 26–32, Springer, Berlin, 1992.

[27] R. Fuhrmann, A. Garcia, and F. Torres, On maximal curves, *J. Number Theory* **67**, 29–51 (1997).

[28] R. Fuhrmann and F. Torres, The genus of curves over finite fields with many rational points, *Manuscripta Math.* **89**, 103–106 (1996).

[29] A. Garcia and H. Stichtenoth, Algebraic function fields over finite fields with many rational places, *IEEE Trans. Inform. Theory* **41**, 1548–1563 (1995).

[30] A. Garcia and H. Stichtenoth, A tower of Artin-Schreier extensions of function fields attaining the Drinfeld-Vladut bound, *Invent. Math.* **121**, 211–222 (1995).

[31] A. Garcia and H. Stichtenoth, On the asymptotic behaviour of some towers of function fields over finite fields, *J. Number Theory* **61**, 248–273 (1996).

[32] A. Garcia and H. Stichtenoth, Asymptotically good towers of function fields over finite fields, *C.R. Acad. Sci. Paris Sér. I Math.* **322**, 1067–1070 (1996).

[33] A. Garcia and H. Stichtenoth, A class of polynomials over finite fields, *Finite Fields Appl.* **5**, 424–435 (1999).

[34] A. Garcia and H. Stichtenoth, On Chebyshev polynomials and maximal curves, *Acta Arith.* **90**, 301–311 (1999).

[35] A. Garcia and H. Stichtenoth, Skew pyramids of function fields are asymptotically bad, *Coding Theory, Cryptography and Related Areas* (J. Buchmann *et al.*, eds.), pp. 111–113, Springer, Berlin, 2000.

[36] A. Garcia, H. Stichtenoth, and M. Thomas, On towers and composita of towers of function fields over finite fields, *Finite Fields Appl.* **3**, 257–274 (1997).

[37] A. Garcia, H. Stichtenoth, and C.P. Xing, On subfields of the Hermitian function field, *Compositio Math.* **120**, 137–170 (2000).

[38] V.D. Goppa, Codes that are associated with divisors (Russian), *Problemy Peredači Informacii* **13**, 33–39 (1977).

[39] V.D. Goppa, Codes on algebraic curves (Russian), *Dokl. Akad. Nauk SSSR* **259**, 1289–1290 (1981).

[40] V.D. Goppa, Algebraic-geometric codes (Russian), *Izv. Akad. Nauk SSSR Ser. Mat.* **46**, 762–781 (1982).

[41] V.D. Goppa, *Geometry and Codes*, Kluwer, Dordrecht, 1988.

[42] D. Goss, *Basic Structures of Function Field Arithmetic*, Springer, Berlin, 1996.

[43] F.G. Gustavson, Analysis of the Berlekamp-Massey linear feedback shift-register synthesis algorithm, *IBM J. Res. Develop.* **20**, 204–212 (1976).

[44] G. Haché, Computation in algebraic function fields for effective construction of algebraic-geometric codes, *Applied Algebra, Algebraic Algorithms and Error-Correcting Codes* (G. Cohen, M. Giusti, and T. Mora, eds.), Lecture Notes in Computer Science, Vol. **948**, pp. 262–278, Springer, Berlin, 1995.

[45] R. Hartshorne, *Algebraic Geometry*, Springer, New York, 1977.

[46] D.R. Hayes, Explicit class field theory for rational function fields, *Trans. Amer. Math. Soc.* **189**, 77–91 (1974).

[47] D.R. Hayes, Explicit class field theory in global function fields, *Studies in Algebra and Number Theory*, Advances in Math. Supp. Studies, Vol. **6**, pp. 173–217, Academic Press, New York, 1979.

[48] D.R. Hayes, Stickelberger elements in function fields, *Compositio Math.* **55**, 209–239 (1985).

[49] D.R. Hayes, A brief introduction to Drinfeld modules, *The Arithmetic of Function Fields* (D. Goss, D.R. Hayes, and M.I. Rosen, eds.), pp. 1–32, W. de Gruyter, Berlin, 1992.

[50] T. Helleseth and T. Johansson, Universal hash functions from exponential sums over finite fields and Galois rings, *Advances in Cryptology – CRYPTO '96* (N. Koblitz, ed.), Lecture Notes in Computer Science, Vol. **1109**, pp. 31–44, Springer, Berlin, 1996.

[51] F. Heß, Zur Divisorenklassengruppenberechnung in globalen Funktionenkörpern, Dissertation, TU Berlin, 1999.

[52] T. Høholdt and R. Pellikaan, On the decoding of algebraic-geometric codes, *IEEE Trans. Inform. Theory* **41**, 1589–1614 (1995).

[53] T. Høholdt, J.H. van Lint, and R. Pellikaan, Algebraic geometry codes, *Handbook of Coding Theory* (V.S. Pless and W.C. Huffman, eds.), Vol. 1, pp. 871–961, Elsevier, Amsterdam, 1998.

[54] L.K. Hua and Y. Wang, *Applications of Number Theory to Numerical Analysis*, Springer, Berlin, 1981.

[55] Y. Ihara, Congruence relations and Shimura curves, *Automorphic Forms, Representations and L-Functions* (A. Borel and W. Casselman, eds.), Proc. Symp. Pure Math., Vol. **33**, Part 2, pp. 291–311, American Math. Society, Providence, RI, 1979.

[56] Y. Ihara, Some remarks on the number of rational points of algebraic curves over finite fields, *J. Fac. Sci. Univ. Tokyo Sect. IA Math.* **28**, 721–724 (1981).

[57] Y. Ihara, Shimura curves over finite fields and their rational points, *Applications of Curves over Finite Fields* (M.D. Fried, ed.), Contemporary Math., Vol. **245**, pp. 15–23, American Math. Society, Providence, RI, 1999.

[58] A. Keller, Cyclotomic function fields with many rational places, *Finite Fields and Applications* (D. Jungnickel and H. Niederreiter, eds.), Springer, Berlin, to appear.

[59] N. Koblitz, *Algebraic Aspects of Cryptography*, Springer, Berlin, 1998.

[60] H. Koch, *Algebraic Number Theory*, Springer, Berlin, 1997.

[61] D. Kohel, S. Ling, and C.P. Xing, Explicit sequence expansions, *Sequences and Their Applications* (C. Ding, T. Helleseth, and H. Niederreiter, eds.), pp. 308–317, Springer, London, 1999.

[62] L. Kuipers and H. Niederreiter, *Uniform Distribution of Sequences*, Wiley, New York, 1974.

[63] G. Larcher, Digital point sets: analysis and applications, *Random and Quasi-Random Point Sets* (P. Hellekalek and G. Larcher, eds.), Lecture Notes in Statistics, Vol. **138**, pp. 167–222, Springer, New York, 1998.

[64] G. Larcher, H. Niederreiter, and W.Ch. Schmid, Digital nets and sequences constructed over finite rings and their application to quasi-Monte Carlo integration, *Monatsh. Math.* **121**, 231–253 (1996).

[65] K. Lauter, Ray class field constructions of curves over finite fields with many rational points, *Algorithmic Number Theory* (H. Cohen, ed.), Lecture Notes in Computer Science, Vol. **1122**, pp. 187–195, Springer, Berlin, 1996.

[66] K. Lauter, Deligne-Lusztig curves as ray class fields, *Manuscripta Math.* **98**, 87–96 (1999).

[67] K. Lauter, A formula for constructing curves over finite fields with many rational points, *J. Number Theory* **74**, 56–72 (1999).

[68] K. Lauter, Improved upper bounds for the number of rational points on algebraic curves over finite fields, *C.R. Acad. Sci. Paris Sér. I Math.* **328**, 1181–1185 (1999).

[69] K. Lauter, Non-existence of a curve over $\mathbf{F}_3$ of genus 5 with 14 rational points, *Proc. Amer. Math. Soc.* **128**, 369–374 (2000).

[70] K. Lauter, Zeta functions of curves over finite fields with many rational points, *Coding Theory, Cryptography and Related Areas* (J. Buchmann *et al.*, eds.), pp. 167–174, Springer, Berlin, 2000.

[71] K. Lauter and J.-P. Serre, Geometric methods for improving the upper bounds on the number of rational points on algebraic curves over finite fields, *J. Algebraic Geometry* **10**, 19–36 (2001).

[72] K.M. Lawrence, A. Mahalanabis, G.L. Mullen, and W.Ch. Schmid, Construction of digital (t, m, s)-nets from linear codes, *Finite Fields and Applications* (S. Cohen and H. Niederreiter, eds.), London Math. Soc. Lecture Note Series, Vol. **233**, pp. 189–208, Cambridge University Press, Cambridge, 1996.

[73] C.F. Laywine and G.L. Mullen, *Discrete Mathematics Using Latin Squares*, Wiley, New York, 1998.

[74] D. Le Brigand and J.J. Risler, Algorithme de Brill-Noether et codes de Goppa, *Bull. Soc. Math. France* **116**, 231–253 (1988).

[75] W.-C.W. Li and H. Maharaj, Coverings of curves with asymptotically many rational points, *J. Number Theory*, to appear.

[76] R. Lidl and H. Niederreiter, *Finite Fields*, Addison-Wesley, Reading, MA, 1983; now distributed by Cambridge University Press.

[77] F.J. MacWilliams and N.J.A. Sloane, *The Theory of Error-Correcting Codes*, North-Holland, Amsterdam, 1977.

[78] Yu.I. Manin, What is the maximum number of points on a curve over F_2?, *J. Fac. Sci. Univ. Tokyo Sect. IA Math.* **28**, 715–720 (1981).

[79] Yu.I. Manin and S.G. Vlăduţ, Linear codes and modular curves, *J. Soviet Math.* **30**, 2611–2643 (1985).

[80] R. Matsumoto and S. Miura, Computing a basis of $\mathcal{L}(D)$ on an affine algebraic curve with one rational place at infinity, *Applied Algebra, Algebraic Algorithms and Error-Correcting Codes* (M. Fossorier *et al.*, eds.), Lecture Notes in Computer Science, Vol. **1719**, pp. 271–281, Springer, Berlin, 1999.

[81] A.J. Menezes, *Elliptic Curve Public Key Cryptosystems*, Kluwer, Boston, 1993.

[82] C.J. Moreno, *Algebraic Curves over Finite Fields*, Cambridge Tracts in Math., Vol. **97**, Cambridge University Press, Cambridge, 1991.

[83] G.L. Mullen, A. Mahalanabis, and H. Niederreiter, Tables of (t, m, s)-net and (t, s)-sequence parameters, *Monte Carlo and Quasi-Monte Carlo Methods in Scientific Computing* (H. Niederreiter and P.J.-S. Shiue, eds.), Lecture Notes in Statistics, Vol. **106**, pp. 58–86, Springer, New York, 1995.

[84] J. Neukirch, *Class Field Theory*, Springer, Berlin, 1986.

[85] J. Neukirch, *Algebraic Number Theory*, Springer, Berlin, 1999.

[86] H. Niederreiter, Quasi-Monte Carlo methods and pseudo-random numbers, *Bull. Amer. Math. Soc.* **84**, 957–1041 (1978).

[87] H. Niederreiter, Point sets and sequences with small discrepancy, *Monatsh. Math.* **104**, 273–337 (1987).

[88] H. Niederreiter, Continued fractions for formal power series, pseudorandom numbers, and linear complexity of sequences, *Contributions to General Algebra 5* (Proc. Salzburg Conf., 1986), pp. 221–233, B.G. Teubner, Stuttgart, 1987.

[89] H. Niederreiter, Sequences with almost perfect linear complexity profile, *Advances in Cryptology - EUROCRYPT '87* (D. Chaum and W.L. Price, eds.), Lecture Notes in Computer Science, Vol. **304**, pp. 37–51, Springer, Berlin, 1988.

[90] H. Niederreiter, The probabilistic theory of linear complexity, *Advances in Cryptology - EUROCRYPT '88* (C.G. Günther, ed.), Lecture Notes in Computer Science, Vol. **330**, pp. 191–209, Springer, Berlin, 1988.

[91] H. Niederreiter, Low-discrepancy and low-dispersion sequences, *J. Number Theory* **30**, 51–70 (1988).

[92] H. Niederreiter, A combinatorial approach to probabilistic results on the linear-complexity profile of random sequences, *J. Cryptology* **2**, 105–112 (1990).

[93] H. Niederreiter, The linear complexity profile and the jump complexity of keystream sequences, *Advances in Cryptology - EUROCRYPT '90* (I.B. Damgård, ed.), Lecture Notes in Computer Science, Vol. **473**, pp. 174–188, Springer, Berlin, 1991.

[94] H. Niederreiter, *Random Number Generation and Quasi-Monte Carlo Methods*, SIAM, Philadelphia, 1992.

[95] H. Niederreiter, Finite fields and cryptology, *Finite Fields, Coding Theory, and Advances in Communications and Computing* (G.L. Mullen and P.J.-S. Shiue, eds.), pp. 359–373, Dekker, New York, 1993.

[96] H. Niederreiter, Nets, (t, s)-sequences, and algebraic curves over finite fields with many rational points, *Proc. International Congress of Mathematicians* (Berlin, 1998), Documenta Math. Extra Volume **ICM III**, 377–386 (1998).

[97] H. Niederreiter, Some computable complexity measures for binary sequences, *Sequences and Their Applications* (C. Ding, T. Helleseth, and H. Niederreiter, eds.), pp. 67–78, Springer, London, 1999.

[98] H. Niederreiter, Constructions of (t, m, s)-nets, *Monte Carlo and Quasi-Monte Carlo Methods 1998* (H. Niederreiter and J. Spanier, eds.), pp. 70–85, Springer, Berlin, 2000.

[99] H. Niederreiter and G. Pirsic, Duality for digital nets and its applications, *Acta Arith.* **97**, 173–182 (2001).

[100] H. Niederreiter and M. Vielhaber, Linear complexity profiles: Hausdorff dimensions for almost perfect profiles and measures for general profiles, *J. Complexity* **13**, 353–383 (1997).

[101] H. Niederreiter and C.P. Xing, Low-discrepancy sequences obtained from algebraic function fields over finite fields, *Acta Arith.* **72**, 281–298 (1995).

[102] H. Niederreiter and C.P. Xing, Low-discrepancy sequences and global function fields with many rational places, *Finite Fields Appl.* **2**, 241–273 (1996).

[103] H. Niederreiter and C.P. Xing, Explicit global function fields over the binary field with many rational places, *Acta Arith.* **75**, 383–396 (1996).

[104] H. Niederreiter and C.P. Xing, Quasirandom points and global function fields, *Finite Fields and Applications* (S. Cohen and H. Niederreiter, eds.), London Math. Soc. Lecture Note Series, Vol. **233**, pp. 269–296, Cambridge University Press, Cambridge, 1996.

[105] H. Niederreiter and C.P. Xing, Cyclotomic function fields, Hilbert class fields, and global function fields with many rational places, *Acta Arith.* **79**, 59–76 (1997).

[106] H. Niederreiter and C.P. Xing, Drinfeld modules of rank 1 and algebraic curves with many rational points. II, *Acta Arith.* **81**, 81–100 (1997).

[107] H. Niederreiter and C.P. Xing, Global function fields with many rational places over the quinary field, *Demonstratio Math.* **30**, 919–930 (1997).

[108] H. Niederreiter and C.P. Xing, Algebraic curves over finite fields with many rational points, *Number Theory: Diophantine, Computational and Algebraic Aspects* (K. Györy, A. Pethö, and V.T. Sós, eds.), pp. 423–443, W. de Gruyter, Berlin, 1998.

[109] H. Niederreiter and C.P. Xing, The algebraic-geometry approach to low-discrepancy sequences, *Monte Carlo and Quasi-Monte Carlo Methods 1996* (H. Niederreiter *et al.*, eds.), Lecture Notes in Statistics, Vol. **127**, pp. 139–160, Springer, New York, 1998.

[110] H. Niederreiter and C.P. Xing, Global function fields with many rational places over the ternary field, *Acta Arith.* **83**, 65–86 (1998).

[111] H. Niederreiter and C.P. Xing, Global function fields with many rational places over the quinary field. II, *Acta Arith.* **86**, 277–288 (1998).

[112] H. Niederreiter and C.P. Xing, Towers of global function fields with asymptotically many rational places and an improvement on the Gilbert-Varshamov bound, *Math. Nachr.* **195**, 171–186 (1998).

[113] H. Niederreiter and C.P. Xing, Nets, (t, s)-sequences, and algebraic geometry, *Random and Quasi-Random Point Sets* (P. Hellekalek and G. Larcher, eds.), Lecture Notes in Statistics, Vol. **138**, pp. 267–302, Springer, New York, 1998.

[114] H. Niederreiter and C.P. Xing, A general method of constructing global function fields with many rational places, *Algorithmic Number Theory* (J.P. Buhler, ed.), Lecture Notes in Computer Science, Vol. **1423**, pp. 555–566, Springer, Berlin, 1998.

[115] H. Niederreiter and C.P. Xing, Curve sequences with asymptotically many rational points, *Applications of Curves over Finite Fields* (M.D. Fried, ed.), Contemporary Math., Vol. **245**, pp. 3–14, American Math. Society, Providence, RI, 1999.

[116] H. Niederreiter and C.P. Xing, Algebraic curves with many rational points over finite fields of characteristic 2, *Number Theory in Progress* (K. Györy, H. Iwaniec, and J. Urbanowicz, eds.), pp. 359–380, W. de Gruyter, Berlin, 1999.

[117] H. Niederreiter and C.P. Xing, Global function fields with many rational places and their applications, *Finite Fields: Theory, Applications, and Algorithms* (R.C. Mullin and G.L. Mullen, eds.), Contemporary Math., Vol. **225**, pp. 87–111, American Math. Society, Providence, RI, 1999.

[118] H. Niederreiter and C.P. Xing, A counterexample to Perret's conjecture on infinite class field towers for global function fields, *Finite Fields Appl.* **5**, 240–245 (1999).

[119] H. Niederreiter and C.P. Xing, Algebraic curves over finite fields with many rational points and their applications, *Number Theory* (R.P. Bambah, V.C. Dumir, and R.J. Hans-Gill, eds.), pp. 287–300, Birkhäuser, Basel, 2000.

[120] H. Niederreiter and C.P. Xing, A propagation rule for linear codes, *Applicable Algebra Engrg. Comm. Comput.* **10**, 425–432 (2000).

[121] H. Niederreiter, C.P. Xing, and K.Y. Lam, A new construction of algebraic-geometry codes, *Applicable Algebra Engrg. Comm. Comput.* **9**, 373–381 (1999).

[122] F. Özbudak and H. Stichtenoth, Curves with many points and configurations of hyperplanes over finite fields, *Finite Fields Appl.* **5**, 436–449 (1999).

[123] F. Özbudak and H. Stichtenoth, Constructing codes from algebraic curves, *IEEE Trans. Inform. Theory* **45**, 2502–2505 (1999).

[124] F. Özbudak and M. Thomas, A note on towers of function fields over finite fields, *Comm. Algebra* **26**, 3737–3741 (1998).

[125] R. Pellikaan, B.-Z. Shen, and G.J.M. van Wee, Which linear codes are algebraic-geometric?, *IEEE Trans. Inform. Theory* **37**, 583–602 (1991).

[126] M. Perret, Tours ramifiées infinies de corps de classes, *J. Number Theory* **38**, 300–322 (1991).

[127] O. Pretzel, *Error-Correcting Codes and Finite Fields*, Oxford University Press, Oxford, 1992.

[128] O. Pretzel, *Codes and Algebraic Curves*, Oxford University Press, Oxford, 1998.

[129] H.-G. Quebbemann, Cyclotomic Goppa codes, *IEEE Trans. Inform. Theory* **34**, 1317–1320 (1988).

[130] M. Rosen, S-units and S-class group in algebraic function fields, *J. Algebra* **26**, 98–108 (1973).

[131] M. Rosen, The Hilbert class field in function fields, *Exposition. Math.* **5**, 365–378 (1987).

[132] H.-G. Rück and H. Stichtenoth, A characterization of Hermitian function fields over finite fields, *J. Reine Angew. Math.* **457**, 185–188 (1994).

[133] R.A. Rueppel, *Analysis and Design of Stream Ciphers*, Springer, Berlin, 1986.

[134] R.A. Rueppel, Stream ciphers, *Contemporary Cryptology: The Science of Information Integrity* (G.J. Simmons, ed.), pp. 65–134, IEEE Press, New York, 1992.

[135] W.Ch. Schmid, Improvements and extensions of the "Salzburg tables" by using irreducible polynomials, *Monte Carlo and Quasi-Monte Carlo Methods 1998* (H. Niederreiter and J. Spanier, eds.), pp. 436–447, Springer, Berlin, 2000.

[136] W.Ch. Schmid and R. Wolf, Bounds for digital nets and sequences, *Acta Arith.* **78**, 377–399 (1997).

[137] R. Schoof, Nonsingular plane cubic curves over finite fields, *J. Combin. Theory Ser. A* **46**, 183–211 (1987).

[138] R. Schoof, Algebraic curves over F_2 with many rational points, *J. Number Theory* **41**, 6–14 (1992).

[139] A. Schweizer, On Drinfeld modular curves with many rational points over finite fields, preprint, Academia Sinica, Taipei, 2000.

[140] J.-P. Serre, *Local Fields*, Springer, New York, 1979.

[141] J.-P. Serre, Sur le nombre des points rationnels d'une courbe algébrique sur un corps fini, *C.R. Acad. Sci. Paris Sér. I Math.* **296**, 397–402 (1983).

[142] J.-P. Serre, Nombres de points des courbes algébriques sur F_q, *Sém. Théorie des Nombres 1982–1983*, Exp. 22, Université de Bordeaux I, Talence, 1983.

[143] J.-P. Serre, Résumé des cours de 1983–1984, *Annuaire du Collège de France* **1984**, 79–83.

[144] J.-P. Serre, *Rational Points on Curves over Finite Fields*, Lecture Notes, Harvard University, 1985.

[145] J.-P. Serre, *Galois Cohomology*, Springer, Berlin, 1997.

[146] J.H. Silverman, *The Arithmetic of Elliptic Curves*, Springer, New York, 1986.

[147] G.J. Simmons, A survey of information authentication, *Contemporary Cryptology: The Science of Information Integrity* (G.J. Simmons, ed.), pp. 379–419, IEEE Press, New York, 1992.

[148] I.M. Sobol', The distribution of points in a cube and the approximate evaluation of integrals (Russian), *Zh. Vychisl. Mat. i Mat. Fiz.* **7**, 784–802 (1967).

[149] J. Spanier and E.H. Maize, Quasi-random methods for estimating integrals using relatively small samples, *SIAM Review* **36**, 18–44 (1994).

[150] S.A. Stepanov, *Arithmetic of Algebraic Curves*, Plenum, New York, 1994.

[151] S.A. Stepanov, *Codes on Algebraic Curves*, Kluwer, New York, 1999.

[152] H. Stichtenoth, *Algebraic Function Fields and Codes*, Springer, Berlin, 1993.

[153] D.R. Stinson, Universal hashing and authentication codes, *Advances in Cryptology – CRYPTO '91* (J. Feigenbaum, ed.), Lecture Notes in Computer Science, Vol. **576**, pp. 74–85, Springer, Berlin, 1992.

[154] D.R. Stinson, *Cryptography: Theory and Practice*, CRC Press, Boca Raton, FL, 1995.

[155] A. Temkine, Hilbert class field towers of function fields over finite fields and lower bounds for $A(q)$, *J. Number Theory*, to appear.

[156] M.A. Tsfasman and S.G. Vlăduţ, *Algebraic-Geometric Codes*, Kluwer, Dordrecht, 1991.

[157] M.A. Tsfasman, S.G. Vlăduţ, and T. Zink, Modular curves, Shimura curves, and Goppa codes, better than Varshamov-Gilbert bound, *Math. Nachr.* **109**, 21–28 (1982).

[158] G. van der Geer and M. van der Vlugt, Curves over finite fields of characteristic 2 with many rational points, *C.R. Acad. Sci. Paris Sér. I Math.* **317**, 593–597 (1993).

[159] G. van der Geer and M. van der Vlugt, Generalized Hamming weights of codes and curves over finite fields with many points, *Proc. Conf. on Algebraic Geometry* (Ramat Gan, 1993), Israel Math. Conf. Proc., Vol. **9**, pp. 417–432, Bar-Ilan University, Ramat Gan, 1996.

[160] G. van der Geer and M. van der Vlugt, Quadratic forms, generalized Hamming weights of codes and curves with many points, *J. Number Theory* **59**, 20–36 (1996).

[161] G. van der Geer and M. van der Vlugt, How to construct curves over finite fields with many points, *Arithmetic Geometry* (F. Catanese, ed.), pp. 169–189, Cambridge University Press, Cambridge, 1997.

[162] G. van der Geer and M. van der Vlugt, Generalized Reed-Muller codes and curves with many points, *J. Number Theory* **72**, 257–268 (1998).

[163] G. van der Geer and M. van der Vlugt, Constructing curves over finite fields with many points by solving linear equations, *Applications of Curves over Finite Fields* (M.D. Fried, ed.), Contemporary Math., Vol. **245**, pp. 41–47, American Math. Society, Providence, RI, 1999.

[164] G. van der Geer and M. van der Vlugt, Tables of curves with many points, *Math. Comp.* **69**, 797–810 (2000).

[165] G. van der Geer and M. van der Vlugt, Kummer covers with many points, *Finite Fields Appl.* **6**, 327–341 (2000).

[166] J.H. van Lint, *Introduction to Coding Theory*, Springer, New York, 1982; 3rd ed., Springer, Berlin, 2000.

[167] S.G. Vlăduţ and V.G. Drinfeld, Number of points of an algebraic curve, *Funct. Anal. Appl.* **17**, 53–54 (1983).

[168] C. Voss and T. Høholdt, A family of Kummer extensions of the Hermitian function field, *Comm. Algebra* **23**, 1551–1566 (1995).

[169] C. Voss and T. Høholdt, An explicit construction of a sequence of codes attaining the Tsfasman-Vlăduţ-Zink bound: the first steps, *IEEE Trans. Inform. Theory* **43**, 128–135 (1997).

[170] H.X. Wang and C.P. Xing, Explicit constructions of perfect hash families from algebraic curves over finite fields, *J. Combin. Theory Ser. A*, to appear.

[171] W.C. Waterhouse, Abelian varieties over finite fields, *Ann. Sci. Ecole Norm. Sup.* (4)**2**, 521–560 (1969).

[172] A. Weil, *Basic Number Theory*, 2nd ed., Springer, New York, 1973.

[173] E. Weiss, *Algebraic Number Theory*, McGraw-Hill, New York, 1963.

[174] C.P. Xing, Multiple Kummer extension and the number of prime divisors of degree one in function fields, *J. Pure Appl. Algebra* **84**, 85–93 (1993).

[175] C.P. Xing, Algebraic-geometry codes with asymptotic parameters better than the Gilbert-Varshamov and the Tsfasman-Vlăduţ-Zink bounds, *IEEE Trans. Inform. Theory*, to appear.

[176] C.P. Xing and K.Y. Lam, Sequences with almost perfect linear complexity profiles and curves over finite fields, *IEEE Trans. Inform. Theory* **45**, 1267–1270 (1999).

[177] C.P. Xing and S. Ling, A class of linear codes with good parameters from algebraic curves, *IEEE Trans. Inform. Theory* **46**, 1527–1532 (2000).

[178] C.P. Xing and S. Ling, A class of linear codes with good parameters, *IEEE Trans. Inform. Theory* **46**, 2184–2188 (2000).

[179] C.P. Xing and H. Niederreiter, A construction of low-discrepancy sequences using global function fields, *Acta Arith.* **73**, 87–102 (1995).

[180] C.P. Xing and H. Niederreiter, Modules de Drinfeld et courbes algébriques ayant beaucoup de points rationnels, *C.R. Acad. Sci. Paris Sér. I Math.* **322**, 651–654 (1996).

[181] C.P. Xing and H. Niederreiter, Drinfeld modules of rank 1 and algebraic curves with many rational points, *Monatsh. Math.* **127**, 219–241 (1999).

[182] C.P. Xing and H. Niederreiter, Applications of algebraic curves to constructions of codes and almost perfect sequences, *Finite Fields and Applications* (D. Jungnickel and H. Niederreiter, eds.), Springer, Berlin, to appear.

[183] C.P. Xing, H. Niederreiter, and K.Y. Lam, Constructions of algebraic-geometry codes, *IEEE Trans. Inform. Theory* **45**, 1186–1193 (1999).

[184] C.P. Xing, H. Niederreiter, and K.Y. Lam, A generalization of algebraic-geometry codes, *IEEE Trans. Inform. Theory* **45**, 2498–2501 (1999).

[185] C.P. Xing, H. Niederreiter, K.Y. Lam, and C.S. Ding, Constructions of sequences with almost perfect linear complexity profile from curves over finite fields, *Finite Fields Appl.* **5**, 301–313 (1999).

[186] C.P. Xing and H. Stichtenoth, The genus of maximal function fields over finite fields, *Manuscripta Math.* **86**, 217–224 (1995).

[187] C.P. Xing, H.X. Wang, and K.Y. Lam, Constructions of authentication codes from algebraic curves over finite fields, *IEEE Trans. Inform. Theory* **46**, 886–892 (2000).

[188] T. Zink, Degeneration of Shimura surfaces and a problem in coding theory, *Fundamentals of Computation Theory* (L. Budach, ed.), Lecture Notes in Computer Science, Vol. **199**, pp. 503–511, Springer, Berlin, 1985.

Index

Printed in the United States
By Bookmasters